First RILEM International Conference on Concrete and Digital Fabrication – Digital Concrete 2018

RILEM BOOKSERIES

Volume 19

RILEM, The International Union of Laboratories and Experts in Construction Materials, Systems and Structures, founded in 1947, is a non-governmental scientific association whose goal is to contribute to progress in the construction sciences, techniques and industries, essentially by means of the communication it fosters between research and practice. RILEM's focus is on construction materials and their use in building and civil engineering structures, covering all phases of the building process from manufacture to use and recycling of materials. More information on RILEM and its previous publications can be found on www.RILEM.net.

More information about this series at http://www.springer.com/series/8781

Timothy Wangler · Robert J. Flatt
Editors

First RILEM International Conference on Concrete and Digital Fabrication – Digital Concrete 2018

Springer

Editors
Timothy Wangler
Physical Chemistry of Building Materials
ETH Zürich
Zürich, Switzerland

Robert J. Flatt
Physical Chemistry of Building Materials
ETH Zürich
Zürich, Switzerland

ISSN 2211-0844 ISSN 2211-0852 (electronic)
RILEM Bookseries
ISBN 978-3-030-07603-0 ISBN 978-3-319-99519-9 (eBook)
https://doi.org/10.1007/978-3-319-99519-9

This Springer imprint is published by the registered company Springer Nature Switzerland AG
The registered company address is: Gewerbestrasse 11, 6330 Cham, Switzerland

Preface

Dear Colleagues,

We are pleased to organize the RILEM 1st International Conference on Concrete and Digital Fabrication, also known as Digital Concrete 2018, held from 10 to 12 September 2018 in Zurich, Switzerland. This is the first conference in this series, marking the beginning of a new research field in digital fabrication with concrete.

The conference arose as a result of work initiated by the RILEM Technical Committee on Digital Fabrication with Cement-based Materials (RILEM TC 276-DFC), which was kicked off in 2016 in Washington, DC, due to a burgeoning interest in the techniques of digital fabrication and their application to cement-based materials such as concrete. The interest in this topic intensified to the point that at the second meeting of the RILEM TC 276-DFC, in Zurich, Switzerland, an international workshop was hosted in which some of the pioneers of the field were invited to speak. The attendance of the workshop was overwhelming compared to what was expected, and what was particularly striking was the very large industry presence at the workshop. At this point, the decision to host the 1st International Conference on Concrete and Digital Fabrication was made, with the host being ETH Zurich, and with the support of RILEM and the National Centre for Competence in Research for Digital Fabrication in Architecture, a Swiss National Science Foundation research funding initiative to explore the possibilities of digital fabrication for construction and architecture. A key part of the decision for Zurich was that the ROB|ARCH Conference, the largest conference on robotics in architecture, was taking place at the same time, and the expected interactions between the two conferences would likely be a fertile ground to produce new collaborations and new ideas.

For the conference, over 100 extended abstracts from around the world were submitted and reviewed, with 65 invited for oral presentations, and 45 invited to write a proceedings paper. Of these, 30 authors submitted papers that were also peer reviewed. All reviews were performed by members of the scientific committee, which consisted of the membership of the TC 276-DFC. These papers reflect closely the content of the submissions to the conference, which can be broadly divided into three major themes: (1) Materials and Processing, (2) Mechanics and

Structure and (3) Applications and Other Topics. These themes are also highly reflective of the researchers and practitioners who are most interested in this field: experts in fresh state properties and their control and evolution (rheology), experts in hardened properties and structural performance (structural engineering and durability) and experts in utilization, design and realization of these budding technologies (design and architecture). All extended abstracts will be available electronically to conference participants.

During the conference, two awards were given to recognize two pioneers in the field of digital fabrication with concrete: Prof. Berokh Khoshnevis of the University of Southern California and Contour Crafting Corporation and pioneer of the Contour Crafting process, and Enrico Dini, founder of D-Shape, a pioneer in the field of large-scale particle bed printing.

In addition to these proceedings, a special issue of Cement and Concrete Research has been published on the topic of digital fabrication with concrete, with invited papers from groups of experts all contributing on various topics, again closely correlated with the above-named themes, and the editors of these proceedings invite the participants to read these papers as well, which we feel will join these proceedings as keystone scientific works that will help to guide future researchers with interest in this topic.

The organizers would like to reiterate their special thanks to the RILEM Technical Committee on Digital Fabrication with Cement-based Materials for playing their role as Scientific Committee in reviewing the extended abstracts and proceedings.

The excellent contributions and expertise of the authors and conference participants are acknowledged here, as well as the direct support of the industrial sponsors, listed here in alphabetical order: Akzo Nobel, Arup, Autodesk, BASF, Basler Hofmann, Boston Consulting Group, Dow Chemical, Kerneos, LafargeHolcim, Laticrete, Omya, Sika AG and Voxeljet.

The editors also offer special thanks to all the institutional support in organization of this seminal event: RILEM, ETH Zurich, the Department of Civil, Environmental and Geomatic Engineering, the Institute for Building Materials, and the National Centre for Competence in Research in Digital Fabrication in Architecture in particular, whose management team, being directly involved in organization of ROB|ARCH, provided much direct and behind the scenes support throughout the organization of this conference. We finally offer our deepest thanks to the local Organizing Committee, for their tireless support throughout the organization and execution of this new event.

September 2018

Timothy Wangler
Robert J. Flatt

Organization

Conference Committees

Conference Chair

Robert J. Flatt	ETH Zurich, Switzerland

Conference Vice Chair

Timothy Wangler	ETH Zurich, Switzerland

International Scientific Committee

Nicolas Roussel (Chair)	IFSTTAR, France
Domenico Asprone	University of Naples Federico II, Italy
Miguel Azenha	University of Minho, Portugal
Hela Bessaies-Bey	IFSTTAR, France
Freek Bos	TU Eindhoven, Netherlands
Richard Buswell	Loughborough University, UK
Rolands Cepuritis	NORCEM, Norway
Vitor Cunha	University of Minho, Portugal
Geert De Schutter	Ghent University, Belgium
Fabrice Decroix	Sika, France
Vivien Esnault	LafargeHolcim, France
Liberato Ferrara	Politecnico di Milano, Italy
Asko Fromm	Hochschule Wismar, Germany
Steffen Grünewald	TU Delft, Netherlands

Scott Jones	National Institute of Standards and Technology, USA
Shiho Kawashima	Columbia University, USA
Jae Hong Kim	Ulsan National Institute of Science and Technology, South Korea
Ulrich Knaack	TU Darmstadt, Germany
Wilson Ricardo Leal da Silva	Danish Technological Institute, Denmark
Karel Lesage	Ghent University, Belgium
Ena Lloret-Fritschi	ETH Zurich, Switzerland
Dirk Lowke	TU Braunschweig, Germany
Delphine Marchon	University of California, Berkeley, USA
Jaime Mata-Falcón	ETH Zurich, Switzerland
Viktor Mechtcherine	TU Dresden, Germany
Costantino Menna	University of Naples Federico II, Italy
Sandro Moro	BASF, Italy
Berta Mota	BAM Berlin, Germany
Sandra Nunes	University of Porto, Portugal
Lars Nyholm Thrane	Danish Technological Institute, Denmark
Aurélie Papon	INSA Toulouse, France
Marco Pepe	University of Salerno, Italy
Arnaud Perrot	Université de Bretagne Sud, Lorient, France
Bertrand Pouteau	Eurovia, France
Milan Radosavljevic	University of the West of Scotland, UK
Theo Salet	TU Eindhoven, Netherlands
Jay Sanjayan	Swinburne University, Australia
Roel Schipper	TU Delft, Netherlands
Surendra Shah	Northwestern University, USA
Mohammed Sonebi	Queen's University Belfast, UK
Stephan Dietmar	TU Berlin, Germany
Toledo Filho Romildo	COPPE/UFRJ, Brazil
Fabrice Toussaint	LafargeHolcim, France
Kim Van Tittelboom	Ghent University, Belgium
Ksenjia Vasilic	BAM Berlin, Germany
Giovanni Volpatti	Cemex, Switzerland
Zhendi Wang	China Building Materials Academy, China
Xiangming Zhou	Brunel University, UK

Local Organizing Committee

Robert J. Flatt (Chair)	ETH Zurich, Switzerland
Timothy Wangler (Vice Chair)	ETH Zurich, Switzerland
Andrea Louys	ETH Zurich, Switzerland

Contents

RILEM Publications

The following list is presenting the global offer of RILEM Publications, sorted by series. Each publication is available in printed version and/or in online version.

RILEM Proceedings (PRO)

PRO 1: Durability of High Performance Concrete (ISBN: 2-912143-03-9; e-ISBN: 2-351580-12-5; e-ISBN: 2351580125); *Ed. H. Sommer*
PRO 2: Chloride Penetration into Concrete (ISBN: 2-912143-00-04; e-ISBN: 2912143454); *Eds. L.-O. Nilsson and J.-P. Ollivier*
PRO 3: Evaluation and Strengthening of Existing Masonry Structures (ISBN: 2-912143-02-0; e-ISBN: 2351580141); *Eds. L. Binda and C. Modena*
PRO 4: Concrete: From Material to Structure (ISBN: 2-912143-04-7; e-ISBN: 2351580206); *Eds. J.-P. Bournazel and Y. Malier*
PRO 5: The Role of Admixtures in High Performance Concrete (ISBN: 2-912143-05-5; e-ISBN: 2351580214); *Eds. J. G. Cabrera and R. Rivera-Villarreal*
PRO 6: High Performance Fiber Reinforced Cement Composites - HPFRCC 3 (ISBN: 2-912143-06-3; e-ISBN: 2351580222); *Eds. H. W. Reinhardt and A. E. Naaman*
PRO 7: 1st International RILEM Symposium on Self-Compacting Concrete (ISBN: 2-912143-09-8; e-ISBN: 2912143721); *Eds. Å. Skarendahl and Ö. Petersson*
PRO 8: International RILEM Symposium on Timber Engineering (ISBN: 2-912143-10-1; e-ISBN: 2351580230); *Ed. L. Boström*
PRO 9: 2nd International RILEM Symposium on Adhesion between Polymers and Concrete ISAP '99 (ISBN: 2-912143-11-X; e-ISBN: 2351580249); *Eds. Y. Ohama and M. Puterman*
PRO 10: 3rd International RILEM Symposium on Durability of Building and Construction Sealants (ISBN: 2-912143-13-6; e-ISBN: 2351580257); *Eds. A. T. Wolf*

PRO 11: 4th International RILEM Conference on Reflective Cracking in Pavements (ISBN: 2-912143-14-4; e-ISBN: 2351580265); *Eds. A. O. Abd El Halim, D. A. Taylor and El H. H. Mohamed*

PRO 12: International RILEM Workshop on Historic Mortars: Characteristics and Tests (ISBN: 2-912143-15-2; e-ISBN: 2351580273); *Eds. P. Bartos, C. Groot and J. J. Hughes*

PRO 13: 2nd International RILEM Symposium on Hydration and Setting (ISBN: 2-912143-16-0; e-ISBN: 2351580281); *Ed. A. Nonat*

PRO 14: Integrated Life-Cycle Design of Materials and Structures - ILCDES 2000 (ISBN: 951-758-408-3; e-ISBN: 235158029X); (ISSN: 0356-9403); *Ed. S. Sarja*

PRO 15: Fifth RILEM Symposium on Fibre-Reinforced Concretes (FRC) - BEFIB'2000 (ISBN: 2-912143-18-7; e-ISBN: 291214373X); *Eds. P. Rossi and G. Chanvillard*

PRO 16: Life Prediction and Management of Concrete Structures (ISBN: 2-912143-19-5; e-ISBN: 2351580303); *Ed. D. Naus*

PRO 17: Shrinkage of Concrete – Shrinkage 2000 (ISBN: 2-912143-20-9; e-ISBN: 2351580311); *Eds. V. Baroghel-Bouny and P.-C. Aïtcin*

PRO 18: Measurement and Interpretation of the On-Site Corrosion Rate (ISBN: 2-912143-21-7; e-ISBN: 235158032X); *Eds. C. Andrade, C. Alonso, J. Fullea, J. Polimon and J. Rodriguez*

PRO 19: Testing and Modelling the Chloride Ingress into Concrete (ISBN: 2-912143-22-5; e-ISBN: 2351580338); *Eds. C. Andrade and J. Kropp*

PRO 20: 1st International RILEM Workshop on Microbial Impacts on Building Materials (CD 02) (e-ISBN 978-2-35158-013-4); *Ed. M. Ribas Silva*

PRO 21: International RILEM Symposium on Connections between Steel and Concrete (ISBN: 2-912143-25-X; e-ISBN: 2351580346); *Ed. R. Eligehausen*

PRO 22: International RILEM Symposium on Joints in Timber Structures (ISBN: 2-912143-28-4; e-ISBN: 2351580354); *Eds. S. Aicher and H.-W. Reinhardt*

PRO 23: International RILEM Conference on Early Age Cracking in Cementitious Systems (ISBN: 2-912143-29-2; e-ISBN: 2351580362); *Eds. K. Kovler and A. Bentur*

PRO 24: 2nd International RILEM Workshop on Frost Resistance of Concrete (ISBN: 2-912143-30-6; e-ISBN: 2351580370); *Eds. M. J. Setzer, R. Auberg and H.-J. Keck*

PRO 25: International RILEM Workshop on Frost Damage in Concrete (ISBN: 2-912143-31-4; e-ISBN: 2351580389); *Eds. D. J. Janssen, M. J. Setzer and M. B. Snyder*

PRO 26: International RILEM Workshop on On-Site Control and Evaluation of Masonry Structures (ISBN: 2-912143-34-9; e-ISBN: 2351580141); *Eds. L. Binda and R. C. de Vekey*

PRO 27: International RILEM Symposium on Building Joint Sealants (CD03; e-ISBN: 235158015X); *Ed. A. T. Wolf*

PRO 28: 6th International RILEM Symposium on Performance Testing and Evaluation of Bituminous Materials - PTEBM'03 (ISBN: 2-912143-35-7; e-ISBN: 978-2-912143-77-8); *Ed. M. N. Partl*

PRO 29: 2nd International RILEM Workshop on Life Prediction and Ageing Management of Concrete Structures (ISBN: 2-912143-36-5; e-ISBN: 2912143780); *Ed. D. J. Naus*

PRO 30: 4th International RILEM Workshop on High Performance Fiber Reinforced Cement Composites - HPFRCC 4 (ISBN: 2-912143-37-3; e-ISBN: 2912143799); *Eds. A. E. Naaman and H. W. Reinhardt*

PRO 31: International RILEM Workshop on Test and Design Methods for Steel Fibre Reinforced Concrete: Background and Experiences (ISBN: 2-912143-38-1; e-ISBN: 2351580168); *Eds. B. Schnütgen and L. Vandewalle*

PRO 32: International Conference on Advances in Concrete and Structures 2 vol. (ISBN (set): 2-912143-41-1; e-ISBN: 2351580176); *Eds. Ying-shu Yuan, Surendra P. Shah and Heng-lin Lü*

PRO 33: 3rd International Symposium on Self-Compacting Concrete (ISBN: 2-912143-42-X; e-ISBN: 2912143713); *Eds. Ó. Wallevik and I. Níelsson*

PRO 34: International RILEM Conference on Microbial Impact on Building Materials (ISBN: 2-912143-43-8; e-ISBN: 2351580184); *Ed. M. Ribas Silva*

PRO 35: International RILEM TC 186-ISA on Internal Sulfate Attack and Delayed Ettringite Formation (ISBN: 2-912143-44-6; e-ISBN: 2912143802); *Eds. K. Scrivener and J. Skalny*

PRO 36: International RILEM Symposium on Concrete Science and Engineering – A Tribute to Arnon Bentur (ISBN: 2-912143-46-2; e-ISBN: 2912143586); *Eds. K. Kovler, J. Marchand, S. Mindess and J. Weiss*

PRO 37: 5th International RILEM Conference on Cracking in Pavements – Mitigation, Risk Assessment and Prevention (ISBN: 2-912143-47-0; e-ISBN: 2912143764); *Eds. C. Petit, I. Al-Qadi and A. Millien*

PRO 38: 3rd International RILEM Workshop on Testing and Modelling the Chloride Ingress into Concrete (ISBN: 2-912143-48-9; e-ISBN: 2912143578); *Eds. C. Andrade and J. Kropp*

PRO 39: 6th International RILEM Symposium on Fibre-Reinforced Concretes - BEFIB 2004 (ISBN: 2-912143-51-9; e-ISBN: 2912143748); *Eds. M. Di Prisco, R. Felicetti and G. A. Plizzari*

PRO 40: International RILEM Conference on the Use of Recycled Materials in Buildings and Structures (ISBN: 2-912143-52-7; e-ISBN: 2912143756); *Eds. E. Vázquez, Ch. F. Hendriks and G. M. T. Janssen*

PRO 41: RILEM International Symposium on Environment-Conscious Materials and Systems for Sustainable Development (ISBN: 2-912143-55-1; e-ISBN: 2912143640); *Eds. N. Kashino and Y. Ohama*

PRO 42: SCC'2005 - China: 1st International Symposium on Design, Performance and Use of Self-Consolidating Concrete (ISBN: 2-912143-61-6; e-ISBN: 2912143624); *Eds. Zhiwu Yu, Caijun Shi, Kamal Henri Khayat and Youjun Xie*

PRO 43: International RILEM Workshop on Bonded Concrete Overlays (e-ISBN: 2-912143-83-7); *Eds. J. L. Granju and J. Silfwerbrand*

PRO 44: 2nd International RILEM Workshop on Microbial Impacts on Building Materials (CD11) (e-ISBN: 2-912143-84-5); *Ed. M. Ribas Silva*

PRO 45: 2nd International Symposium on Nanotechnology in Construction, Bilbao (ISBN: 2-912143-87-X; e-ISBN: 2912143888); *Eds. Peter J. M. Bartos, Yolanda de Miguel and Antonio Porro*

PRO 46: ConcreteLife'06 - International RILEM-JCI Seminar on Concrete Durability and Service Life Planning: Curing, Crack Control, Performance in Harsh Environments (ISBN: 2-912143-89-6; e-ISBN: 291214390X); *Ed. K. Kovler*

PRO 47: International RILEM Workshop on Performance Based Evaluation and Indicators for Concrete Durability (ISBN: 978-2-912143-95-2; e-ISBN: 9782912143969); *Eds. V. Baroghel-Bouny, C. Andrade, R. Torrent and K. Scrivener*

PRO 48: 1st International RILEM Symposium on Advances in Concrete through Science and Engineering (e-ISBN: 2-912143-92-6); *Eds. J. Weiss, K. Kovler, J. Marchand, and S. Mindess*

PRO 49: International RILEM Workshop on High Performance Fiber Reinforced Cementitious Composites in Structural Applications (ISBN: 2-912143-93-4; e-ISBN: 2912143942); *Eds. G. Fischer and V. C. Li*

PRO 50: 1st International RILEM Symposium on Textile Reinforced Concrete (ISBN: 2-912143-97-7; e-ISBN: 2351580087); *Eds. Josef Hegger, Wolfgang Brameshuber and Norbert Will*

PRO 51: 2nd International Symposium on Advances in Concrete through Science and Engineering (ISBN: 2-35158-003-6; e-ISBN: 2-35158-002-8); *Eds. J. Marchand, B. Bissonnette, R. Gagné, M. Jolin and F. Paradis*

PRO 52: Volume Changes of Hardening Concrete: Testing and Mitigation (ISBN: 2-35158-004-4; e-ISBN: 2-35158-005-2); *Eds. O. M. Jensen, P. Lura and K. Kovler*

PRO 53: High Performance Fiber Reinforced Cement Composites - HPFRCC5 (ISBN: 978-2-35158-046-2; e-ISBN: 978-2-35158-089-9); *Eds. H. W. Reinhardt and A. E. Naaman*

PRO 54: 5th International RILEM Symposium on Self-Compacting Concrete (ISBN: 978-2-35158-047-9; e-ISBN: 978-2-35158-088-2); *Eds. G. De Schutter and V. Boel*

PRO 55: International RILEM Symposium Photocatalysis, Environment and Construction Materials (ISBN: 978-2-35158-056-1; e-ISBN: 978-2-35158-057-8); *Eds. P. Baglioni and L. Cassar*

PRO 56: International RILEM Workshop on Integral Service Life Modelling of Concrete Structures (ISBN 978-2-35158-058-5; e-ISBN: 978-2-35158-090-5); *Eds. R. M. Ferreira, J. Gulikers and C. Andrade*

PRO 57: RILEM Workshop on Performance of cement-based materials in aggressive aqueous environments (e-ISBN: 978-2-35158-059-2); *Ed. N. De Belie*

PRO 58: International RILEM Symposium on Concrete Modelling - CONMOD'08 (ISBN: 978-2-35158-060-8; e-ISBN: 978-2-35158-076-9); *Eds. E. Schlangen and G. De Schutter*

PRO 59: International RILEM Conference on Site Assessment of Concrete, Masonry and Timber Structures - SACoMaTiS 2008 (ISBN set: 978-2-35158-061-5; e-ISBN: 978-2-35158-075-2); *Eds. L. Binda, M. di Prisco and R. Felicetti*
PRO 60: Seventh RILEM International Symposium on Fibre Reinforced Concrete: Design and Applications - BEFIB 2008 (ISBN: 978-2-35158-064-6; e-ISBN: 978-2-35158-086-8); *Ed. R. Gettu*
PRO 61: 1st International Conference on Microstructure Related Durability of Cementitious Composites 2 vol., (ISBN: 978-2-35158-065-3; e-ISBN: 978-2-35158-084-4); *Eds. W. Sun, K. van Breugel, C. Miao, G. Ye and H. Chen*
PRO 62: NSF/RILEM Workshop: In-situ Evaluation of Historic Wood and Masonry Structures (e-ISBN: 978-2-35158-068-4); *Eds. B. Kasal, R. Anthony and M. Drdácký*
PRO 63: Concrete in Aggressive Aqueous Environments: Performance, Testing and Modelling, 2 vol., (ISBN: 978-2-35158-071-4; e-ISBN: 978-2-35158-082-0); *Eds. M. G. Alexander and A. Bertron*
PRO 64: Long Term Performance of Cementitious Barriers and Reinforced Concrete in Nuclear Power Plants and Waste Management - NUCPERF 2009 (ISBN: 978-2-35158-072-1; e-ISBN: 978-2-35158-087-5); *Eds. V. L'Hostis, R. Gens, C. Gallé*
PRO 65: Design Performance and Use of Self-consolidating Concrete - SCC'2009 (ISBN: 978-2-35158-073-8; e-ISBN: 978-2-35158-093-6); *Eds. C. Shi, Z. Yu, K. H. Khayat and P. Yan*
PRO 66: 2nd International RILEM Workshop on Concrete Durability and Service Life Planning - ConcreteLife'09 (ISBN: 978-2-35158-074-5; ISBN: 978-2-35158-074-5); *Ed. K. Kovler*
PRO 67: Repairs Mortars for Historic Masonry (e-ISBN: 978-2-35158-083-7); *Ed. C. Groot*
PRO 68: Proceedings of the 3rd International RILEM Symposium on 'Rheology of Cement Suspensions such as Fresh Concrete (ISBN 978-2-35158-091-2; e-ISBN: 978-2-35158-092-9); *Eds. O. H. Wallevik, S. Kubens and S. Oesterheld*
PRO 69: 3rd International PhD Student Workshop on 'Modelling the Durability of Reinforced Concrete (ISBN: 978-2-35158-095-0); *Eds. R. M. Ferreira, J. Gulikers and C. Andrade*
PRO 70: 2nd International Conference on 'Service Life Design for Infrastructure' (ISBN set: 978-2-35158-096-7, e-ISBN: 978-2-35158-097-4); *Ed. K. van Breugel, G. Ye and Y. Yuan*
PRO 71: Advances in Civil Engineering Materials - The 50-year Teaching Anniversary of Prof. Sun Wei' (ISBN: 978-2-35158-098-1; e-ISBN: 978-2-35158-099-8); *Eds. C. Miao, G. Ye, and H. Chen*
PRO 72: First International Conference on 'Advances in Chemically-Activated Materials – CAM'2010' (2010), 264 pp, ISBN: 978-2-35158-101-8; e-ISBN: 978-2-35158-115-5, *Eds. Caijun Shi and Xiaodong Shen*

PRO 73: 2nd International Conference on 'Waste Engineering and Management - ICWEM 2010' (2010), 894 pp, ISBN: 978-2-35158-102-5; e-ISBN: 978-2-35158-103-2, *Eds. J. Zh. Xiao, Y. Zhang, M. S. Cheung and R. Chu*
PRO 74: International RILEM Conference on 'Use of Superabsorsorbent Polymers and Other New Addditives in Concrete' (2010) 374 pp., ISBN: 978-2-35158-104-9; e-ISBN: 978-2-35158-105-6; *Eds. O. M. Jensen, M. T. Hasholt, and S. Laustsen*
PRO 75: International Conference on 'Material Science - 2nd ICTRC - Textile Reinforced Concrete - Theme 1' (2010) 436 pp., ISBN: 978-2-35158-106-3; e-ISBN: 978-2-35158-107-0; *Ed. W. Brameshuber*
PRO 76: International Conference on 'Material Science - HetMat - Modelling of Heterogeneous Materials - Theme 2' (2010) 255 pp., ISBN: 978-2-35158-108-7; e-ISBN: 978-2-35158-109-4; *Ed. W. Brameshuber*
PRO 77: International Conference on 'Material Science - AdIPoC - Additions Improving Properties of Concrete - Theme 3' (2010) 459 pp., ISBN: 978-2-35158-110-0; e-ISBN: 978-2-35158-111-7; *Ed. W. Brameshuber*
PRO 78: 2nd Historic Mortars Conference and RILEM TC 203-RHM Final Workshop – HMC2010 (2010) 1416 pp., e-ISBN: 978-2-35158-112-4; *Eds. J. Válek, C. Groot, and J. J. Hughes*
PRO 79: International RILEM Conference on Advances in Construction Materials Through Science and Engineering (2011) 213 pp., ISBN: 978-2-35158-116-2, e-ISBN: 978-2-35158-117-9; *Eds. Christopher Leung and K. T. Wan*
PRO 80: 2nd International RILEM Conference on Concrete Spalling due to Fire Exposure (2011) 453 pp., ISBN: 978-2-35158-118-6, e-ISBN: 978-2-35158-119-3; *Eds. E. A. B. Koenders and F. Dehn*
PRO 81: 2nd International RILEM Conference on Strain Hardening Cementitious Composites (SHCC2-Rio) (2011) 451 pp., ISBN: 978-2-35158-120-9, e-ISBN: 978-2-35158-121-6; *Eds. R. D. Toledo Filho, F. A. Silva, E. A. B. Koenders and E. M. R. Fairbairn*
PRO 82: 2nd International RILEM Conference on Progress of Recycling in the Built Environment (2011) 507 pp., e-ISBN: 978-2-35158-122-3; *Eds. V. M. John, E. Vazquez, S. C. Angulo and C. Ulsen*
PRO 83: 2nd International Conference on Microstructural-related Durability of Cementitious Composites (2012) 250 pp., ISBN: 978-2-35158-129-2; e-ISBN: 978-2-35158-123-0; *Eds. G. Ye, K. van Breugel, W. Sun and C. Miao*
PRO 84: CONSEC13 - Seventh International Conference on Concrete under Severe Conditions – Environment and Loading (2013) 1930 pp., ISBN: 978-2-35158-124-7; e-ISBN: 978-2- 35158-134-6; *Eds. Z. J. Li, W. Sun, C. W. Miao, K. Sakai, O. E. Gjorv and N. Banthia*
PRO 85: RILEM-JCI International Workshop on Crack Control of Mass Concrete and Related issues concerning Early-Age of Concrete Structures – ConCrack 3 – Control of Cracking in Concrete Structures 3 (2012) 237 pp., ISBN: 978-2-35158-125-4; e-ISBN: 978-2-35158-126-1; *Eds. F. Toutlemonde and J.-M. Torrenti*

PRO 86: International Symposium on Life Cycle Assessment and Construction (2012) 414 pp., ISBN: 978-2-35158-127-8, e-ISBN: 978-2-35158-128-5; *Eds. A. Ventura and C. de la Roche*

PRO 87: UHPFRC 2013 – RILEM-fib-AFGC International Symposium on Ultra-High Performance Fibre-Reinforced Concrete (2013), ISBN: 978-2-35158-130-8, e-ISBN: 978-2-35158-131-5; *Eds. F. Toutlemonde*

PRO 88: 8th RILEM International Symposium on Fibre Reinforced Concrete (2012) 344 pp., ISBN: 978-2-35158-132-2, e-ISBN: 978-2-35158-133-9; *Eds. Joaquim A. O. Barros*

PRO 89: RILEM International workshop on performance-based specification and control of concrete durability (2014) 678 pp, ISBN: 978-2-35158-135-3, e-ISBN: 978-2-35158-136-0; *Eds. D. Bjegović, H. Beushausen and M. Serdar*

PRO 90: 7th RILEM International Conference on Self-Compacting Concrete and of the 1st RILEM International Conference on Rheology and Processing of Construction Materials (2013) 396 pp, ISBN: 978-2-35158-137-7, e-ISBN: 978-2-35158-138-4; *Eds. Nicolas Roussel and Hela Bessaies-Bey*

PRO 91: CONMOD 2014 - RILEM International Symposium on Concrete Modelling (2014), ISBN: 978-2-35158-139-1; e-ISBN: 978-2-35158-140-7; *Eds. Kefei Li, Peiyu Yan and Rongwei Yang*

PRO 92: CAM 2014 - 2nd International Conference on advances in chemically-activated materials (2014) 392 pp., ISBN: 978-2-35158-141-4; e-ISBN: 978-2-35158-142-1; *Eds. Caijun Shi and Xiadong Shen*

PRO 93: SCC 2014 - 3rd International Symposium on Design, Performance and Use of Self-Consolidating Concrete (2014) 438 pp., ISBN: 978-2-35158-143-8; e-ISBN: 978-2-35158-144-5; *Eds. Caijun Shi, Zhihua Ou, Kamal H. Khayat*

PRO 94 (online version): HPFRCC-7 - 7th RILEM conference on High performance fiber reinforced cement composites (2015), e-ISBN: 978-2-35158-146-9; *Eds. H. W. Reinhardt, G. J. Parra-Montesinos, H. Garrecht*

PRO 95: International RILEM Conference on Application of superabsorbent polymers and other new admixtures in concrete construction (2014), ISBN: 978-2-35158-147-6; e-ISBN: 978-2-35158-148-3; *Eds. Viktor Mechtcherine, Christof Schroefl*

PRO 96 (online version): XIII DBMC: XIII International Conference on Durability of Building Materials and Components (2015), e-ISBN: 978-2-35158-149-0; *Eds. M. Quattrone, V. M. John*

PRO 97: SHCC3 – 3rd International RILEM Conference on Strain Hardening Cementitious Composites (2014), ISBN: 978-2-35158-150-6; e-ISBN: 978-2-35158-151-3; *Eds. E. Schlangen, M. G. Sierra Beltran, M. Lukovic, G. Ye*

PRO 98: FERRO-11 – 11th International Symposium on Ferrocement and 3rd ICTRC - International Conference on Textile Reinforced Concrete (2015), ISBN: 978-2-35158-152-0; e-ISBN: 978-2-35158-153-7; *Ed. W. Brameshuber*

PRO 99 (online version): ICBBM 2015 - 1st International Conference on Bio-Based Building Materials (2015), e-ISBN: 978-2-35158-154-4; *Eds. S. Amziane, M. Sonebi*

PRO 100: SCC16 - RILEM Self-Consolidating Concrete Conference (2016), ISBN: 978-2-35158-156-8; e-ISBN: 978-2-35158-157-5; *Ed. Kamal H. Kayat*
PRO 101 (online version): III Progress of Recycling in the Built Environment (2015), e-ISBN: 978-2-35158-158-2; *Eds. I. Martins, C. Ulsen and S. C. Angulo*
PRO 102 (online version): RILEM Conference on Microorganisms-Cementitious Materials Interactions (2016), e-ISBN: 978-2-35158-160-5; *Eds. Alexandra Bertron, Henk Jonkers, Virginie Wiktor*
PRO 103 (online version): ACESC'16 - Advances in Civil Engineering and Sustainable Construction (2016), e-ISBN: 978-2-35158-161-2; *Eds. T. Ch. Madhavi, G. Prabhakar, Santhosh Ram and P. M. Rameshwaran*
PRO 104 (online version): SSCS'2015 - Numerical Modeling - Strategies for Sustainable Concrete Structures (2015), e-ISBN: 978-2-35158-162-9
PRO 105: 1st International Conference on UHPC Materials and Structures (2016), ISBN: 978-2-35158-164-3, e-ISBN: 978-2-35158-165-0
PRO 106: AFGC-ACI-fib-RILEM International Conference on Ultra-High-Performance Fibre-Reinforced Concrete – UHPFRC 2017 (2017), ISBN: 978-2-35158-166-7, e-ISBN: 978-2-35158-167-4; *Eds. François Toutlemonde and Jacques Resplendino*
PRO 107 (online version): XIV DBMC – 14th International Conference on Durability of Building Materials and Components (2017), e-ISBN: 978-2-35158-159-9; *Eds. Geert De Schutter, Nele De Belie, Arnold Janssens, Nathan Van Den Bossche*
PRO 108: MSSCE 2016 - Innovation of Teaching in Materials and Structures (2016), ISBN: 978-2-35158-178-0, e-ISBN: 978-2-35158-179-7; *Ed. Per Goltermann*
PRO 109 (2 volumes): MSSCE 2016 - Service Life of Cement-Based Materials and Structures (2016), ISBN Vol. 1: 978-2-35158-170-4, Vol. 2: 978-2-35158-171-4, Set Vol. 1&2: 978-2-35158-172-8, e-ISBN : 978-2-35158-173-5; *Eds. Miguel Azenha, Ivan Gabrijel, Dirk Schlicke, Terje Kanstad and Ole Mejlhede Jensen*
PRO 110: MSSCE 2016 - Historical Masonry (2016), ISBN: 978-2-35158-178-0, e-ISBN: 978-2-35158-179-7; *Eds. Inge Rörig-Dalgaard and Ioannis Ioannou*
PRO 111: MSSCE 2016 - Electrochemistry in Civil Engineering (2016), ISBN: 978-2-35158-176-6, e-ISBN: 978-2-35158-177-3; *Ed. Lisbeth M. Ottosen*
PRO 112: MSSCE 2016 - Moisture in Materials and Structures (2016), ISBN: 978-2-35158-178-0, e-ISBN: 978-2-35158-179-7; *Eds. Kurt Kielsgaard Hansen, Carsten Rode and Lars-Olof Nilsson*
PRO 113: MSSCE 2016 - Concrete with Supplementary Cementitious Materials (2016), ISBN: 978-2-35158-178-0, e-ISBN: 978-2-35158-179-7; *Eds. Ole Mejlhede Jensen, Konstantin Kovler and Nele De Belie*
PRO 114: MSSCE 2016 - Frost Action in Concrete (2016), ISBN: 978-2-35158-182-7, e-ISBN: 978-2-35158-183-4; *Eds. Marianne Tange Hasholt, Katja Fridh and R. Doug Hooton*

PRO 115: MSSCE 2016 - Fresh Concrete (2016), ISBN: 978-2-35158-184-1, e-ISBN: 978-2-35158-185-8; *Eds. Lars N. Thrane, Claus Pade, Oldrich Svec and Nicolas Roussel*

PRO 116: BEFIB 2016 – 9th RILEM International Symposium on Fiber Reinforced Concrete (2016), ISBN: 978-2-35158-187-2, e-ISBN: 978-2-35158-186-5; *Eds. N. Banthia, M. di Prisco and S. Soleimani-Dashtaki*

PRO 117: 3rd International RILEM Conference on Microstructure Related Durability of Cementitious Composites (2016), ISBN: 978-2-35158-188-9, e-ISBN: 978-2-35158-189-6; *Eds. Changwen Miao, Wei Sun, Jiaping Liu, Huisu Chen, Guang Ye and Klaas van Breugel*

PRO 118 (4 volumes): International Conference on Advances in Construction Materials and Systems (2017), ISBN Set: 978-2-35158-190-2, Vol. 1: 978-2-35158-193-3, Vol. 2: 978-2-35158-194-0, Vol. 3: ISBN:978-2-35158-195-7, Vol. 4: ISBN:978-2-35158-196-4, e-ISBN: 978-2-35158-191-9; *Ed. Manu Santhanam*

PRO 119 (online version): ICBBM 2017 - Second International RILEM Conference on Bio-based Building Materials, (2017), e-ISBN: 978-2-35158-192-6; *Ed. Sofiane Amziane*

PRO 120 (2 volumes): EAC-02 - 2nd International RILEM/COST Conference on Early Age Cracking and Serviceability in Cement-based Materials and Structures, (2017), Vol. 1: 978-2-35158-199-5, Vol. 2: 978-2-35158-200-8, Set: 978-2-35158-197-1, e-ISBN: 978-2-35158-198-8; *Eds. Stéphanie Staquet and Dimitrios Aggelis*

PRO 121 (2 volumes): SynerCrete18: Interdisciplinary Approaches for Cement-based Materials and Structural Concrete: Synergizing Expertise and Bridging Scales of Space and Time, (2018), Set: 978-2-35158-202-2, Vol.1: 978-2-35158-211-4, Vol.2: 978-2-35158-212-1, e-ISBN: 978-2-35158-203-9; *Ed. Miguel Azenha, Dirk Schlicke, Farid Benboudjema, Agnieszka Knoppik*

PRO 122: SCC'2018 China - Fourth International Symposium on Design, Performance and Use of Self-Consolidating Concrete, (2018), ISBN: 978-2-35158-204-6, e-ISBN: 978-2-35158-205-3

PRO 123: Final Conference of RILEM TC 253-MCI: Microorganisms-Cementitious Materials Interactions (2018), Set: 978-2-35158-207-7, Vol.1: 978-2-35158-209-1, Vol.2: 978-2-35158-210-7, e-ISBN: 978-2-35158-206-0; *Ed. Alexandra Bertron*

PRO 124 (online version): Fourth International Conference Progress of Recycling in the Built Environment (2018), e-ISBN: 978-2-35158-208-4; *Eds. Isabel M. Martins, Carina Ulsen, Yury Villagran*

PRO 125 (online version): SLD4 - 4th International Conference on Service Life Design for Infrastructures (2018), e-ISBN: 978-2-35158-213-8; *Eds. Guang Ye, Yong Yuan, Claudia Romero Rodriguez, Hongzhi Zhang, Branko Savija*

PRO 126: Workshop on Concrete Modelling and Material Behaviour in honor of Professor Klaas van Breugel (2018), ISBN: 978-2-35158-214-5, e-ISBN: 978-2-35158-215-2; *Ed. Guang Ye*

PRO 127 (online version): CONMOD2018 - Symposium on Concrete Modelling (2018), e-ISBN: 978-2-35158-216-9; *Eds. Erik Schlangen, Geert de Schutter, Branko Savija, Hongzhi Zhang, Claudia Romero Rodriguez*
PRO 128: SMSS2019 - International Conference on Sustainable Materials, Systems and Structures (2019), ISBN: 978-2-35158-217-6, e-ISBN: 978-2-35158-218-3

RILEM Reports (REP)

Report 19: Considerations for Use in Managing the Aging of Nuclear Power Plant Concrete Structures (ISBN: 2-912143-07-1); *Ed. D. J. Naus*
Report 20: Engineering and Transport Properties of the Interfacial Transition Zone in Cementitious Composites (ISBN: 2-912143-08-X); *Eds. M. G. Alexander, G. Arliguie, G. Ballivy, A. Bentur and J. Marchand*
Report 21: Durability of Building Sealants (ISBN: 2-912143-12-8); *Ed. A. T. Wolf*
Report 22: Sustainable Raw Materials - Construction and Demolition Waste (ISBN: 2-912143-17-9); *Eds. C. F. Hendriks and H. S. Pietersen*
Report 23: Self-Compacting Concrete state-of-the-art report (ISBN: 2-912143-23-3); *Eds. Å. Skarendahl and Ö. Petersson*
Report 24: Workability and Rheology of Fresh Concrete: Compendium of Tests (ISBN: 2-912143-32-2); *Eds. P. J. M. Bartos, M. Sonebi and A. K. Tamimi*
Report 25: Early Age Cracking in Cementitious Systems (ISBN: 2-912143-33-0); *Ed. A. Bentur*
Report 26: Towards Sustainable Roofing (Joint Committee CIB/RILEM) (CD 07) (e-ISBN 978-2-912143-65-5); *Eds. Thomas W. Hutchinson and Keith Roberts*
Report 27: Condition Assessment of Roofs (Joint Committee CIB/RILEM) (CD 08) (e-ISBN 978-2-912143-66-2); *Ed. CIB W 83/RILEM TC166-RMS*
Report 28: Final report of RILEM TC 167-COM 'Characterisation of Old Mortars with Respect to Their Repair (ISBN: 978-2-912143-56-3); *Eds. C. Groot, G. Ashall and J. Hughes*
Report 29: Pavement Performance Prediction and Evaluation (PPPE): Interlaboratory Tests (e-ISBN: 2-912143-68-3); *Eds. M. Partl and H. Piber*
Report 30: Final Report of RILEM TC 198-URM 'Use of Recycled Materials' (ISBN: 2-912143-82-9; e-ISBN: 2-912143-69-1); *Eds. Ch. F. Hendriks, G. M. T. Janssen and E. Vázquez*
Report 31: Final Report of RILEM TC 185-ATC 'Advanced testing of cement-based materials during setting and hardening' (ISBN: 2-912143-81-0; e-ISBN: 2-912143-70-5); *Eds. H. W. Reinhardt and C. U. Grosse*
Report 32: Probabilistic Assessment of Existing Structures. A JCSS publication (ISBN 2-912143-24-1); *Ed. D. Diamantidis*
Report 33: State-of-the-Art Report of RILEM Technical Committee TC 184-IFE 'Industrial Floors' (ISBN 2-35158-006-0); *Ed. P. Seidler*

Report 34: Report of RILEM Technical Committee TC 147-FMB 'Fracture mechanics applications to anchorage and bond' Tension of Reinforced Concrete Prisms – Round Robin Analysis and Tests on Bond (e-ISBN 2-912143-91-8); *Eds. L. Elfgren and K. Noghabai*

Report 35: Final Report of RILEM Technical Committee TC 188-CSC 'Casting of Self Compacting Concrete' (ISBN 2-35158-001-X; e-ISBN: 2-912143-98-5); *Eds. Å. Skarendahl and P. Billberg*

Report 36: State-of-the-Art Report of RILEM Technical Committee TC 201-TRC 'Textile Reinforced Concrete' (ISBN 2-912143-99-3); *Ed. W. Brameshuber*

Report 37: State-of-the-Art Report of RILEM Technical Committee TC 192-ECM 'Environment-conscious construction materials and systems' (ISBN: 978-2-35158-053-0); *Eds. N. Kashino, D. Van Gemert and K. Imamoto*

Report 38: State-of-the-Art Report of RILEM Technical Committee TC 205-DSC 'Durability of Self-Compacting Concrete' (ISBN: 978-2-35158-048-6); *Eds. G. De Schutter and K. Audenaert*

Report 39: Final Report of RILEM Technical Committee TC 187-SOC 'Experimental determination of the stress-crack opening curve for concrete in tension' (ISBN 978-2-35158-049-3); *Ed. J. Planas*

Report 40: State-of-the-Art Report of RILEM Technical Committee TC 189-NEC 'Non-Destructive Evaluation of the Penetrability and Thickness of the Concrete Cover' (ISBN 978-2-35158-054-7); *Eds. R. Torrent and L. Fernández Luco*

Report 41: State-of-the-Art Report of RILEM Technical Committee TC 196-ICC 'Internal Curing of Concrete' (ISBN 978-2-35158-009-7); *Eds. K. Kovler and O. M. Jensen*

Report 42: 'Acoustic Emission and Related Non-destructive Evaluation Techniques for Crack Detection and Damage Evaluation in Concrete' - Final Report of RILEM Technical Committee 212-ACD (e-ISBN: 978-2-35158-100-1); *Ed. M. Ohtsu*

Report 45: Repair Mortars for Historic Masonry - State-of-the-Art Report of RILEM Technical Committee TC 203-RHM (e-ISBN: 978-2-35158-163-6); *Eds. Paul Maurenbrecher and Caspar Groot*

Report 46: Surface delamination of concrete industrial floors and other durability related aspects guide - Report of RILEM Technical Committee TC 268-SIF (e-ISBN: 978-2-35158-201-5); *Ed. Valerie Pollet*

Materials and Processing

Fresh and Hardened Properties of 3D Printable Geopolymer Cured in Ambient Temperature

Shin Hau Bong(✉), Behzad Nematollahi, Ali Nazari, Ming Xia, and Jay G. Sanjayan

Centre for Sustainable Infrastructure, Faculty of Science, Engineering and Technology, Swinburne University of Technology, Hawthorn, VIC 3122, Australia
sbong@swin.edu.au

Abstract. This paper reports the fresh and hardened properties of an ambient temperature cured 3D printable geopolymer suitable for extrusion-based 3D concrete printing process. Effects of several key geopolymer synthesis parameters including type of alkaline activator (sodium (Na)-based versus potassium (K)-based), mass ratio of silicate to hydroxide solutions, viscosity and SiO_2/M_2O ratio (where M = Na or K) of silicate solution on extrudability, open time, shape retention ability and compressive strength of the 3D printable geopolymers were investigated. The results revealed that the type of alkaline activator solution and SiO_2/Na_2O ratio of the silicate solution had a significant influence on the open time and shape retention ability of the mixtures. The parameters investigated in this study did not have significant effect on the extrudability of the mixtures. The Na-based activators resulted in higher compressive strength of 3D printed geopolymer than the K-based activators. The 3D printable geopolymer mixture made by 8.0 M NaOH solution (25% w/w) and Na_2SiO_3 solution (75% w/w) with a SiO_2/Na_2O ratio = 2.0 exhibited the highest compressive strength of 16.6 MPa when cured for only 3 days in the ambient temperature.

Keywords: Geopolymer · 3D concrete printing · Extrusion · Alkaline activator Compressive strength · Ambient temperature curing

1 Introduction

Additive Manufacturing (AM), also known as three-dimensional (3D) printing, is a manufacturing process that fabricates complex 3D objects directly from a digital 3D model in a layer-by-layer deposition approach. Although the AM has been widely adopted in many industries such as aerospace, automotive and medicine [1], the construction industry has been rather slow in its adopting. The application of AM in the concrete construction industry may offer several advantages over conventional approach of casting concrete into a formwork, including enhanced geometrical freedom and reduction in construction cost and time [2].

T. Wangler and R. J. Flatt (Eds.): DC 2018, RILEM Bookseries 19, pp. 3–11, 2019.
https://doi.org/10.1007/978-3-319-99519-9_1

One of the main challenges of the current 3D concrete printing (3DCP) technology is the very limited scope of printable concretes. To tackle this limitation, the researchers at Swinburne University of Technology, Australia have been working on developing innovative methodologies for formulating geopolymers for 3DCP for construction applications. Geopolymer is a sustainable alternative binder to ordinary Portland cement (OPC). It is made by alkaline activation of fly ash and/or slag, being industrial by-products of coal power stations and iron manufacture, respectively [3]. Different geopolymer-based formulations have recently been developed by the authors of this study for the requirements and demands of commercially available powder-based 3D printers [4, 5]. Recently, the researchers at Nanyang Technological University, Singapore have reported the properties of a 3D printable fly ash and slag blended geopolymer mortar. However, the selection of material ingredients and mixture proportions were based on trial-and-error [6]. In other words, no study has been undertaken to investigate the effect of material ingredients and mixture proportions on the properties of the 3D printed geopolymer. Therefore, the authors of this study has recently conducted a comprehensive study to investigate the effect of several parameters such as type of activator, type of sodium silicate solution, mass ratio of sodium silicate to sodium hydroxide solutions, and activator to fly ash ratio on the fresh and hardened properties of a 3D printable fly ash based geopolymer for extrusion-based 3DCP [7]. Nevertheless, the developed 3D printable fly ash based geopolymer requires heat curing (24 h at 60 °C) which may limit its in-situ application.

To tackle this limitation, this study aims to develop an ambient temperature cured 3D printable geopolymer to expand the application of this ecologically friendly material in 3DCP. The effect of several parameters such as type of alkaline activators (sodium (Na)-based versus potassium (K)-based), mass ratio of the silicate solution to hydroxide solution, SiO_2/M_2O ratio (where M = Na or K) of the silicate solutions and viscosity of the silicate solutions on the fresh and hardened properties of the ambient temperature cured 3D printable geopolymers were investigated.

2 Materials and Methods

2.1 Materials and Mixture Proportions

Low calcium (Class F) fly ash and granulated ground blast furnace slag (slag) were supplied by Gladstone power station located in Queensland, Australia and Building Products Supplies Pty Ltd., Australia, respectively. Two types of sands were used in this study. The fine sand supplied by TGS Industrial Sands Pty Ltd., Australia had an average particle size of 172 μm. The relatively coarser sand supplied by Sibelco Australia Ltd. had an average particle size of 898 μm.

Three types of liquid alkaline activators including two Na-based and one K-based were used in this study. Two sodium silicate (Na_2SiO_3) solutions with two different SiO_2/Na_2O mass ratios and one potassium silicate (K_2OSi_3) solution with SiO_2/K_2O mass ratio of 2.02 supplied by PQ Australia Pty Ltd., Australia were used as the silicate solutions. Table 1 presents the viscosity and SiO_2/M_2O ratio (M = Na or K) of the silicate solutions. 8.0 M of sodium/potassium hydroxide solutions were prepared by dissolving sodium hydroxide (NaOH) and potassium hydroxide (KOH) pellets in tap water, respectively. The NaOH and KOH pellets were supplied by Sigma Aldrich Pty Ltd., Australia and Merck Pty Ltd., Australia, respectively.

Table 1. The viscosity and SiO_2/M_2O ratio (M = Na or K) of the silicate solutions.

Grade of Silicate Solution	SiO_2/M_2O ratio (M = Na or K)	Viscosity, cps @ 20 °C
D Grade Na_2SiO_3	2.00	250–450
N Grade Na_2SiO_3	3.22	100–300
KASIL 2040 K_2SiO_3	2.02	100–300

Table 2 presents the mixture proportions of the ambient temperature cured 3D printable geopolymers invedstigated in this study. Two silicate solution to hydroxide solution ratios of 1.5 and 3.0 were used in this study. In all mixtures a combination of fly ash and slag was used as the binder with fly ash/slag ratio of 3.0. A constant activator solution to binder ratio of 0.4 was used in all mixtures. Sodium carboxymethyl cellulose (CMC) powder supplied by DKS Co. Ltd, Japan was used as a viscosity modifying agent (VMA) in this study to improve the rheological properties of the mixtures. Anhydrous borax was used as a retarder to prevent setting of the mixture during mixing and printing processes. The dosages of these admixtures in each mixture were adjusted accordingly to achieve similar rheology, as different types and ratios of alkaline activators have different viscosities and setting time.

2.2 3D Printing Apparatus

A small scale custom-made 3D printing apparatus was used to simulate the extrusion-based 3D printing process (see Fig. 1(a)). A 3D printed 45° angle nozzle (see Fig. 1(b)) with an opening of 15 mm × 30 mm was attached to the end of the extruder.

2.3 Printing of 3D Printable Geopolymer

The fresh geopolymer mortar was filled into the 3D printing apparatus and extruded through the nozzle by moving the extrusion unit along the longitudinal direction at a constant speed. The printed filaments were left in the laboratory environment for ambient temperature curing until the day of testing. For each mixture, 9 prismatic specimens with the dimension of 60 mm (L) × 30 mm (W) × 15 mm (H) were cut from the printed filaments one day before testing.

2.4 Characterization Methods

The extrudability of fresh 3D printable geopolymer was evaluated by extruding four 250 mm long single-layer filaments through the nozzle. A mixture was evaluated as good extrudability if four 250 mm long single-layer filaments were extruded successfully without any blockage or separation.

Table 2. The mixture proportions of the ambient temperature cured 3D printable geopolymers.

Mix ID	Binder[a]	Activator	Sand		Retarder[h]	VMA[i]
			Coarse	Fine		
Na-D15	1.0	0.4[b]	1.0	0.5	0.002	0.015
Na-D30	1.0	0.4[c]	1.0	0.5	0.002	0.020
Na-N15	1.0	0.4[d]	1.0	0.5	0.002	0.011
Na-N30	1.0	0.4[e]	1.0	0.5	0.002	0.010
K-15	1.0	0.4[f]	1.0	0.5	0.002	0.010
K-30	1.0	0.4[g]	1.0	0.5	0.005	0.007

Note: All numbers are mass ratios of the binder weight.
[a]Mixture of fly ash and slag with fly ash/slag ratio of 3.0.
[b]Mixture of 8.0 M NaOH and D Grade Na_2SiO_3 solutions with Na_2SiO_3/NaOH mass ratio of 1.5.
[c]Mixture of 8.0 M NaOH and D Grade Na_2SiO_3 solutions with Na_2SiO_3/NaOH mass ratio of 3.0.
[d]Mixture of 8.0 M NaOH and N Grade Na_2SiO_3 solutions with Na_2SiO_3/NaOH mass ratio of 1.5.
[e]Mixture of 8.0 M NaOH and N Grade Na_2SiO_3 solutions with Na_2SiO_3/NaOH mass ratio of 3.0.
[f]Mixture of 8.0 M KOH and KASIL 2040 K_2SiO_3 solutions with K_2SiO_3/KOH mass ratio of 1.5.
[g]Mixture of 8.0 M KOH and KASIL 2040 K_2SiO_3 solutions with K_2SiO_3/KOH mass ratio of 3.0.
[h]Anhydrous borax.
[i]Sodium carboxymethyl cellulose.

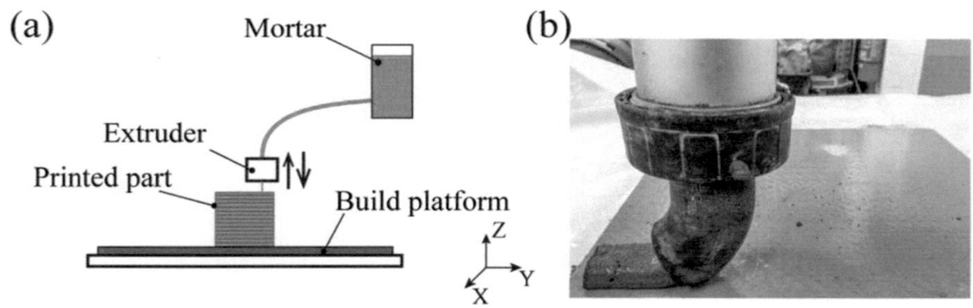

Fig. 1. (a) Schematic illustration of the extrusion-based 3D printing process, (b) 3D printed 45° angle nozzle with an opening of 15 mm × 30 mm.

The open time of each mixture was defined as the period from the time of addition of the liquid alkaline activator to the mixture until the material showed poor extrudability [8]. 100 mm long filaments were extruded with time interval of 1 min to investigate the open time of each mixture. The end of open time was determined once blockage of the material or separation of the filament occurs.

The shape retention ability of each mixture was quantified using a shape retention ratio (SRR) calculated according to Eq. (1).

$$SRR = \frac{W_{Top}}{W_{Bottom}} \tag{1}$$

where W_{Top} and W_{Bottom} are the top and bottom widths of the printed prismatic specimen, respectively measured using a digital Vernier caliper with an accuracy of up to 0.01 mm.

Compression test was conducted to determine the compressive strength of printed specimens. All specimens were tested in longitudinal direction (i.e. the extrusion direction) by using a universal testing machine under a loading rate of 0.33 MPa/s.

3 Results and Discussions

3.1 Extrudability

A nozzle with an opening of 15 mm × 30 mm was used to evaluate the extrudability of fresh geopolymer mortars. Four 250 mm long single-layer filaments were extruded from each batch of fresh mortar prepared as shown in Fig. 2. From the test results obtained, all types of mixtures exhibited good extrudability where no blockage or separation was observed during the printing process. This indicates that the extrudability of the mixtures is not affected by the parameters studied in this paper.

Fig. 2. The filament was extruded to evaluate the extrudability.

3.2 Open Time

The measured open times of the fresh 3D printable geopolymer mortars are tabulated in Table 3. As can be seen from the table, mixtures with K-based activators had shorter open time as compared to the ones with Na-based activator. According to Le et al. [9], the open time of cementitious materials has some correlation with its setting time. In addition, previous study has reported that geopolymer activated with K-based activator have faster setting time than that with Na-based activator [10]. Therefore, the shorter open time of mixture with K-based activator indicates that the setting time of the mixtures is faster than that of the Na-based mixtures. Besides that, it was also found that the open times of "Na-D15" and "Na-D30" mixtures were almost double as compared to other mixtures. This is most likely due to the lower SiO_2/Na_2O ratio of D Grade Na_2SiO_3 solution which results in lower rate of geopolymerisation reaction and longer setting time [11, 12]. Although "Na-30" mixture contained the highest dosage of CMC, authors noticed that the extruded filaments failed to retain its shape when the time was less than 18 min as shown in Fig. 3. This is mostly due to the slower geopolymerisation reaction in the system as a result of higher silicate to hydroxide solutions ratio [11].

Table 3. Open time of 3D printable geopolymers.

Mix ID	Open time (min)
Na-D15	26
Na-D30	29
Na-N15	17
Na-N30	18
K-15	16
K-30	15

3.3 Shape Retention Ability

The SRR of mixtures with different types and ratios of alkaline activators are shown in Fig. 4. As shown in the figure, the variations of SRR among the mixtures were very small as the values were in the range of 0.89–0.94, which can be considered as consistent. By comparing the dosages of CMC and grades of Na_2SiO_3 solution used in different mixtures (Table 2), it was obvious that the mixtures containing D Grade Na_2SiO_3 solution, regardless of silicate to hydroxide solutions ratio, required the higher amount of CMC to obtain similar SSR with the one with N Grade Na_2SiO_3 solution. It was also noted that the mixtures with K-based alkaline activators exhibited better shape retention ability (less CMC was added) than the one with N Grade Na_2SiO_3 solution, despite both silicate solutions had identical viscosity. This shows that the rate of geopolymerisation reaction not only influences the open time but also affecting the shape retention ability of the 3D printable geopolymer mortars. Moreover, it was found that more CMC was added to the mixture when silicate to hydroxide solutions ratio increased, irrespective of type of activator used. This was expected as hydroxide solution has lower viscosity than silicate solution. However, opposite trend was found in the mixtures with D Grade Na_2SiO_3 solution.

3.4 Compressive Strength

The compression test results of printed prism specimens after 3 days of ambient temperature curing are shown in Fig. 5. The compressive strengths of specimens were in the range of 8.5–16.6 MPa. By comparing the specimens between different types of activator, it was found that the compressive strengths of the printed specimens activated

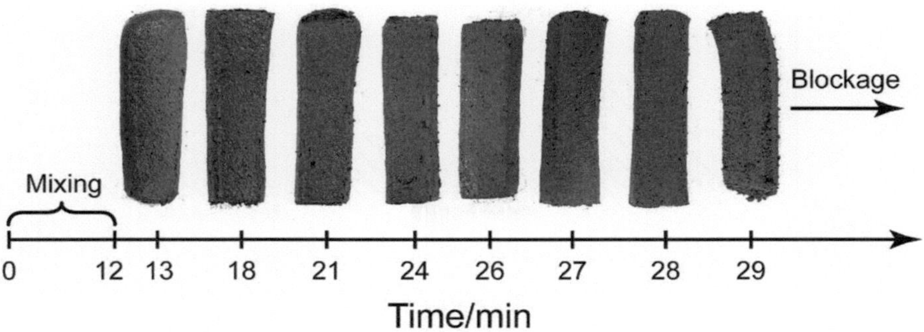

Fig. 3. Filaments extruded from "Na-D30" mixture at different rest times.

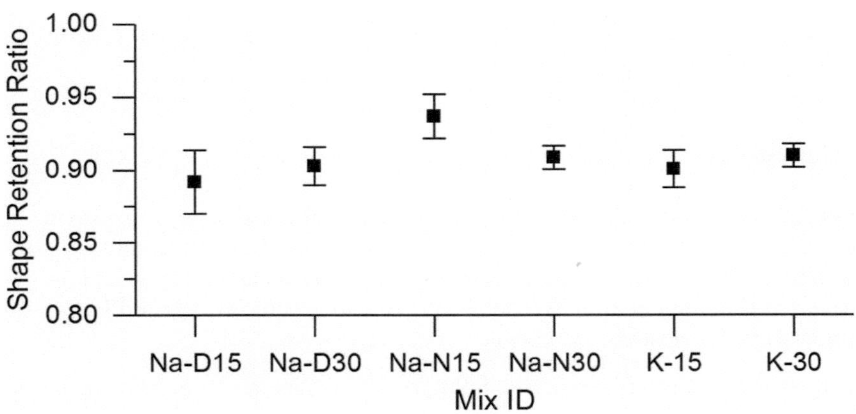

Fig. 4. Shape retention ratio of 3D printable geopolymer.

by Na-based activators were higher than K-based ones. A similar result was also obtained by Palomo et al. [13] for cast geopolymer specimens. As can be seen in Fig. 5, the increase in silicate to hydroxide solutions ratio from 1.5 to 3.0, regardless type of activator, resulted in the increase of the compressive strength of specimens. This is because silicate solution provides additional soluble silicate into the geopolymeric system which promotes the geopolymerisation reaction and results in increasing the compressive strength [12, 14]. However, a decreasing trend in compressive strength of specimens with Na-based activators was observed when silicate solution with higher SiO_2/Na_2O ratio was used. This can be well explained as previous study showed that the compressive strength of geopolymer increased when the silicon content increased, but further increase of silicon content in the system either led to a small raise of compressive strength or, in some cases, a decrease in compressive strength [14, 15].

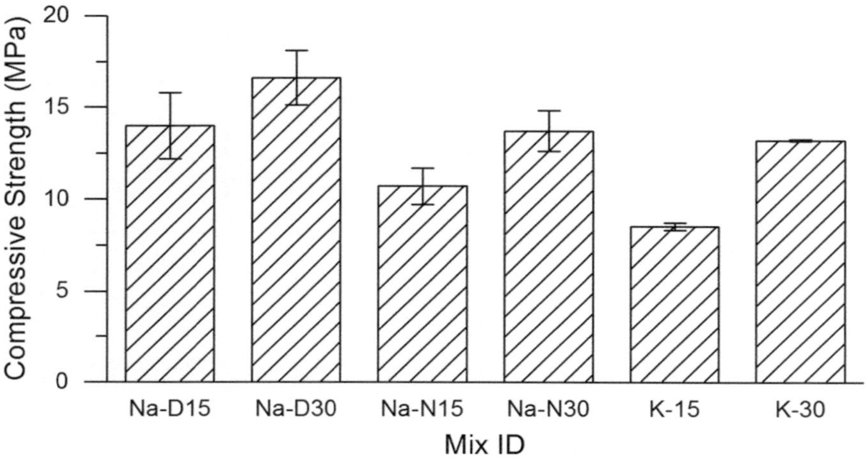

Fig. 5. Compressive strength of 3D printed geopolymers.

4 Conclusion

The effects of type of alkaline activator, mass ratio of the silicate to hydroxide solutions, viscosity and SiO_2/M_2O ratio (M = Na or K) of the silicate solutions on the extrudability, open time, shape retention ability and compressive strengths of the ambient temperature cured 3D printable geopolymers were investigated in this study. The following conclusions are drawn:

1. Among the parameters investigated, the type of alkaline activator solution and SiO_2/Na_2O ratio of the silicate solution were the most influential parameters in regards to the open time and shape retention ability of the 3D printable geopolymers. This is because these two parameters govern the rate of geopolymerisation reaction and consequently the setting time.
2. The compressive strength of the 3D printed geopolymers synthesised with the Na-based activators was generally higher than the K-based activators. The highest compressive strength of the specimens with the Na-based activators was 16.6 MPa after 3-days of ambient temperature curing.
3. The parameters studied in this paper had no significant influence on the extrudability of the 3D printable geopolymers.

Future work will focus on optimising geopolymer mixture formulation and investigating the effects of testing directions on the mechanical properties of the optimum 3D printed geopolymer mix.

References

1. Gao, W., Zhang, Y., Ramanujan, D., Ramani, K., Chen, Y., Williams, C.B., Wang, C.C., Shin, Y.C., Zhang, S., Zavattieri, P.D.: The status, challenges, and future of additive manufacturing in engineering. Comput. Aided Des. **69**, 65–89 (2015)
2. Nematollahi, B., Xia, M., Sanjayan, J.: Current progress of 3D concrete printing technologies. In: 34th International Symposium on Automation and Robotics in Construction (ISARC 2017), Taiwan, pp. 260–267 (2017)
3. Nematollahi, B., Sanjayan, J., Shaikh, F.U.A.: Synthesis of heat and ambient cured one-part geopolymer mixes with different grades of sodium silicate. Ceram. Int. **41**, 5696–5704 (2015)
4. Xia, M., Nematollahi, B., Sanjayan, J.: Printability, accuracy and strength of fly ash/slag geopolymer made using powder-based 3D printing for construction applications. Autom. Constr.
5. Xia, M., Nematollahi, B., Sanjayan, J.: Influence of binder saturation level on compressive strength and dimensional accuracy of powder-based 3D printed geopolymer. In: Proceedings of the 2nd International Conference on Advanced Manufacturing and Materials (ICAMM 2018), Tokyo (2018)
6. Panda, B., Paul, S.C., Hui, L.J., Tay, Y.W.D., Tan, M.J.: Additive manufacturing of geopolymer for sustainable built environment. J. Clean. Prod. **167**, 281–288 (2017)
7. Nematollahi, B., Xia, M., Sanjayan, J.: Effect of type of fiber on inter-layer bond and flexural strengths of extrusion-based 3D printed geopolymer. In: Proceedings of the 2nd International Conference on Advanced Manufacturing and Materials (ICAMM), Tokyo (2018)
8. Paul, S.C., Tay, Y.W.D., Panda, B., Tan, M.J.: Fresh and hardened properties of 3D printable cementitious materials for building and construction. Arch. Civ. Mech. Eng. **18**, 311–319 (2018)
9. Le, T.T., Austin, S.A., Lim, S., Buswell, R.A., Gibb, A.G., Thorpe, T.: Mix design and fresh properties for high-performance printing concrete. Mater. Struct. **45**, 1221–1232 (2012)
10. Xu, H., Van Deventer, J., Lukey, G.: Effect of alkali metals on the preferential geopolymerization of stilbite/kaolinite mixtures. Ind. Eng. Chem. Res. **1**(40), 3749–3756 (2001)
11. Panias, D., Giannopoulou, I.P., Perraki, T.: Effect of synthesis parameters on the mechanical properties of fly ash-based geopolymers. Colloids Surf. A Physicochem. Eng. Asp. **301**, 246–254 (2007)
12. Nadoushan, M.J., Ramezanianpour, A.A.: The effect of type and concentration of activators on flowability and compressive strength of natural pozzolan and slag-based geopolymers. Constr. Build. Mater. **111**, 337–347 (2016)
13. Palomo, A., Grutzeck, M., Blanco, M.: Alkali-activated fly ashes: a cement for the future. Cem. Concr. Res. **29**, 1323–1329 (1999)
14. Komljenović, M., Baščarević, Z., Bradić, V.: Mechanical and microstructural properties of alkali-activated fly ash geopolymers. J. Hazard. Mater. **181**, 35–42 (2010)
15. Law, D.W., Adam, A.A., Molyneaux, T.K., Patnaikuni, I., Wardhono, A.: Long term durability properties of class F fly ash geopolymer concrete. Mater. Struct. **48**, 721–731 (2015)

Evolution of Concrete/Formwork Interface in Slipforming Process

T. Craipeau[1,2(✉)], T. Lecompte[1], F. Toussaint[2], and A. Perrot[1]

[1] Institut de Recherche Dupuy de Lôme, Centre de Recherche Christian Huygens, Lorient, France
typhanie.craipeau@lafargeholcim.com
[2] LafargeHolcim R&D Center, Saint Quentin Fallavier, France

Abstract. In a slipforming process, concrete is continuously poured and the formwork simultaneously raised so that the older concrete at the bottom supports the fresher at the top after a few hours of hydration. Such a complex process has to be industrially optimized to ensure the quality surface of concrete. The objective of the present work is to study the evolution of the concrete/formwork interface at very early age.

An experimental device has been designed in order to measure the evolution of the friction at this interface during the first hydration period. With cement hydration and thus water consumption, pore water pressure decreases and creates suction which could increase granular stress on the formwork. The role of pore water pressure on formwork friction is investigated. The first hydrates may also bond to the surface and create a strong adhesion at the same time.

Keywords: Interface · Friction · Slipforming

1 Introduction

Concrete can be formed using very different processes. Their common denominator is that the fresh and early age concrete properties have to be optimized regarding the way of forming. For instance, in order to be pumped and then cast inside a formwork, the concrete must be fluid enough to flow in the pipe but firm enough to resist bleeding or segregation.

An interesting process is the slipforming, where concrete is poured and the formwork raised continuously during hardening. "Smart Dynamic Casting" [1] is a 3D digital evolution of the slipforming process. The problem of optimizing concrete properties is very similar in both processes. Many questions, concerning fresh concrete, are linked with the interface conditions between the concrete and the moving formwork.

In a slipforming process, the formwork is in contact with concrete at different states of hydration. Concrete behavior changes drastically at the beginning of cement hydration. This technique, see Fig. 1, is a widely used construction methodology for high rise structures such as skyscrapers, pylons, silos or marine foundations that benefit from its high construction speed and the absence of cold joint that may affect the durability of the structure.

T. Wangler and R. J. Flatt (Eds.): DC 2018, RILEM Bookseries 19, pp. 12–23, 2019.
https://doi.org/10.1007/978-3-319-99519-9_2

Fig. 1. Slipforming illustration

The concrete is directly in contact with the formwork from its fresh state to a set state without any demolding oil. During sliding, possible micro-cracking/lump formation can happen on the wall surface. These defects are most probably related to the concrete adhesion on the formwork [2]. These issues should be reduced and prevented to avoid a reduction of the durability of the structures that could turn into strong damages if not properly treated.

The aim of this study is to improve our knowledge of the physical phenomena which govern friction and adhesion between concrete and various supports. The particularity of slipforming technique is that the walls are composed of a material exhibiting a gradient of hydration state with concrete shifting from fresh state to early age (Fig. 2). Usually, the formwork is about 1.2 m length, the concrete has to be set at the bottom of the formwork. The average speed is around 15–25 cm per hour. It has been shown that water consumption caused by cement hydration induces capillary suction which is assumed to play a major role on the interface [3]. It first induces pore water depression that increases granular stresses as the material behavior may become frictional with hydration [4]. Moreover, hydrates formation is expected to create a physical bond with the surface (Fig. 3). In order to appreciate the overall behavior, we attempt to evaluate the contribution of each phenomenon separately. An experimental device was built in order to measure the evolution of the friction behaviour of the concrete during the first hydration period. It is described later in this paper. Thereby, a multi-scale approach linking interfacial phenomena and cement hydration is proposed.

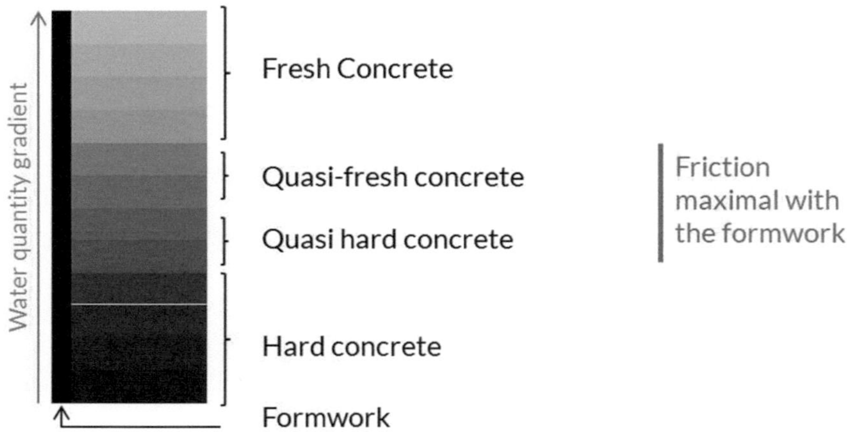

Fig. 2. Concrete evolution inside a slipform formwork

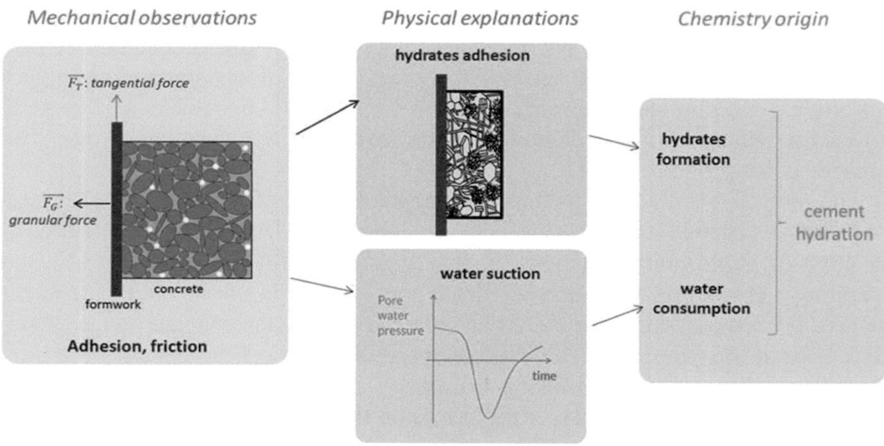

Fig. 3. Summary of granular forces origin

2 Interfacial Behavior of Fresh Concrete: Physical Phenomena

2.1 Chemistry Origin and Hydration Consequences

The origin of adhesion is the cement hydration. It firstly induces water consumption thus suction, (because of Le Chatelier contraction, see below) and hydrates formation which may bond to the surface as well.

The main hydration reactions of the Portland cement from the most reactive hydrate to the less one are presented below:

$3CaO.Al_2O_3(C_3A) + 3CaSO_4.2H_2O + H_2O \rightarrow C_3A.3CaSO_4.32H_2O$ (ettringite)
$3CaO.SiO_2(C_3S$ alite$) + 5.3\,H_2O \rightarrow C_{1.7}SH_4$ (CSH) $+ 1.3\,Ca(OH)_2$ (Portlandite CH)
$2CaO.SiO_2$ (C_2S belite) $+ 5.3\,H_2O \rightarrow C_{1.7}SH_4$ (CSH) $+ 0.3\,Ca(OH)_2$ (Portlandite CH)
$4CaO.Al_2O_3Fe_2O_3$ (C_4AF) reaction (with gypse $CaSO_4$) is less reactive.

Those equations are important to keep in mind for the following results.

The sum of the initial volumes of water and cement is larger than the volume of the formed hydrates; this is called Le Chatelier contraction or selfdessication. Before Vicat setting time, the material can be considered as plastic matter: with contraction and hydrates formation, the grains will come closer together.

When the sample becomes harder, there is a percolation of the grains and the air will occupy the difference of water volume. Thus, the pressure of water will be in depressure compared to the atmospheric pressure. The capillary desiccation induces material shrinkage called as well endogenous shrinkage [5].

2.2 Pore Water Pressure and Granular Stress (Terzaghi Equation)

Thanks to Terzaghi equation, suction can be related to the normal force. Terzaghi [8] describes that the total stress, σ (kPa) is equal to the sum of the effective stress σ′ (granular stress, kPa) and the pore water pressure, uw (kPa):

$$\sigma = \sigma' + u_w \quad (1)$$

The total stress of the concrete is considered in this study almost constant. Therefore, the effective pressure of the concrete only varies with concrete pore pressure. Terzaghi equation is illustrated in Fig. 4. At constant total constant, negative pore water pressure increases the stress σ′ acting on the granular skeleton (which is the case in the early stage of hydration).

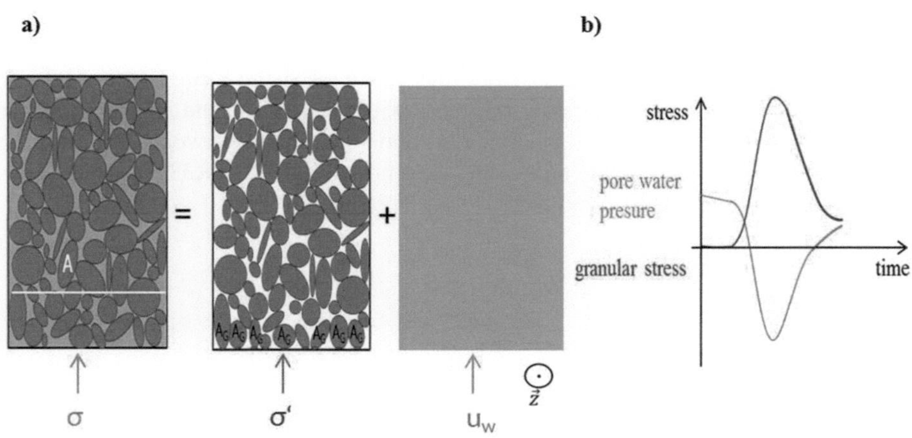

Fig. 4. Terzaghi equation illustration

3 Experimental Procedures

3.1 Materials and Mixing Procedure

The aim of the study is to understand the physical phenomena involved in the lifting resistance exerted by the concrete during the slipforming; therefore we chose usual materials: a Portland cement (CEM I 52.5 N PM-ES CP2 NF), with 37v% of a fine sand (PE2LS 0/0.315 μm) to help the mixing (also to reduce heating during mixing). A low Water/Cement ratio (here 0.3) is used to reduce strongly the pore water pressure. In order to obtain a low W/C ratio and a good rheology, a common polycarboxylate plasticiser admixture was used: Glenium 27 from BASF (0.063 wt%/cement).

In the fresh state, the mortar is fluid (slump ASTM is 260 mm), and its air volume fraction is 2.7%.

The designed mortar was prepared with a Perrier mixer with the same mixing protocol in order to ensure tests reproducibility. The sand was first mixed with part of the water at low speed (140 rpm) during 1 min. Then, after a waiting period of 4 min, cement is introduced and mixed at low speed during 1 min. The rest of water is introduced during 30 s at low speed. Finally, the material is mixed at high speed (180 rpm) during 2 min.

3.2 Device and Procedure

A specific device was developed. It consists of a parallelepiped box, filled with mortar. One face of the box, attached on a traction machine, can be moved and is made of the material that represents the formwork interface. A load cell permits to measure the global interfacial shear load (Fig. 5). It is also fitted with pore pressure sensors and force transducers to estimate those critical physical parameters. It allows to improve the understanding of the impact of suction on granular/interfacial friction and to assess its relative influence on the total stress. The advantage of this apparatus is to benefit from a "long run" of 400 mm enabling to approach slipforming kinematics (frequency and amplitude of lifts). For each lifting step, the lifting speed is 1 $mm \cdot s^{-1}$ for a move of 5 mm. The time at rest is 565 s between each step. The average lifting speed is then 0.5 $mm \cdot min^{-1}$. There are 67 lifting: the total test lasts around 11 h.

Pore water pressure is measured along the formwork using a pressure transducer device: a porous membrane is directly in contact with the concrete and is connected to a saturated chamber which allows obtaining a continuous water network to a pressure transducer [6] (Fig. 6). A relative pressure transmitter (Keller® PR33X) allows measuring and recording positive and negative pressure up to 1 bar.

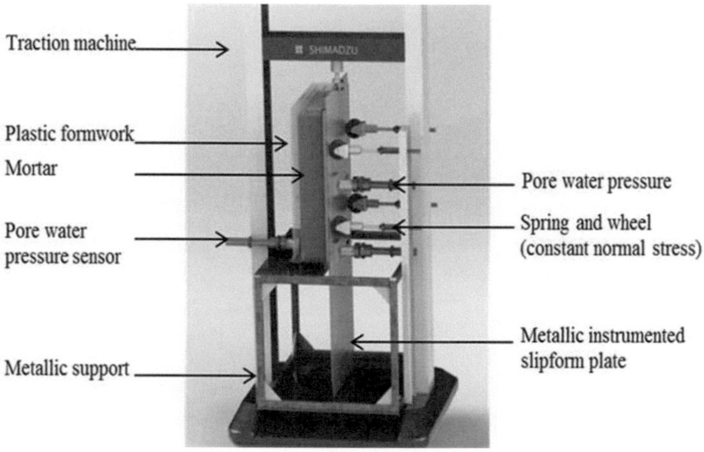

Fig. 5. Vertical shear stress device developed to study slipforming operations.

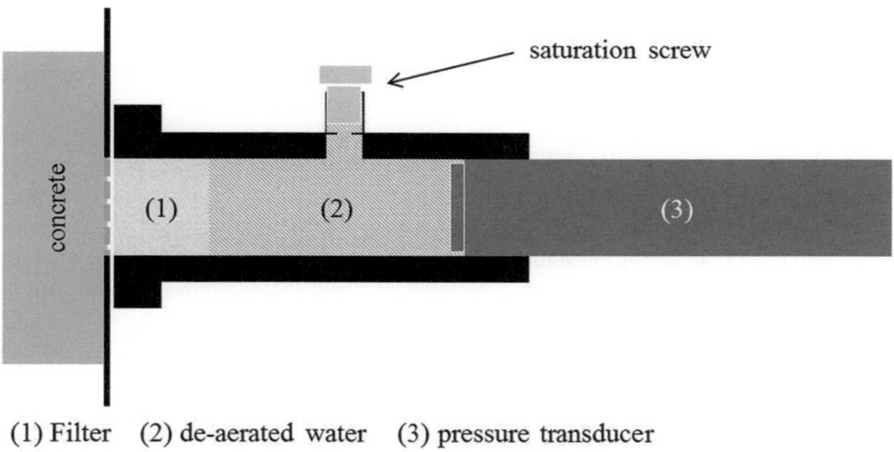

Fig. 6. Pore water pressure measurement system

4 Experimental Results and Discussions

4.1 Cement Hydration Characterization

Pore water pressure (u_w), free water quantity evolution (determined with NMR relaxometry) and the volume variations (measured by hydrostatic weighting) are plotted versus the degree of hydration on Fig. 7. The degree of hydration is calculated from calorimetry measurements: it is defined as the ratio between the cumulated heat flow (which is the surfaces under the curve of the heat flow evolution over time) and the total theoretical enthalpy of the hydrates.

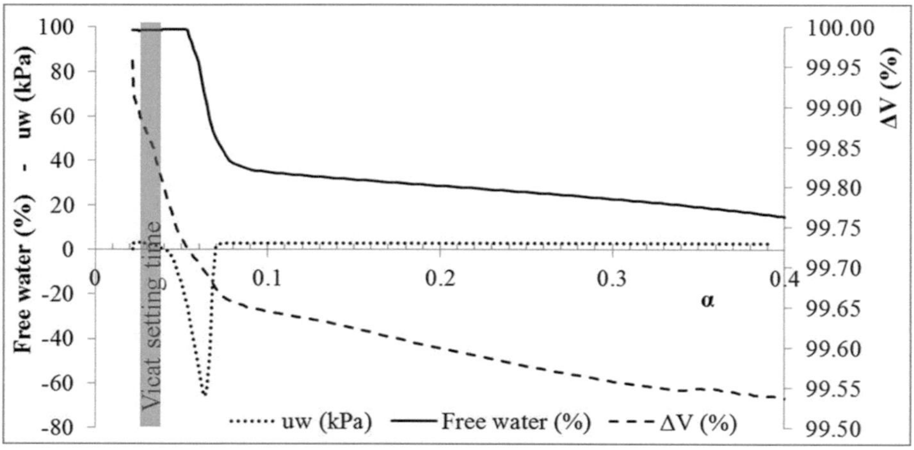

Fig. 7. Evolution of free water content, pore water pressure and volume variation during early hydration of the studied mortar.

Pore water pressure (dotted curve) is decreasing until the air network becomes continuous and connected with ambient atmospheric pressure, if an air drop comes in contact with the sensor the pressure comes back to positive values.

The volume (dashed curve) decreases faster when pore water pressure is decreasing. After that, it continues to decrease but the slope is lower. Free water quantity (continuous line)-obtained thanks to RMN relaxometry- is constant at the beginning and then water is consumed faster [7]. This means that a small amount of consumed water will induce a strong structuration of the material [4]. Thus, we see that the evolution of the three parameters is correlated and can be separated in two periods. Below $\alpha = 0.07$ the hydration and water consumption is low. Above $\alpha = 0.07$, CSH precipitation begins significantly. The system becomes rigid, pore water pressure and the quantity of water drop suddenly. We can see that pore water pressure, free water quantity, volume variations are strongly linked to hydration degree of cement.

4.2 Tangential Stress

The measured shear stress as a function of the lifting height is plotted (Fig. 8) between 100 and 300 mm of lifting.

The behaviour of the mortar is changing over hydration time. In the first period, it has a viscoplastic behaviour: the shear stress decreases during the rest period indicating material relaxation. Then after a transition phase, its behaviour changes to an elastic solid state: between two liftings the shear stress increases.

Mortar yield stress and the interface tangential stress are plotted on the same graph to evaluate the influence of the structural build-up of the mortar (i.e. evolution of the yield stress before setting time in dotted line) on the tangential stress (continuous line) (Fig. 9). The tangential stress evolution is obtained with our experimental device (Sect. 3.2). The yield stress is measured thanks to a Vane test.

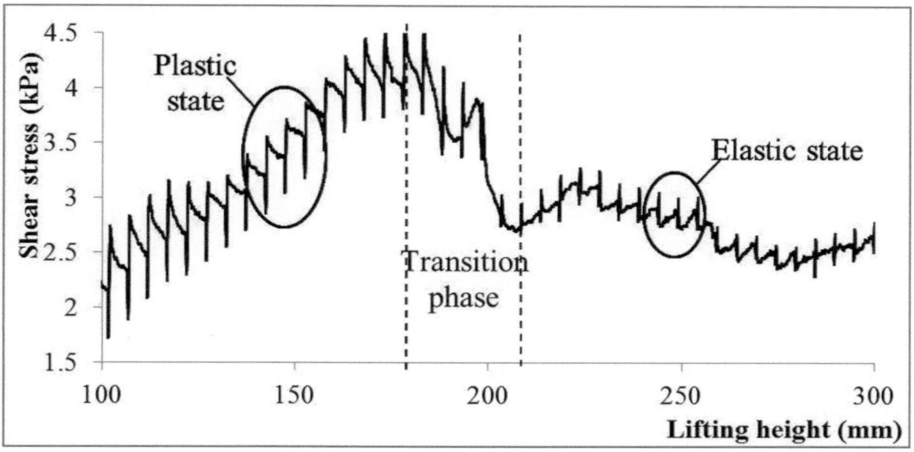

Fig. 8. Mortar behavior evolution illustration: change in behavior type

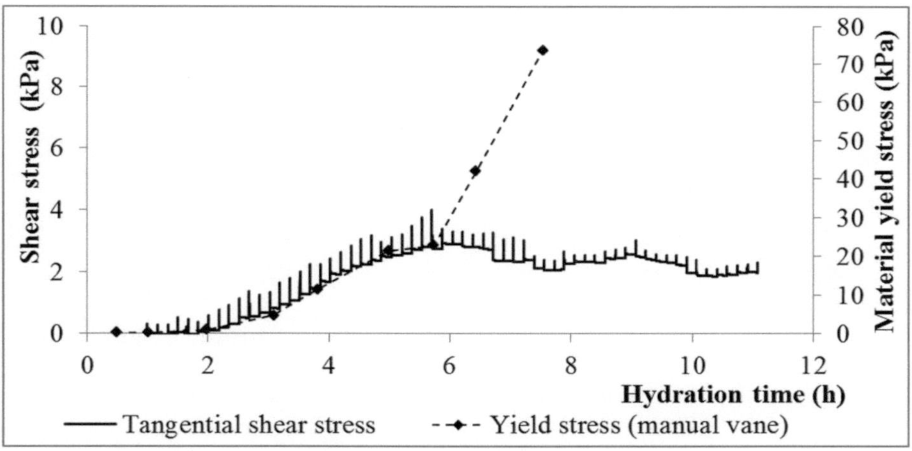

Fig. 9. Evolution of mortar yield stress evolution and tangential stress evolution measured during lifting operation using the developed device

Figure 9 shows that there is a scaling between mortar yield stress and the average interfacial tangential stress with a factor 8 (comparing the y axis scales) until 6 h of hydration time which is 1 h after the end of Vicat setting time. We can observe that after 6 h the yied stress increases drastically whereas tangential decreases a little. This indicates that after 6 h solid-solid friction is measured.

4.3 Friction Law

The lifting operation of slipforming is carried out with mortar running from fresh state where it presents a viscoplastic fluid-like behavior to a set state where it has reached an almost elastic solid-type behavior. It can be noted that between those two extreme states, the mortar exhibits frictional behavior. The lifting force must overcome the adhesion and the friction on the metallic panel before the sliding starts. This force peak is called static lifting stress. After the first movement, the force required to lift the plate is decreasing as adhesion is cut off and only friction should be overcome (Fig. 10). The minimum friction that occurs during sliding is called sliding lifting stress. The difference between the static lifting force and the sliding lifting force will be called adhesion and the sliding lifting force resulting is only friction [8].

During one cycle (Fig. 10), it seems that the formwork movement induces a shear stress reduction.

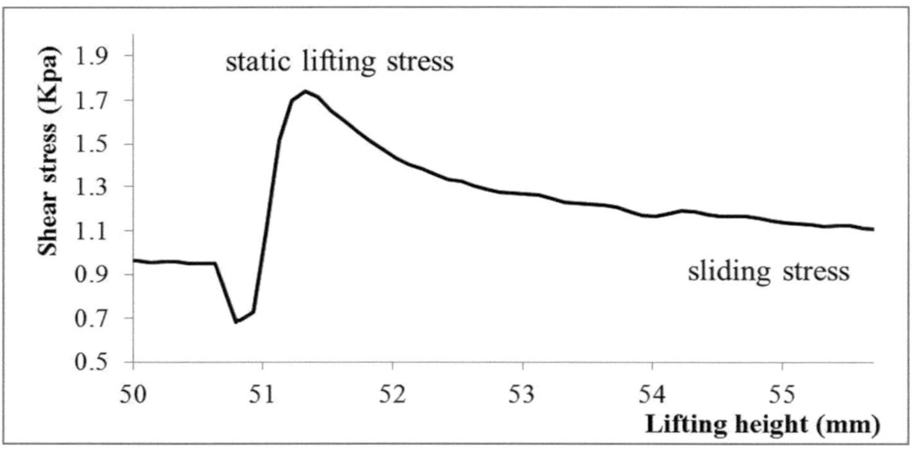

Fig. 10. Illustration of static lifting stress and sliding stress

It has been shown in [8] that the lifting force is proportional to the effective stress of the concrete on the panels. The relation is described by the adhesion and friction laws:

$$\mu_{static} = \frac{F_{Static}}{\sigma'} \tag{2}$$

$$\mu_{sliding} = \frac{F_{sliding}}{\sigma'} \tag{3}$$

With: F_{static} = static friction stress [kPa]; $F_{sliding}$ = sliding friction stress [kPa]; μ_{static} = static friction coefficient [–]; $\mu_{sliding}$ = sliding friction coefficient [–]; σ' = effective stress [kPa].

The effective stress represents the stress between the solid particles and the slipform panels.

4.4 Hypothesis of Mechanisms

The evolution of pore water pressure and tangential shear stress obtained with the design device are plotted in Fig. 10. We observe that the order of magnitude of the pore water pressure (dotted line) is really higher than the tangential shear stress (continuous line). For a first interpretation of the results, to see if there is a correlation between those two distinct measurements, we zoomed between 0 and 8 h of hydration and we plotted the opposite of the addition of pore water pressure and hydrostatic pressure of the mortar (Fig. 11).

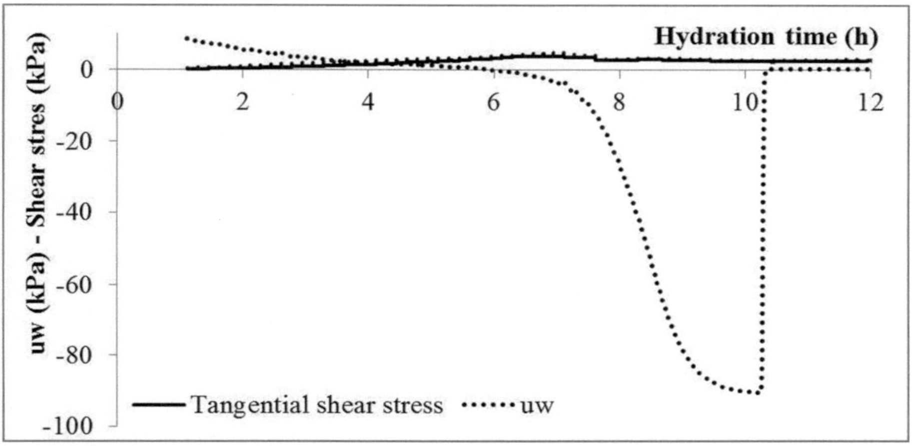

Fig. 11. Pore water pressure and shear stress evolution measured with the experimental device.

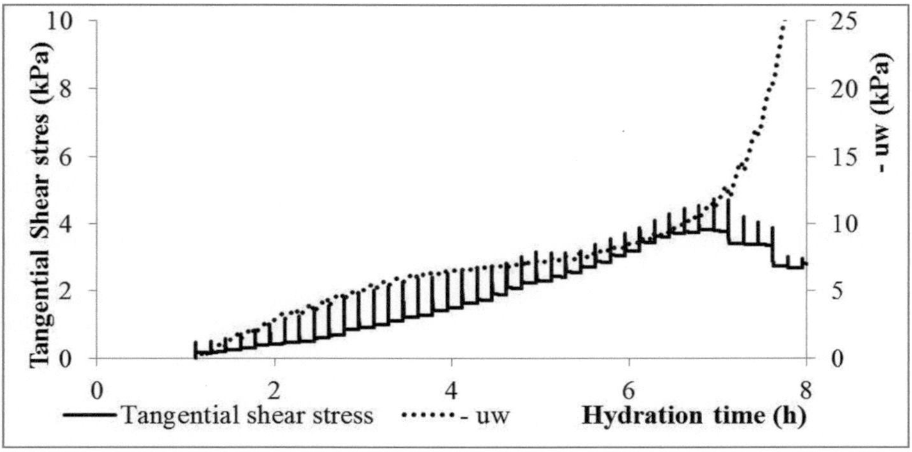

Fig. 12. Zoom of calibrated pore water pressure and shear stress evolution measured with the experimental device.

It seems that those two measures are related: the calibrated pore water pressure seems to increase with the maximum of the peaks of the tangential shear stress. Comparing the two different y-scales of Fig. 12, a friction coefficient of about 0.4 can be computed. Considering a constant total shear stress, as proposed by Terzaghi theory, the granular stress should fluctuate with the opposite of the pore water pressure. This is in agreement with our measurements until 7 h when the water network becomes discontinuous (thus Terzaghi equation cannot be apply because we are not in saturated condition). Those first results have to be confirmed.

5 Conclusion and Perspectives

We characterized water consumption, caused by cement hydration which induces hydrate bonding to the surface on one hand and material shrinkage which drives the pore water depressure on the other hand. As a result, there is an increase of the granular strength on the formwork. Thus, we estimated a friction coefficient by the tangential shear stress.

Our first results show an interesting correlation between structuration of the material and the transition state between liquid to solid-type behaviour.

Terzaghi equation seems to be valuable during the first state when water network is continuous. After that, the air void quantity is increasing, water network is not continuous anymore, pore water pressure and friction increase, and the granular shear stress decreases as we have solid-solid friction with the evolution of the structuration of the material.

Acknowledgements. The authors are thankful to Agata M. Gajewicz and Prof. Peter McDonald (University of Surrey) for allowing us measuring the water consumption with relaxometry NMR and there useful feedbacks.

The authors also wish to thank Alexandre Menguy which made the drawings of the experimental device.

References

1. Lloret, E., et al.: Complex concrete structures. Comput.-Aided Des. **60**, 40–49 (2015)
2. Internal document J. N. Rivoal Low adhesion caisson concrete
3. Hammer, T.A.: The use of pore water presure to follow the evolution from fresh to hardened concrete, 2006 (e-ISBN: 2351580028) RILEM Publications SARL
4. Mettler, L.K., Wittel, F.K., Flatt, R.J., Herrmann, H.J.: Evolution of strength and failure of SCC during early hydration. Cem. Concr. Res. **89**, 288–296 (2016)
5. Barcelo, L.: Influence des caractéristiques des ciments sur la stucturation et le comportement dimensionnel des matériaux cimentaires au jeune âge. Ecole normale supérieure de Cachan (2001)
6. Lecompte, T., Perrot, A., Picandet, V., Bellegou, H., Amziane, S.: Cement-based mixes: shearing properties and pore pressure. Cem. Concr. Res. **42**(1), 139–147 (2012)

7. Gajewicz, A.M., Gartner, E., Kang, K., McDonald, P.J., Yermakou, V.: A 1H NMR relaxometry investigation of gel-pore drying shrinkage in cement pastes. Cem. Concr. Res. **86**, 12–19 (2016)
8. Fossaa, K.T.: Slipforming of Vertical Concrete Structures. Friction between Concrete and Slipform Panel. Fakultet for ingeniørvitenskap og teknologi (2001)

Experience in Online Modification of Rheology and Strength Acquisition of 3D Printable Mortars

V. Esnault(✉), A. Labyad, M. Chantin, and F. Toussaint

Lafarge Centre de Recherche, 38290 Saint Quentin Fallavier, France
vivien.esnault@lafargeholcim.com

Abstract. This study focus on the early age properties of two mortar formulations designed for a 3D printing extrusion process. They follow a new design and process strategy, which consists in formulating a mortar to be self-levelling, to optimize pumpability, and then incorporating an additive in the extrusion nozzle to modify rheology properties and setting properties to adapt it to the requirement of the printing process (self-sustaining as soon as the material exits the nozzle, and fast strength acquisition). Two types of additives are considered: an alkali-free shotcrete accelerator and a starch ether based VMA. Compression and shear strength measurements from 2 min to 4 h after the incorporation of the additive demonstrate the capacity of the method to create mortars with strength acquisition vastly superior to results from the literature. Lab-scale extrusion and operational feedback from 3D printing customers demonstrate the feasibility at operational scale. The variety of properties obtainable by playing with different types of additives is also discussed.

Keywords: 3D printing · Formulation · Extrusion process

1 Introduction

Extrusion 3D printing of concrete, initially invented Pr. Berokh Koshnevis [1], is now gaining increasing attention as an alternative to traditional casting techniques. The method consists in the extrusion of a ribbon of cementitious material (usually a mortar, due to the difficulty to process big aggregates), which are disposed one upon another by a robotic system. It might allow both drastic gain in productivity (due to the use of robotic systems), and increase freedom of conception for designers, no longer limited by the limitations imposed by formworks [2–4].

3D printable mortar or concrete formulations are subjected to contradictory requirements [4]. They need to be pumpable, in order to be easily conveyed from a reservoir or a mixer to a nozzle head. But the 3D printing application also requires some mechanical properties on the extruded product, in order to sustain first its own weight, then the weight of the subsequent layers. General ease of the process also requires the formula to present a sufficient "open time" for the transport and the application of the concrete on the structure, without early setting of the product in the system which could prove catastrophic for the installation.

T. Wangler and R. J. Flatt (Eds.): DC 2018, RILEM Bookseries 19, pp. 24–38, 2019.
https://doi.org/10.1007/978-3-319-99519-9_3

The most common approach [5–7] consists in finding a compromise between those contradictory requirements, so the mortar is able to perform adequately at all steps. We present in this paper a competing method, where the properties of the mortar are modified during the printing process. In this method, the mortar is initially self-levelling, to optimize the pumpability through the system, and has a long open time. An additive is injected and incorporated into the mix in the extrusion head. This addition can drastically modify the properties: accelerating setting time, and "thicken" instantly the rheology of the mortar so it has the capacity to sustain its own weight. This method initially developed by the Lafarge Centre de Recherche for the needs of an academic project [8], is now routinely used by the company XtreeE for the realization of operational projects [9]. Based on published information, the technique has also been developed independently by teams from the Sika Group [10].

Whatever the process and formulation method adopted, early age performance and their development with time are critical to the performance of the overall 3D printing process. As explained in [6], the 3D printed mortar is put under stress just after deposition by the continuation of the 3D printing process. If the concrete does not acquire enough strength sufficiently quickly, either the structure might deform (or collapse catastrophically), or the process speed needs to be drastically limited to adapt to the concrete limitations, leading to a drop of productivity for the system.

Studies on the fresh properties of 3D printable materials are relatively rare [6, 11, 12], and all concerns materials which are not modified online. This work provides the first study concerning the early age properties of online modified 3D printable mortars. It aims at demonstrating that this new approach allows for drastic improvement in performance, possibly by several orders of magnitude.

2 Materials and Methods

2.1 3D Printable Formulations and Preparation

In the course of this project, we tested two methods to modify the properties of the mortar:

- Injection of an accelerator, in order to modify both the rheology and the hydration kinetics in a short timeframe.
- Injection of a viscosity modifying admixtures (VMA), in order to selectively modify the rheology without altering to much the hydration kinetics.

A variety of fast-action accelerators has been developed for shotcrete applications since at least the 1970s [13]. The shotcrete process presents similarities with our strategy for 3D printing formulations: a fast change of properties is needed just before the product leaves the system, and adding the additive in the nozzle head is a possible strategy. We settled for this study for a family of alkali-free shotcrete accelerators, where the main active agent is aluminum sulfate salts [14, 15]. Those accelerators are safer to operate due to their low alkalinity, and are known to affect less final strength then their alkaline counterparts, probably because they stimulate the reaction of the aluminum phases of cement without perturbating too much the reaction of the silicates [16]. For this study, we used Sika 40 AF, commercialized by Sika Group.

VMAs are a class of additives, typically constituted of long organic molecules, such as cellulose ethers, which are added to cementitious materials to increase its viscosity or yield stress [17, 18]. They work following a variety mechanism, from raising the viscosity of the interstitial fluid phase, bridging flocculation between grains, polymer to polymer interaction through association or entanglement [19, 20]. We selected for this study a type of starch ether. The mechanism of action of this product is not fully clear. The product increases viscosity without affecting yield stress at relatively low concentration and high water content [21], while an important increase of yield stress is reported at higher concentration and lower water content [22]. Schmidt et al. [23] explains the phenomenon by an important interaction between the starch ether and the grains in the paste. An explanation somehow contradicted by Palacios et al. [24] who observes limited adsorption on the cement grains and proposes a mechanism they call "depletion flocculation". Whatever the mechanism, the fact that the molecule provides important yield stress while being soluble in water with little influence on its viscosity is interesting for our application, as it allows for easy pumping and incorporation. The effect on setting time and strength acquisition is also reported to be limited [25]. For this study we used Foxcrete S200, commercialized by Avebe.

Formulations tested are mortar which were developed for the operational needs of 3D printing customers:

- The "NAG3" formulation is derived from a UHPC dry mix (Ductal®), with slightly modified additives and water content to maximize pumpability and open time. Maximum grain size is 300 μm. It is modified online using the accelerator additive at a dosage of 31 g/L of mortar. A similarly designed product is routinely supplied to XtreeE in its operations [9].
- The "C60" formulation is a mortar mix with a lower level of target performance (Rc > 60 MPa at 28 days). Maximum grain size is 1.6 mm. It is modified online using the starch-based VMA at a dosage of 0.38 g/L mortar. This product has not yet been deployed operationally on a printing system, but it has tested on our experimental extrusion system.

We present results appealing to each of the addition strategy, but we insist that the starting formulations are not the same so direct comparisons should be made carefully, and be the object of future works.

The properties of both formulation are listed in Table 1, before and after addition of the accelerator or VMA. Flow properties were assessed using the Marsh cone method (ASTM C230), and a non-standard funnel flow (which results are not presented). Those properties correspond to self-levelling mortars. Beside this, both formulations retains good flow properties for an estimated open time of 2 h, so the material can be easily pumped and use in substantial batches.

Setting time of the formulations were measured using the Vicat method (norm EN 196-1.) and by following the heat released by the setting reaction through a calorimetry device. The Vicat setting time is considerably accelerated by both additives, however this measurement can be considered misleading. Due to the heavy modifications implemented in the system, the Vicat method might detect a "setting" while the

Table 1. Properties for the modified and unmodified NAG3 and C60 formulations

Formula	Marsh cone slump flow (mm)	Density (kg/m^3)	Vicat setting time (min)		Peak hydration (h)	Rc (MPa)		
			Initial	Final		1 d.	7 d.	28 d.
NAG3 (no acc.)	140	2330	235	305	10.5	59.0	93.1	103
NAG3 (acc.)	–		40	80	9.5	51.5	84.0	102
C60 (no VMA)	210	2230	120	150	9	44.1	–	71.3
C60 (VMA)	–		85	115	9	27.7	–	62.6

chemical reactions associated to this in an unperturbed system had not taken place. This is confirmed by the calorimetry values, where the peak in heat release can be associated to the silicate reaction. The peak time is not changed by the addition of the VMA, and only slightly reduced by the accelerator (which does not stimulate directly the silicate reaction). We therefore argue that the Vicat setting time does not correspond here to the formation of a C-S-H network, but either to a particularly intense thixotropy of the system (C60), or a controlled form of flash setting ((NAG3).

Although the focus of this study is specifically early age properties, Table 1 also display the compression resistance values between 1 day and 28 days on the formulations considered, with and without the addition of accelerator or VMA. It demonstrates it is possible to attain high level of final strength with this approach, and that the final properties are only modestly impacted by the late addition process.

2.2 Lab-Scale Extrusion Testing

Formulations developed were extruded using an experimental setup which simulate conditions which are encountered on a 3D printing device, though at a limited scale: throughput is fixed at 2 L/min. The schematic of the system and a photo are presented in Figs. 1 and 2. The self-levelling mortar is conveyed through a pump and a piping system. A secondary pump conveys the liquid additive to a dynamic mixing device which incorporates the additive within the mortar flow using a rotative blade. Pressures are monitored on both circuits to detect any clogging and pressure rise that could accidentally occur. Due to the long open time and high fluidity of the mix used, it has been calculated that the mortar could be pumped without difficulties on several dozens of meters, though of course the pumping distance in our tests was shortened for practical reasons. Average residency time of the mortar in the mixing device is on the order of magnitude of 30 s.

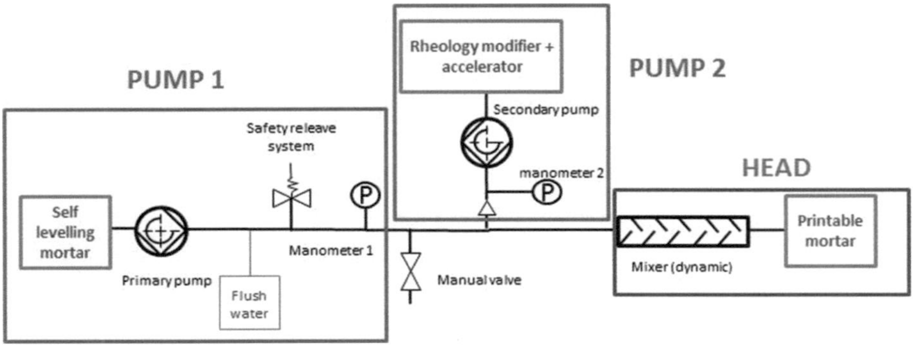

Fig. 1. Schematic of the lab-scale extrusion system

Fig. 2. Lab-scale extrusion system layout. Main pump can be seen on the left hand side, front. Secondary pump: left hand side, back. Mixer system in the middle. Material in buckets: C60

2.3 Early Age Characterization Methods

For early age characterization, the materials were not extruded but prepared in a batch of 2 L on a mortar mixer. First the self-levelling mortar were prepared. Then the modifying additive (accelerator or VMA) were added, always exactly 10 min after cement went into contact with water. Incorporation of the additive in the mortar is made using a standardized mixing procedure (30 s of mixing at moderate speed) designed to incorporate homogeneously the additive without risking of damaging the rheology modifying effect by over-mixing.

Early-age characterization were done using two equipments: a scissometer and a texturometer:

- The scissometer (Fig. 3 left) is a cross-shaped blade linked to an equipment able to measure torque. The blade is carefully inserted inside the mortar, and rotated slowly until the blades captures a cylinder of mortar and make it slip against the rest of the mortar. The recipient containing the mortar has to be chosen adequately so the bulk of the sample is not rotated with the blade. The maximum torque exerted when the material start slipping is measured, and it is possible to calculate simply the yield shear stress from the geometry of the system.

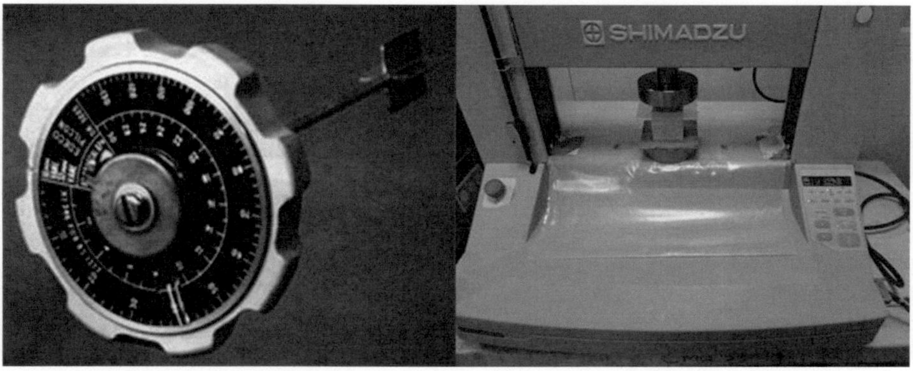

Fig. 3. Equipments used for early age characterization : scissometer (left), texturometer (right)

- The texturometer (Fig. 3 right) is essentially a press designed to operate in a low range of stress. Samples were prepared as cubes of 5 × 5 × 5 cm. They were casted just after the addition of the VMA or accelerator, and systematically demolded 10 min after casting. A special procedure was put into place to manipulate the cubes without deforming them even at very early ages, and we estimate geometrical defects never exceeded 2 mm. The force sensor in use had a capacity of 5000 kN with a precision of 10 N Which with the geometry adopted means we could measure stresses with a precision of 4 kPa. Samples were loaded at a fixed speed of mm/min. Contact detection was set at 10 N, or 4 kPa. Loading speed is set at 2 mm/min.

The two equipments cannot be used on exactly the same timing. The scissometer equipment can be used very fast, but not after the material hardens too much. The texturometer can be used for longer time range (eventually, until the mortar fully hardens), but preparation of sample requires some minimum time (15 min), and that the mortar had reached some level of consistency.

In practice, scissometry was performed at 2, 5, 10, 15, 20 and 30 min for the NAG3 formulation, at 5, 10, 15, 20, 30, 49 and 60 min for the slower evolving C60. Texturometer measurements were performed at 15, 30, 45, 60, 90 and 120 min for the NAG3, 30, 60, 90, 120, 180, 240 min for the C60. All times take the addition of the VMA or accelerator as t = 0 which means 10 min after the contact between cement and water. All points were repeated 3 times on 3 different batches (for a total of 9 points), to limit the impact of variability in batch preparation and error in sample preparation and measurement.

As can be seen from this selection of points, there is some overlap between the ranges of the two measurement methods. We exploited this overlap to calculate an average ratio between yield stress measured through the two methods, for instance to extrapolate a compression strength in the very early time range where texturometer data are not available. This assumption that this ratio remains constant through the evolution of the material is not solidly supported, and those calculated values can only be considered as gross estimations.

3 Results

3.1 Texturometry

Figure 4 displays the stress vs strain curve obtained through the texturometer test, for both formulations at the different times. Those are average curves calculated from the nine individual curves obtained through repeated testing. Stress is calculated by dividing the force measured by the equipment by the contact surface. We did not have practical ways to measure the evolution of this contact surface, which is problematic as the sample undergo important deformations and the widening of the contact surfaces cannot be neglected. To remedy this we calculated the contact surface assuming an isotropic linear media with a poisson ratio of 0.3 (in accordance to observation made by Wolfs et al. [12])

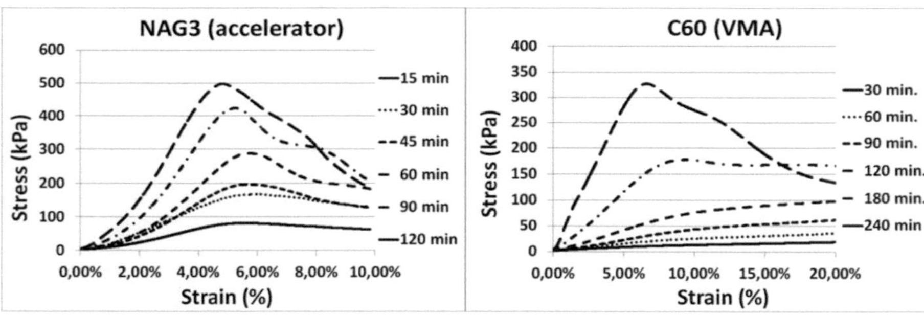

Fig. 4. Stress vs strain curves at differents times, NAG3 formulation (left), and C60 (right)

It can be observed immediately that the curves can present two different type of behaviour:

- For the NAG3 formulation at all times, and the C60 formulation for 180 min and more, the stress goes through a maximum which can be associated with the fragile breakage of the sample.
- For the C60 formulation before 180 min, no such maximum is observed in the range of strains tested (0–20%) and the material keeps on deforming plastically with increasing stress.

This observation is confirmed visually on the loaded sample. Figure 5 shows two samples, one displaying cracks and which did present a stress maximum associated with brittle behavior, one which obviously deformed continuously through the test with little or no crack formation.

Fig. 5. Representative aspect of texturometer samples (C60). From left to right: unloaded, loaded at 30 min, loaded at 4 h.

To analyze the curves, we choose to extract two characteristic data:

- Maximum stress, which corresponds to failure resistance of the material. When the material does not exhibit a maximum, we arbitrarily chose to consider the stress for a strain of 10% as the maximum. The choice is motivated by the fact that all samples which presented a fragile behavior were broken before a strain of 10% was reached, and that in all likelihood the printed structure would be either ruined or severely deformed if any of its parts were to be subjected to such intense deformation.
- Rigidity modulus, which is calculated similarly to Young modulus using a ratio between uniaxial stress and strain. More precisely, the ratio is calculated between 25% and 75% of maximum stress (or stress at 10% in the absence of maximum), where most curves seem to display a relatively linear behavior, and results are not perturbed by contact artifacts or crack propagation before breakage. We do not want to adopt directly the term "Young modulus" such as in [6] or [18] as the material clearly deform plastically under load, and not elastically.

We defend that both data are of key importance for the resistance of the printed element at very early ages. As demonstrated by Suiker [26], Printed elements are likely to collapse following two possible mechanisms: "plastic collapse", which correspond to the material locally exceeding its yield strength, but also "global buckling", where the structure prove insufficiently rigid to sustain a given geometry. Buckling failures appears to be very common when slender elements are printed, so a stiffer product would prove much desirable to maintain such shapes.

Figure 6 and Table 2 display the evolution of maximum stress and rigidity modulus for both formulations (average values and Relative Standard Deviation (RSD)). Results were no real maximum stress could be identified are marked with an asterisk.

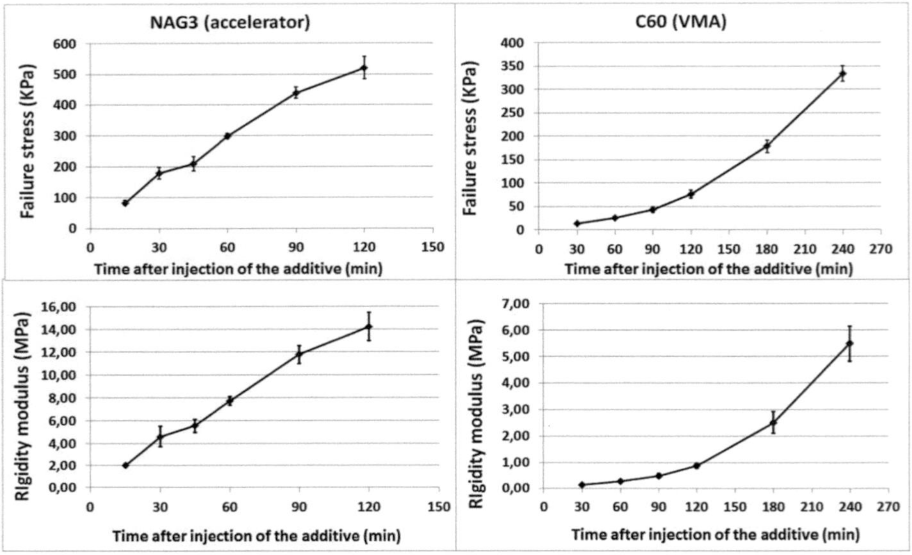

Fig. 6. Maximum stress and rigidity as a function of time, for the NAG3 and C60 formulation. Error bars are established for 95% probability. Scales vary between the two formulations due to important disparities in performance.

Table 2. Maximum stress and rigidity modulus at different times, for the NAG3 and C60 formulations. "*" means no true maximum stress could be defined.

Formulation	Time (min)	Maximum stress		Rigidity modulus	
		Average (kPa)	RSD	Average (MPa)	RSD
NAG3 (accelerator)	15	82.7	12.4%	2.03	11.4%
	30	179	14.2%	4.59	25.1%
	45	210	14.0%	5.52	13.8%
	60	298	3.8%	7.71	6.1%
	90	439	5.5%	11.8	8.7%
	120	520	9.0%	14.2	11.4%
C60 (VMA)	30	13.1*	16.8%	0.14*	17.7%
	60	25.3*	18.8%	0.27*	20.4%
	90	43.3*	18.4%	0.49*	19.2%
	120	75.8*	14.9%	0.88*	14.7%
	180	178	9.7%	2.51	21.2%
	240	334	6.4%	5.49	15.8%

For both materials, Maximum stress and rigidity modulus appears to be following a similar increasing trends. Error levels are kept for all points to acceptable levels, comparable to what can be found in [18]. Strength acquisition is much faster for the NAG3 formulation, and appears to be globally linear with time. In the case of C60, the strength acquisition is much slower, but accelerating with time. It is interesting to notice that the acceleration of strength acquisition coincides with the appearance of a brittle behaviour (around 3 h), probably following the formation of a rigid hydrate structure in the maturing material.

3.2 Scissometry

Table 3 and Fig. 7 display the average and standard deviation of shear resistance for the two formulations, measured with the scissometry equipment. Both formulations are calibrated to display a shear resistance around 2 kPa almost immediately after the additive is inserted, a value we associate with printability based from empirical experience.

Table 3. Early age properties of NAG3 and C60 formulations, measured through scissometry

Formulation	Time (min)	Shear stress (nominal)		Shear stress (additive/2)		Shear stress (additive x2)	
		Mean (kPa)	RSD	Mean (kPa)	RSD	Mean (kPa)	RSD
NAG3 (acc.)	2	1.62	48.0%	–	–	–	–
	5	4.18	41.6%	0.046	11,3%	35.3	5.6%
	10	13.9	23.9%	0.061	17,7%	67.5	6.5%
	15	19.1	16.1%	–	–	–	–
	22	30.2	17.0%	0.133	32.4%	138	4.0%
	30	38.9	9.2%	0.178	35.6%	Not measurable	
C60 (VMA)	5	1.87	15,8%	0.342	26.8%	8.13	6.6%
	10	2.42	17.4%	0.708	12.9%	10.2	14.8%
	15	3.65	19.0%	–	–	–	–
	30	5.48	20.3%	1.06	12.8%	13.9	7.7%
	49	8.35	18.7%	–	–	–	–
	60	13.8	7.8%	–	–	–	–
	90	16.4	9.5%	14.8	5.6%	24.2	3.4%

Both formulations displays an increase of shear resistance with time in those first few minutes after the addition of the product, although once again the resistance increase is much more important in the case of the accelerated NAG3 formulation.

Repeatability appears to be correct for both formulations and for the different times, with one notable exception: the earliest times (2 and 5 min) for the NAG3 formulation. Closer analysis of the results shows that this variability is due to important variations between the three batches measured (and not by variability of measurements within a

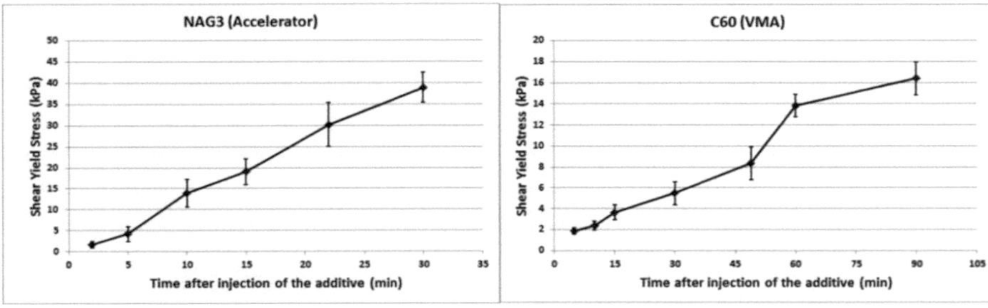

Fig. 7. Shear yield stress measured by scissometry, for NAG3 and C60 formulation

same batch). We think this variation is caused by a strong sensitivity of the NAG3 formulations to preparation conditions, and in particular to mechanical solicitation. The observation is confirmed by empirical observation of the material and experience in extruding this product. Beyond 10 min, the reproducibility goes back to acceptable levels, a possible sign that the sensitivity to preparation conditions is only relevant in the first few minutes.

Table 3 also displays the results of a rough sensitivity analysis to the dosage of additive: scissometer values were taken at three times for both formulations, with a dosage of VMA or accelerator of x0.5 and x2. Both products display as expected a strong sensitivity to additive concentration (especially the accelerated NAG3 formulation), illustrating that the rheology modification we observe can be tuned to needs depending on additive dosage. However, due to the strong variation in dosage adopted, the products were not printable in those test cases (either too firm or not consistent enough). A finer study with more reasonable dosage variations would be needed to assess the operational impact of those variations.

3.3 Lab Scale Extrusion

Unfortunately, difficulties in the availability of experimental equipments prevented us at the time we write this report to conduct systematic extrusion testings, that would systematically confirm the observations presented in the preceding chapter.

However, we were in position to confirm the technical feasibility of extruding both formulations using our experimental system. And in the case of the NAG3 formulation, this claim is supported by the extensive experience of XtreeE with a formulation following a similar concept [9].

Those early extrusion testing also brought interesting empirical observations. The NAG3 formulation appears remarkably fluid when it exists the nozzle, before solidifying very quickly. This behavior is interesting if the material need to deformed to be printed, but prevent from shaping the material precisely using the nozzle section (Fig. 8, left). It can be attributed either (or both) to a delayed effect of the additive, or to a high sensitivity to mixing conditions. In opposition, the C60 formulation appears to be more instantly firm, and keep remarkably well the shape given by the nozzle (Fig. 8, right).

Fig. 8. Extruded samples of NAG3 formulation (left), and C60 formulation (right)

In those early testings, the NAG3 formulations displayed a considerable sensitivity to process parameters such as mixing speed and duration. The C60 formulation also displayed such sensitivity, but to a lesser extent. The question of this sensitivity to process parameters is absolutely key to the operational user, as it could prove a limitation to the predictability of the 3D printable material. For instance, does this impact can still be felt long after deposition, or is it limited to the immediate instants after deposition? We lack for now the scientific evidence for a definitive answer, but there are signs pleading for a relatively short term impact (such as the important variability of the scissometer results, but only for the shorter times).

4 Discussion

Data on texturometry (Table 2) and scissometry (Table 3) overlaps on part of the time range, so we can calculate a average ratio between compression resistance and shear resistance (of 4.47 for the NAG3 and 2.27 for the C60). Assuming this ratio is constant, we can make a gross estimation of the product resistance at very early age: 7.2 kPa for the NAG3 at 2 min, and 4.2 kPa for the C60 at 5 min. Given the densities given in Table 1, this is enough to sustain respectively 32 cm and 19 cm of their own weight, more than enough to assure that the material will not collapse during the disposition of the first layers.

The strength evolution is considerably faster then anything which has been reported in the literature. Di Carlo [6] reports compression yield stress at 5.5 kPa at 11 min, 8.2 kPa at 22 min, 11.7 kPa at 92 min. The same study also report rigidity modulus (0.077 MPa at 11 min, 0.12 MPa at 22 min, 0.13 MPa at 90 min). Perrot et al. [11] reports compression yield stress around 4 kPa initially, 5 kPa at 30 min, and 16 kPa after 90 min. Wolfs et al. [12] reports 6.4 kPa initially, 10 kPa at 30 min, and 19 kPa at 90 min. The same study also report rigidity modulus (0.074 MPa initially, 0.12 MPa at 30 min, 0.19 MPa at 90 min). Results reported in this study are above by at least an order of magnitude for the accelerated NAG3 formulation. The slower C60 formulation presents comparable level of performance before 30 min, but still outperform the benchmark products at longer times (90 min). Those impressive levels of performance are delivered not only in terms of resistance, but also in terms of rigidity, a key factor for structural stability in those early ages. We believe this difference is only made possible by the considerable liberty provided by the opportunity to modify the product online.

This fast strength evolution is of course of great interest to maximize printing speed and process efficiency. If only compression failure is taken into account, the NAG3 is able to sustain 3.6 m of its own weight within 15 min. The C60 formulation would require slower speed, but can still sustain (in compression) a height of 60 cm within 30 min and 1.2 m within an hour.

The spectacular gains in resistance obtained with the NAG3 formulation are not surprising. The aluminum based accelerator generates a fast precipitation of ettringite which leads to a quick consolidation of the system. However, we observe several hints that this consolidation is highly sensitive to mechanical perturbation: the NAG3 formulation gets out of the nozzle very fluid, but consolidate almost instantly when leaving the nozzle, and we have strong reproducibility issues at very young ages, indicating a strong impact of preparation conditions. Following a mechanism presented by Roussel et al. [27], we suggest that the consolidation of the NAG3 formulation is mostly linked to the precipitation of hydrates selectively at the interfaces between grains. Mechanical solicitation can break those bonds and move the grains from one another, effectively destroying the consolidation effect. This induces a considerable sensitivity to process conditions, depending whether the process let the consolidation mechanism take place or not. The structure formed by this mechanism also has limited capacity to deform plastically, so the material displays a fragile behavior at very early ages. The C60 formulation, which early rheology properties are less linked to this mechanism, is less sensitive to mechanical perturbation, and can deform plastically at early ages.

On a broader perspective, we identify two very different patterns of behavior between the formulations. The accelerator modified NAG3 formulation display spectacular strength evolution, from a very early age. However the materials is still relatively fluid when it exists the nozzle, which can allow to operate the printing system smoothly, but might be problematic if retaining a specific shape is required instantly. The material become stiffer quickly, but will present faster a brittle behavior. This strength evolution is also more sensitive to mechanical perturbation in the process. We also expect a rather strong impact of factors such as cement nature, ambient temperature… Finally, such a fast evolution of the material must be taken into account regarding several aspects of the industrial operation: maintenance and cleaning, dealing with stoppage, curing of the printed elements…

The VMA modified C60 formulation displays instantly a robust level of resistance, which is likely less sensitive to a variety of perturbation. The slow, unperturbed setting reaction can be an asset for operations where the material is not subjected to critical stress too rapidly. It also presents the capacity to deform plastically long after deposition. However, the clear weak point of this formulation is the slow performance gain, especially in the first hour, when compared to the accelerated formulation. We insist this problem can be partially corrected by the used of accelerators (with a slower mode of action then the shotcrete accelerator used in this study), which allow to stimulate the strength acquisition in this time range, but keeping the robust initial rheology and causing less operational problems then the purely accelerated system.

We believe it is in fact possible to design a full array of printable systems between the two extreme scenario presented in this study: ultrafast, purely accelerated systems, and unaccelerated VMA based systems.

There remains a considerable amount of work to fully understand the behavior of those systems. The first glaring point is the impact of process, with the acquisition of results on extrusion systems, especially for the accelerated formulations which appears to be very sensitive to such effects. This will clarify the impact of processing conditions on early age properties, and especially for how long such an impact is relevant on the material behavior. Results dispersion indicates this effect could only be sensible in the first few minutes, but this limited impact should be systematically demonstrated. Another key factor, ignored in this study, is the evolution of the batch prior to the addition of the additive. We systematically considered a situation where the batch is modified and used very soon after being prepared, but it is unlikely to be always the case in operational situations. Finally, key environmental factors such as temperature, curing conditions are likely to have a decisive impact on property evolution, especially if the 3D printable mortar is used outside of the controlled conditions of a factory floor and directly on the worksite.

Acknowledgements. The resulted presented in this study are originated in the research project HINDCON (Hybrid INDustrial CONstruction) funded by the European Commission (Grant Agreement n°723611).

References

1. Koshnevis, B.: Innovative rapid prototyping process making large size, smooth surface complex shapes in a wide variety of materials. Mater. Technol. **13**, 52–63 (1998)
2. Koshnevis, B., et al.: Mega-scale fabrication by contour crafting. Int. J. Ind. Syst. Eng. **1**, 301–320 (2006)
3. Buswell, R.A., et al.: Freeform construction: mega-scale rapid manufacturing for construction. Autom. Constr. **16**, 224–231 (2007)
4. Bos, F., et al.: Additive manufacturing of concrete in construction: potentials and challenges of 3D concrete printing. Virtual Phys. Prototyp. **11**, 209–225 (2016)
5. Le, T.T., et al.: Mix design and fresh properties for high performance printing concrete. Mater. Struct. **45**, 1221–1232 (2012)
6. Di Carlo, T.: Experimental and numerical techniques to characterize structural properties of fresh concrete relevant to contour crafting. Ph.D. thesis, University of Southern California (2012)
7. Paul, S.C., et al.: Fresh and hardened properties of 3D printable cementitious materials for building and construction. Arch. Civ. Mech. Eng. **18**, 311–319 (2018)
8. Gosselin, C., et al.: Large scale 3D printing of ultra-high performance concrete – a new processing route for architect and builder. Mater. Des. **100**, 102–109 (2016)
9. XtreeE communication.: http://www.xtreee.eu/projects/. Accessed 26 Feb 2018
10. Sika Group publication.: https://www.sika.com/en/group/Publications/ambitions/ambitions18/concrete_team.html. Accessed 23 Feb 2018

11. Perrot, A., et al.: Structural build-up of cement based materials used for 3D printing extrusion techniques. Mater. Struct. **49**, 1213–1220 (2016)
12. Wolfs, R.J.M., et al.: Early age mechanical behaviour of 3D printed concrete: numerical modelling and experimental testing. Cem. Concr. Res. **106**, 103–116 (2018)
13. Schutz, R.J.: Properties of shotcrete admixtures. In: Shotcrete for Ground Support, Proceedings of the Engineering Foundation Conference, Easton (1977)
14. Paglia, C.: The influence of calcium sulfo aluminate as accelerating component within cementitious systems. Ph.D., ETH Zürich (2000)
15. Lootens, et al.: Some peculiar chemistry aspects of shotcrete accelerators. In: Proceedings of the 1st International Conference on Microstructure Related Durability of Cementitious Composites, Nanjing pp. 1255–1261. RILEM Publications S.A.R.L. (2008)
16. Eberhardt, A.B., et al.: On the retardation caused by some stabilizers in alkali free accelerators. In: Proceedings of the 17th International Conference on Building Materials, Weimar (2009)
17. Palacios, M., Flatt, R.J.: Working mechanism of viscosity modifying admixtures. In: Science and Technology of Concrete Admixtures, pp. 415–432. Woodhead Publishing (2016)
18. Brumaux, C., et al.: Cellulose ethers and yield stress of cement pastes. Cem. Concr. Res. **55**, 14–21 (2014)
19. Khayat, K.H.: Viscosity-enhancing admixtures for cement-based materials – an overview. Cem. Concr. Compos. **2**, 171–188 (1998)
20. Khayat K.H., Ghezal, A.: Effect of viscosity-modifying admixture-superplasticizer combination on flow properties of SCC equivalent motar. In: 3rd International RILEM Symposium on Self-Compacting Concrete, pp. 369–385 (2003)
21. Simonides, H., Terpstra, J.: Use of innovative starch ethers for paving blocks and other concrete products. Concr. Plant Precast Technol. **9**, 38–45 (2007)
22. Rajayogan, V., Santhanam, M., Sarma, B.S.: Evaluation of hydroxy propyl starch as a viscosity-modifying agent for self compacting concrete. In: 3rd International RILEM Symposium on Self-Compacting Concrete, pp. 386–394 (2003)
23. Schmidt, W., et al.: The working mechanism of starch and Diutan gum in cementitious and limestone dispersions in presence of polycarboxylate ether superplasticizers. Appl. Rheol. **23** (5), 52903 (2013)
24. Palacios, M.. et al.: Compatibility Between Polycarboxylate and Viscosity-Modifying Admixtures in Cement Pastes. American Concrete Institute, ACI Special Publication, pp. 29–42 (2012)
25. Schmidt, W., et al.: Interactions of polysaccharide stabilizing agents with early cement hydration without and in the presence of superplasticizers. Constr. Build. Mater. **139**, 584–593 (2017)
26. Suiker, A.S.J.: Mechanical performance of wall structures in 3D printing processes: theory, desing tools and experiments. Int. J. Mech. Sci. **137**, 145–170 (2018)
27. Roussel, N., et al.: The origin of thixotropy of fresh cement pastes. Cem. Concr. Res. **42**, 148–157 (2012)

A Framework for Performance-Based Testing of Fresh Mixtures for Construction-Scale 3D Printing

Ali Kazemian[1,2,3(✉)], Xiao Yuan[3], Ryan Meier[1], and Behrokh Khoshnevis[1,3,4]

[1] The Sonny Astani Department of Civil and Environmental Engineering, University of Southern California, Los Angeles, CA, USA
akazemia@usc.edu
[2] Department of Computer Science, University of Southern California, Los Angeles, CA, USA
[3] Contour Crafting Corporation, El Segundo, CA, USA
[4] Department of Industrial and Systems Engineering, University of Southern California, Los Angeles, CA, USA

Abstract. A step-by-step procedure for performance-based testing of mixtures for construction-scale 3D printing is proposed. Workability of a fresh "printing mixture" is described in terms of print quality, shape stability, robustness, and printability window. To demonstrate the proposed procedure and test methods, an experimental program is carried out using four different mixtures. The experimental results are used as the basis for discussion and comparison of performance of developed mixtures for use in construction-scale 3D printing. Finally, perspectives on the future research areas as critical steps for advancement of construction-scale 3D printing are provided.

Keywords: 3D printing · Contour Crafting · Cementitious materials Workability

1 Introduction

A recent major attempt toward automated construction is based on the idea of scaling up additive manufacturing techniques. Additive manufacturing is defined as "the process of joining materials to make objects from 3D model data, usually layer upon layer, as opposed to subtractive manufacturing methodologies" [1]. It should be noted that additive manufacturing technologies have previously been used for small-scale concept modeling in architecture [2]. However, use of this technique for full-scale in-situ automated building construction is in the process of emerging in the construction industry. A well-developed automated layer-by-layer construction process would present numerous advantages including design freedom, superior construction speed, minimal waste of construction materials, and higher degree of customization.

Contour Crafting (CC) is a pioneering additive fabrication technology that uses computer control to exploit the surface-forming capability of troweling to create smooth and accurate planar and free-form surfaces out of extruded materials [3]. CC is

T. Wangler and R. J. Flatt (Eds.): DC 2018, RILEM Bookseries 19, pp. 39–52, 2019.
https://doi.org/10.1007/978-3-319-99519-9_4

commonly recognized as the first viable construction-scale additive manufacturing process for building construction [4–6]. The prototype CC machine has work envelope dimensions of 5 m × 8 m × 3 m, corresponding to a 120 m^2 printing zone.

With respect to the construction material, Portland cement concrete has been found to be the most viable option for widespread use in automated construction processes in near future [5, 7]. Concrete is well-understood and has unique fresh and hardened properties as well as an extensive variety of readily available admixtures to customize its performance.

Limited research has been carried out on properties of printing concrete. In 2016, Perrot *et al.* [7] studied the time-dependent structural build-up of cementitious materials in layer-wise construction. The time required to harden is important because during the layer-by-layer construction process, the previously deposited layers need to be able to withstand the load caused by following layers. Based on comparison of vertical stress acting on the first printed layer with the critical stress related to the plastic deformation, a theoretical framework was proposed. Assuming linear evolution of yield stress over time, these researchers defined a critical failure time (t_f) as a function of concrete specific weight, concrete yield stress with no time at rest, structuration rate, construction rate and a geometric factor (α_{geom}). Finally, layer-wise construction of a 70 mm diameter column with building rates of 1.1 to 6.2 m/h was used to validate the findings. Except for the smallest building rate, 1.13 m/h, the experimentally measured t_f values were highly correlated with values calculated based on proposed expressions [7].

While few prior studies have focused on specific properties of printing mixtures such as shape stability [7–11], a comprehensive list of performance requirements and test methods for a printing mixture has not yet been developed. In this paper, a step-by-step procedure for testing of fresh printing concrete is introduced and relevant details are discussed. Then, an experimental program which was carried out to demonstrate the proposed approach is discussed. The proportions of four printing mixtures, designed to study effects of Nano-clay, silica fume and fiber, are also presented along with the results of proposed new tests. Finally, perspectives on the future research areas as the critical steps for advancement of construction-scale 3D printing are provided.

2 Experimental Program

2.1 The Step-by-Step Testing Procedure

The suggested procedure for laboratory testing of printing concrete in fresh state is presented in Fig. 1. The testing procedure is designed based on properties of printed layers (rather than parameters directly related to the employed pumping or extrusion mechanism). As such, the proposed framework is applicable to different concrete 3D printing systems.

Based on experience and other research works [8, 12–14], printing concrete can be characterized by high powder content, no coarse aggregate, increased paste fraction, and use of viscosity modifying admixture (VMA). In fact, reported mixture proportions of successful printing mixtures could be used as a starting point. After designing an initial mixture, fulfilling print quality requirements is the first step in the proposed

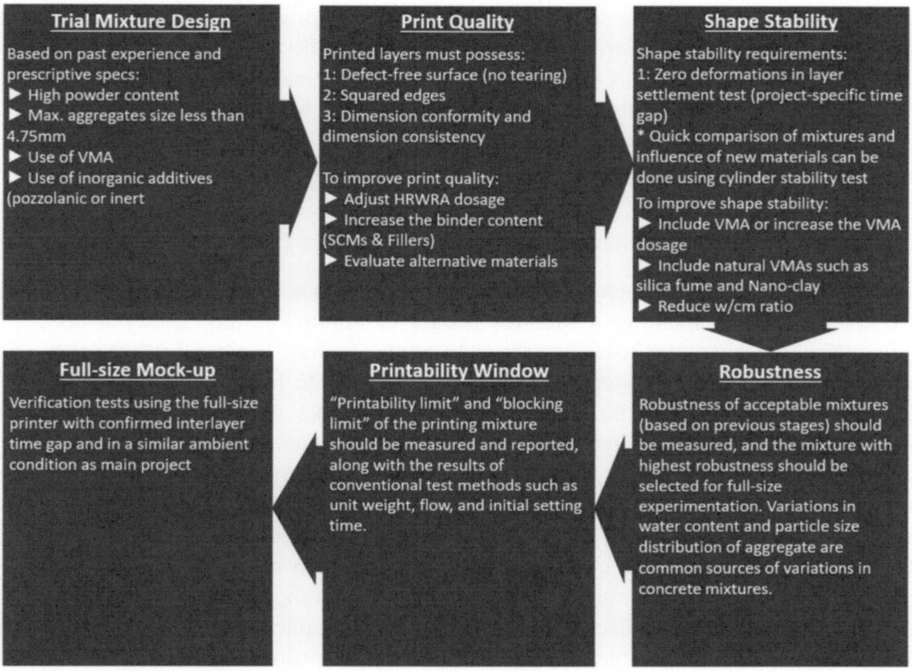

Fig. 1. Suggested procedure for performance-based testing of printing mixture at fresh state

iterative mixture assessment and modification process. With respect to acceptable print quality, three requirements: (a) surface quality, (b) squared edges and (c) dimension conformity and consistency, must all be satisfied.

Next, shape stability of a mixture must be examined and necessary adjustments made. To this end, layer settlement experiment is proposed, where concrete layers are printed on top of each other with the same extrusion mechanism as the full-scale concrete printer. Further, no visible deformations should occur when the target inter-layer time gap is used.

Next, robustness of developed mixtures is evaluated, where the influence of material variations on the fresh properties of the printing mixture is investigated. In this study, "change in water content" is considered as the only source of variation.

The fourth step in laboratory testing of a printing mixture refers to printability window of a mixture. Two important parameters related to the printability window include the printability limit and blockage limit. Printability limit is the longest period during which a mixture can be printed with acceptable print quality. Blockage limit refers to the longest period of time when a mixture can remain in the nozzle before the concrete stiffens and blocks the extrusion. Both limits should be measured and reported for each specific mixture. Finally, when the laboratory testing is aimed at developing a mixture for a specific construction project, a verification test is suggested. This test should be carried out using the full-size printer in a similar ambient temperature and humidity as the intended project. Also, the use of the same concrete batching, mixing, and transporting equipment as the actual project is highly recommended. The main function of the proposed verification test is to subject the designed mixture to actual jobsite-based assessment. In the following sections, the experimental program which was carried out based on the proposed framework will be presented. In addition, the details of proposed requirements and test methods for print quality, shape stability, robustness, and printability window will be discussed.

2.2 Concrete Printing Setup, and Printing Mixtures Proportions

A linear concrete printing machine was constructed and used for extrusion experiments. Compared to implementing a full-size printer (Contour Crafting gantry robot), the advantages of using such a laboratory-scale printer include savings in time, cost, and material. The developed system is capable of printing up to 10 concrete layers, 1.2 m in length. The nozzle uses an extrusion mechanism like the CC machine to print 38.1 mm × 25.4 mm concrete layers. The control system was developed using a combination of Arduino Mega (based on ATmega1280) and Arduino Uno (based on ATmega328) microcontrollers. The machine can print concrete at different linear speeds and deposition rates. The implemented feedback control system (closed-loop) for the extruder ensures the consistent material deposition rate.

In addition, Bluetooth communication with the printing setup enables the user to conveniently control the machine using an Android application on a smartphone or tablet. This application enables the user to select the extrusion parameters (linear speed and deposition rate). For the purposes of this study, a linear printing speed of 60 mm/s was used in all experiments.

With respect to materials, ASTM C150 Type II Portland cement was used to produce printing mixtures. A commercially available manufactured sand with nominal maximum aggregate size of 2.36 mm and fineness modulus of 2.9 was used as fine aggregate. A polycarboxylate-based high-range water reducing admixture (HRWRA) was used to achieve the required flowability for the mixtures. In addition, to increase the plastic viscosity and cohesion of printing mixtures, a commercially available VMA for anti-washout concrete was used. Polypropylene fiber was also used as a shrinkage reinforcement for a printing mixture, as it inhibits and controls the formation of plastic and drying shrinkage cracking in concrete [15]. This is highly important for printing concrete since a higher rate of water evaporation is anticipated for a printed structure, given there is no formwork covering the surface of freshly printed elements. The length of fiber used in this study is 6 mm and tensile strength is 415 MPa. Furthermore, densified silica fume was used as a supplementary cementitious material (SCM). SCMs are well known for causing improvement in various properties of concrete in short and long term [16–18]. A highly-purified attapulgite clay with average particle length of 1.75 μm and average particle diameter of 3 nm was also used in this study. The specific gravity of this clay is 2.29 and the average length divided by average diameter of clay particles is 583, indicating a high aspect ratio [19]. Therefore, they may form a highly entangled gel even at a small addition rate, provided proper dispersion. This Nano-clay is commercially available and several studies [19–21] have investigated the influence of its addition on different properties of other special concretes, such as formwork pressure of self-consolidating concrete.

Four mixtures were used in this study. Total cementitious materials content and water/cementitious materials ratio were kept constant in all mixtures at 600 kg/m^3 and 0.43, respectively. The details of mixtures are presented in Table 1. For NCPM mixture preparation, Nano-clay was initially mixed with water and introduced as a suspension to the mixer.

Table 1. Mixture proportions of printing mixtures [22]

Mixture ID	Fine aggregate (SSD)	Portland cement	Free Water	Silica fume	Fiber	Nano-clay	HRWRA	VMA
	kg/m^3	kg/m^3	kg/m^3	kg/m^3	kg/m^3	%[a]	%[a]	%[a]
PPM	1379	600	259	0	0	0	0.05	0.11
SFPM	1357	540	259	60	0	0	0.16	0
FRPM	1379	600	259	0	1.18	0	0.06	0.10
NCPM	1379	600	259	0	0	0.30	0.15	0

[a]The percentages are reported by cementitious materials mass.

3 Testing and Results

Based on the suggested procedure, workability of printing mixtures is described and evaluated in terms of print quality, shape stability, robustness and printability window. The details and proposed evaluation methods are discussed in this section.

Print Quality

"Print quality" is related to the properties of a printed layer such as surface quality and dimensional conformity/consistency when using a specific printing mixture. A printing mixture could be considered acceptable when the three following requirements are satisfied: (a) The printed layer must be free of surface defects, including any discontinuity due to excessive stiffness and inadequate cohesion; (b) The layer edges must be visible and squared (versus round edges); and (c) Dimension conformity and dimension consistency must be satisfied by printed layer. Based on three proposed criteria, the print quality of a mixture could be evaluated and acceptance decision could be made.

Figure 2 presents a case where a mixture with poor workability is rejected by the first print quality requirement, due to observation of discontinuity. With respect to the third print quality requirement, "dimension conformity" guarantees the dimensions of the printed layers are within an acceptable range of the target dimensions (Fig. 3a), while "dimension consistency" refers to changes in width of a printed layer and acceptable variations (Fig. 3b). It should be noted these dimensional limitations are set for a fresh concrete layer and do not consider variations caused by shrinkage over time. In this study, a width of 38.1 mm was designed for each layer. After running a large number of experiments, it was concluded a 10% error in the target width is a reasonable bound for accepting or rejecting printed layers. In other words, the width of printed layers using all the accepted printing mixtures was in the range of 38.1–42 mm. It should be noted five measurements were done along each printed layer to assess the dimension conformity, and for each mixture, the experiment was carried out four times (four replicates of each mixture). The print quality of a mixture was considered "acceptable" only if the three requirements were satisfied by all four mixture replicates. The printing mixtures presented in Table 1 were selected based on aforementioned print quality requirements. A trial and error approach was adopted for this purpose, as currently there is no guideline or suggested procedure for design and testing of printing mixtures.

Fig. 2. Low print quality (tearing) of extruded layer caused by inappropriate workability

Shape Stability

Shape stability is a critical property of fresh printing concrete, which refers to the concrete ability to resist deformations during layer-wise concrete construction. More specifically, there are three main sources of deformation which apply to a deposited layer: (a) self-weight, (b) weight of following layer(s) which will be printed on top of it, and (c) the extrusion pressure. Based on observations, a mixture with acceptable print quality produces a layer without visible deformations due to self-weight. However, the two latter parameters could possibly lead to undesirable deformations when the following layer(s) are printed. This highlights the importance of laboratory testing of shape stability of printing concrete during the mixture design stage.

In this study, to obtain a realistic notion of the time gap required between deposition of consecutive layers, the layer-by-layer construction of a one-story 108 m^2 house with two bedrooms and one bathroom designed by an architectural and design company [23] is considered. Based on the structure's plan, the nozzle traveling distance for each layer was measured as 67 m. Considering the linear printing speed of 60 mm/s, an interlayer time gap of 19 min was obtained. The calculated 19-min time gap, as well as the worst-case scenario of a zero time gap, were used for shape stability assessment of the four developed mixtures. "layer settlement" test was developed and carried out to study the shape stability of printing concrete. In this test, two concrete layers were printed on top of each other with a specific time gap. A camera was placed in front of printed layers, a ruler was placed next to the layers as a scale and photos were taken both before and after the second layer was printed. Then, ImageJ software, a public domain Java-based image processing program [24], was used to analyze the photos and to measure layer settlement. The average of five readings for a printed layer

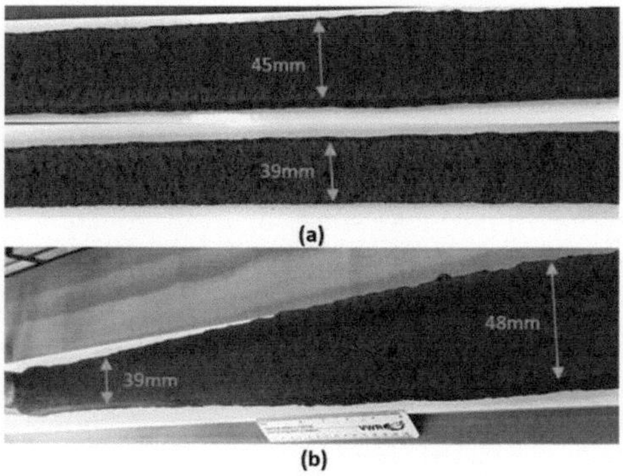

Fig. 3. (a) Variations in width of printed layer for different mixtures at the same printing speed (dimension conformity). (b) Variations in width of a single layer (dimension consistency) [22]

was reported as a test result, while the average of three tests (three printed layers) was used as the final result for a printing mixture. For each experiment, the bottom layer was printed as soon as the mixing procedure was complete.

The layer settlement test results for both scenarios, namely, zero and 19-min time gaps, are presented in Table 2. As anticipated, shape stability of a printed layer improves over time and smaller deformations were measured for all mixtures when the time gap was 19 min. The results indicate all mixtures except PPM possess high shape stability when there is a 19-min time gap between layers, denoted by no visible deformations of bottom layer. An average 1.5 mm deformation, equivalent to 5.9% of layer height, was measured for the layers printed with PPM.

Table 2. Layer settlement test results (mm) [22]

Mix ID	Time gap: 0				Time gap: 19 min			
	Test 1	Test 2	Test 3	Average reading	Test 1	Test 2	Test 3	Average reading
PPM	Collapse	Collapse	Collapse	–	1.9	1.1	1.6	1.5 [0.3]
SFPM	2.2	1.8	1.5	1.8 [0.3][a]	0	0	0	0
FRPM	2.8	3.3	2.5	2.9 [0.3]	0	0	0	0
NCPM	2.0	1.1	1.6	1.6 [0.4]	0	0	0	0

[a]Values in brackets are standard deviations of the deformation measurements for each set of 3 layers (mm)

For PPM mixture with zero time gap, considerable deformations occurred after the second layer was deposited. Considering the significant changes in both width and height of the layer in this case, the result of testing was reported as “collapse”. For the zero time gap, the lowest deformations were measured for SFPM and NCPM mixtures, where the average layer height reduced by approximately 6.7%. However, considering the standard deviation of the obtained results, there is no statistically significant difference between shape stability of these two mixtures. Acceptable print quality, as defined in this study, was not found to guarantee high shape stability, as the four printing mixtures with acceptable print quality showed different levels of shape stability. As such, this property must be separately evaluated during mixture design.

It should be mentioned that yield stress of fresh concrete is the main parameter determining the shape stability before setting. Yield stress increases over time in the absence of agitation and shear stress [7, 25]. This is due to the nucleation of cement grains at their contact point by C-S-H formation during the dormant period before the setting time [26]. Several researchers have reported a linear increase in yield stress during the dormant period [7, 25, 27], which suggests a corresponding linear increase of shape stability with time. Another important consideration for a fundamental study of shape stability is thixotropy, defined as build-up and breakdown of internal 3D structure within cementitious paste. This phenomenon happens due to flocculation or coagulation and dispersion of cement particles which, in turn, result from inter-particle forces and chemical connections [28, 29]. Build-up and breakdown of the internal structure causes an increase and reduction in viscosity of fresh paste, respectively [30]. Considering the process of concrete 3D printing, where the cementitious mixture undergoes considerable shear stress and agitation before being deposited as a layer, these changes in the internal structure and the consequent influence on shape stability need to be considered. For example, Kawashima et al. [20] reported a higher build-up rate of the internal structure after shear-induced breakdown for Nano-clay incorporated cementitious mixtures, especially at early ages. This faster "structuration at rest" of Nano-clay included mixtures could explain the enhanced shape stability of NCPM mixture in this study.

Robustness

There are various chemical admixtures and materials (variety of supplementary cementitious materials, for example) to be used for developing printable mixtures. It is recommended that several mixtures with acceptable print quality and shape stability are developed and the mixture with highest robustness against variations is selected. According to RILEM TC 288MPS [31], the concrete robustness (as a general concept) is the characteristic of a mixture representing its tolerance to variations in constituent characteristics and quantities, variations during concrete mixing, transport, and placement, as well as environmental conditions [32]. Considering the high importance of fresh properties of printing concrete, robustness of printing mixture can be defined as the capacity of mixture to retain its printability when small variations in the properties or quantities of the constituent materials occur. This is very important from practical standpoint, since the concrete extrusion process needs to continue for one or few days uninterrupted and as planned, in order to realize the promise of fast building construction made by construction-scale 3D printing. Higher robustness of the printing mixture could minimize the problems caused by usual variations in the material. Changes in aggregate moisture content (which, in turn, changes the free water content) and variations in particle size distribution of aggregates are considered the most common sources of variation in concrete mixtures. The focus of current research is on the robustness of printing mixture against changes in the water content. ± 10 l/m^3

change in water content was selected as the level of variation used for assessment. The basis for selection of this value is EFNARC recommendations for self-consolidating concrete (SCC): "A well designed and robust SCC can typically accept a 5 to 10 l/m^3 change in water content without falling outside the specified classes of performance when fresh" [33].

Based on shape stability and print quality test results, NCPM and SFPM were selected for robustness evaluation. As such, 4 layers were printed using each mixture and each layer was measured at 5 different points using a digital caliper (similar to print quality test) in order to measure the average width of the layer. Change in the width of printed layers was used as a measure of mixture sensitivity to variations. Table 3 presents the average width of the two reference mixtures as well as deliberately altered mixtures.

Table 3. Average width of layers made with reference and altered mixtures

Mix ID	Average layer width (mm)
SFPM	39.4 [0.9][a]
SFPM+10	47.8 [2.7]
SFPM-10	Not printable (tearing)
NCPM	40.1 [1.2]
NCPM+10	44.3 [1.8]
NCPM-10	Not printable (tearing)

[a]Values in brackets are standard deviations of the test results (mm)

As a reminder, based on print quality experiments, the width of an acceptable printing mixture must be in the range of 38.1–42 mm. Therefore, the results show that none of the two mixtures (NCPM and SFPM) were able to retain acceptable print quality when the water content changed by $\pm$10 l/m^3. Both mixtures show similar behavior when the amount of free water is reduced by 10 l/m^3, that is the resulting mixture is not printable. On the other hand, when the free water is increased by this amount the layer width increases. The average increase for NCPM is 10.7%, while for SFPM it is 21.3%. This implies relatively higher robustness of NCPM mixture. In other words, when considering print quality, the NCPM mixture shows higher resistance to variations in water content compared to SFPM.

Printability Window

The printability window is the period of time during which the printing mixture could be extruded by the nozzle with an acceptable quality, considering the workability loss that occurs over time. This timing of material delivery to the nozzle is essential to the operation of a full-size building printer such as the CC machine. Two time limits are introduced herein to define the printability window of a mixture, the printability and blockage time limits. The printability limit, refers to the time when the quality of printed layer is affected as a result of workability loss, recognized by triple "print quality" requirement, whereas the blockage limit is the time when the concrete cannot be guided out of the printing nozzle at all, so further delay would result in mixture solidification and damage to the nozzle. It should be noted that the experimental results presented elsewhere [22] proved that conventional testing such as setting time measurements and workability loss measurements cannot replace the direct measurement of printability limit and blockage limit.

4 Future Work

There are numerous unexplored areas in the emerging field of construction-scale 3D printing. Some major topics and unresolved challenges that need to be further investigated include curing and early-age strength development of 3D printed concrete, shrinkage of 3D printed concrete, and structural performance and durability of 3D printed structures. Another important area of research which needs to be explored is quality monitoring of concrete 3D printing process and the printed structure.

In construction-scale 3D printing layers of 1–2 in. dimensions are being extruded, and small variations in the extruded layers could possibly lead to process failure. Deviations from specifications could result in extrusion of layers with undesirable properties, which later can cause significant issues such as collapse of freshly printed structure. As such, automated construction processes need real-time, in-process, accurate quality monitoring techniques specially developed and customized for construction-scale 3D printing such that any issue is recognized instantaneously and could be modified in a timely fashion. One of the techniques which is currently investigated is based on computer vision. Figure 4 shows examples of using computer vision for real-time extrusion quality monitoring, where the target layer width is 1.5 in.

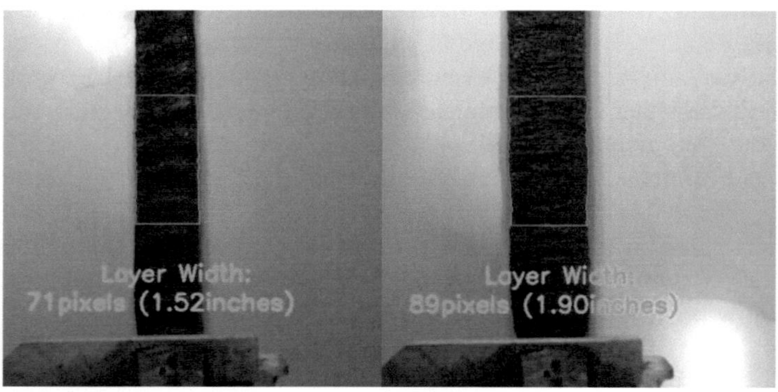

Fig. 4. Real-time extrusion quality monitoring using computer vision

Early age strength monitoring is another topic which needs to be addressed. This is highly important since, compared to conventional construction, it takes much shorter time to print a building and ideally a building will be ready for residence few days after the concrete 3D printing process begins. As such, a reliable strength monitoring system is needed to determine the earliest time when it is safe to start the finishing operations or for residents to use the building. To this end, some ongoing experiments are focused on monitoring electrical resistivity development of freshly printed concrete for early age concrete strength estimation. Some initial results are presented in Fig. 5.

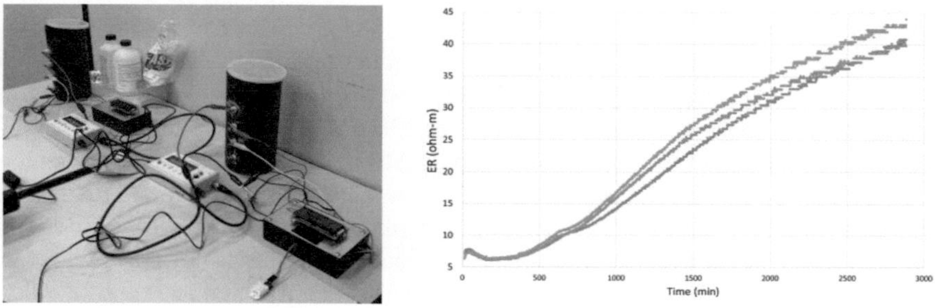

(a) (b)

Fig. 5. (a) Four probe electrical resistivity measurement setup. (b) Initial electrical resistivity results for a printing mixture (repeated 3 times) during first 48 h after mixing

It should be mentioned that industrial construction-scale 3D printers are at early stages of development and still reliability of developed systems is under question. When the current challenges are solved and reliable printers are available, there would be numerous possibilities with respect to the building design and the construction process. In specific, one interesting possibility is use of functionally graded materials (multi-material 3D printing). Based on the structural design and performance requirements, different parts (or elements) of a building could be printed using different materials. Printing mixtures with different values of compressive strength, unit weight, modulus of elasticity, etc. could be used in one project. This concept is visualized using color concrete (iron oxide pigments) in Fig. 6.

Fig. 6. Visualization of graded materials in construction-scale 3D printing using color concrete

5 Conclusions

A step-by-step procedure is proposed for laboratory testing of fresh printing mixtures. The testing procedure includes assessment and modification of the print quality, shape stability, robustness, and printability window of a mixture. It is based only on different properties of already extruded layers; therefore, the procedure is not dependent on extrusion mechanism and could be applied to different concrete 3D printing systems. To demonstrate the suggested procedure and test methods, a laboratory-scale linear concrete printer machine capable of printing 38.1 mm × 25.4 mm layers is constructed. Four different printing mixtures are developed and the results of test methods are reported. Acceptable print quality, as defined in this study, is not found to guarantee high shape stability, because the four printing mixtures with acceptable print quality show different levels of shape stability. Experimental data revealed that inclusion of silica fume and Nano-clay (a highly-purified attapulgite clay) enhance shape stability of fresh printing mixture, while minor improvement is observed from polypropylene fiber addition. The experimental results also suggest that layer settlement test result could be used as an indicator of shape stability. Furthermore, comparison of NCPM and SFPM mixtures in terms of robustness proves the superior performance of NCPM against changes in water content (i.e. higher robustness). Finally, perspectives on research areas which are necessary for advancement of construction-scale 3D printing are provided. In specific, real-time quality monitoring of concrete extrusion process seems to be highly important and a critical step towards reliable automated construction systems.

References

1. ASTM F2792-12a (withdrawn), American Society of Testing and Materials (2012)
2. Buswell, R.A., Soar, R.C., Gibb, A.G., Thorpe, A.: Freeform construction application research. In: Advances in Engineering Structures, Mechanics & Construction, pp. 773–780 (2006)
3. Khoshnevis, B.: Automated construction by contour crafting-related robotics and information technologies. Autom. Constr. **13**(1), 5–19 (2004)
4. Naboni, R., Paoletti, I.: Advanced Customization in Architectural Design and Construction. Springer, Berlin (2015)
5. Wangler, T., Lloret, E., Reiter, L., Hack, N., Gramazio, F., Kohler, M., Bernhard, M., Dillenburger, B., Buchli, J., Rousel, N., Flatt, R.: Digital concrete: opportunities and challenges. RILEM Tech. Lett. **1**, 67–75 (2016)
6. Wolfs, R.: 3D Printing of Concrete Structures. Eindhoven University of Technology (2015)
7. Perrot, A., Rangeard, D., Pierre, A.: Structural built-up of cement-based materials used for 3D-printing extrusion techniques. Mater. Struct. **49**(4), 1213–1220 (2016)
8. Le, T.T., Austin, S.A., Lim, S., Buswell, R.A., Gibb, A.G.F., Thorpe, T.: Mix design and fresh properties for high-performance printing concrete. Mater. Struct. **45**, 1221–1232 (2012)
9. Shah, S.: Design and Application of Low Compaction Energy Concrete for Use in Slip-Form Concrete Paving. Infrastructure Technology Institute, Northwestern University (2008)

10. Voigt, T., Mbele, J., Wang, K., Shah, S.: Using fly ash, clay, and fibers for simultaneous improvement of concrete green strength and consolidatability for slip-form pavement. J. Mater. Civ. Eng. **22**(2), 196–206 (2010)
11. Kazemian, A., Yuan, X., Meier, R., Cochran, E., Khoshnevis, B.: Construction-scale 3D printing: shape stability of fresh printing concrete. In: 12th International Manufacturing Science and Engineering Conference (MSEC 2017), Los Angeles (2017)
12. Nerella, V., Krause, M., Nather, M., Mechtcherine, V.: Studying printability of fresh concrete for formwork free concrete on-site 3D printing technology (CONPrint3D). In: 25th Conference on Rheology of Building Materials, Regensburg (2016)
13. Anell, L.: Concrete 3D printer. Lund University, Sweden (2015)
14. Ma, G., Wang, L.: A critical review of preparation design and workability measurement of concrete material for largescale 3D printing. Front. Struct. Civ. Eng. **12**(3), 382–400 (2018)
15. BASF. https://www.master-builders-solutions.basf.us/en-us/products/masterfiber/1649. Accessed 1 Nov 2016
16. Ramezanianpour, A.A., Zolfagharnasab, A., Zadeh, F.B., Hasanpour, S., Boushehri, R., Pourebrahimi, M.R., Ramezanianpour, A.M.: Effect of supplementary cementing materials on concrete resistance against sulfuric acid attack. In: Hordijk, D., Luković, M. (eds.) High Tech Concrete: Where Technology and Engineering Meet, pp. 2290–2298. Springer, Cham (2018)
17. Ramezanianpour, A., Kazemian, A., Moghaddam, M., Moodi, F., Ramezanianpour, A.: Studying effects of low-reactivity GGBFS on chloride resistance of conventional and high strength concretes. Mater. Struct. **49**(7), 2597–2609 (2016)
18. Ramezanianpour, A., Ghiasvand, E., Nickseresht, I., Moodi, F., Kamel, M.: Engineering properties and durability of concretes containing limestone cements. In: Second International Conference on Sustainable Construction Materials and Technologies, Coventry University and UWM Center for By-Products Utilization (2010)
19. Kawashima, S., Chaouche, M., Corr, D.J., Shah, S.P.: Influence of purified attapulgite clays on the adhesive properties of cement pastes as measured by the tack test. Cem. Concr. Compos. **48**, 35–41 (2014)
20. Kawashima, S., Chaouche, M., Corr, D.J., Shah, S.P.: Rate of thixotropic rebuilding of cement pastes modified with highly purified attapulgite clays. Cem. Concr. Res. **53**, 112–118 (2013)
21. Kim, J., Beacraft, M., Shah, S.P.: Effect of mineral admixtures on formwork pressure of self-consolidating concrete. Cem. Concr. Compos. **32**(9), 665–671 (2010)
22. Kazemian, A., Yuan, X., Cochran, E., Khoshnevis, B.: Cementitious materials for construction-scale 3D printing: laboratory testing of fresh printing mixture. Constr. Build. Mater. **145**, 639–647 (2017)
23. FreeGreen. https://www.houseplans.com/plan/1160-square-feet-2-bedroom-1-bathroom-0-garage-modern-39050. Accessed 15 June 2016
24. ImageJ. https://imagej.nih.gov/ij/
25. Josserand, L., Coussy, O., Larrard, F.D.: Bleeding of concrete as an ageing consolidation process. Cem. Concr. Res. **36**(9), 1603–1608 (2006)
26. Roussel, N., Ovarlez, G., Garrault, S., Brumaud, C.: The origins of thixotropy of fresh cement pastes. Cem. Concr. Res. **42**(1), 148–157 (2012)
27. Roussel, N.: A thixotropy model for fresh fluid concretes: theory, validation and applications. Cem. Concr. Res. **36**, 1797–1806 (2006)
28. Wallevik, J.: Thixotropic investigation on cement paste: experimental and numerical approach. J. Nonnewton. Fluid Mech. **132**, 86–99 (2005)
29. Wallevik, J.: Rheological properties of cement paste: thixotropic behavior and structural breakdown. Cem. Concr. Res. **39**(1), 14–29 (2009)

30. Heirman, G.: Modelling and quantification of the effect of mineral additions on the rheology of fresh powder tupe self-compacting concrete. Arenberg Doctoral School of Science, Engineering and Technology (2011)
31. Khayat, K., De Schutter, G.: Mechanical properties of self-compacting concrete. RILEM Technical Committee TC 228-MPS (2013)
32. Ghoddousi, P., Salehi, A.M.: The robustness of self consolidating concrete due to changes in mixing water. Period. Polytech. Civ. Eng. **61**(2), 216–225 (2017)
33. The European Guidelines for Self-Compacting Concrete: Specification, Production and Use, 1 May 2005. http://www.efnarc.org/pdf/SCCGuidelinesMay2005.pdf

Characterization of 3D Printing Mortars Made with OPC/CSA Mixes

Noura Khalil[1(✉)], Sébastien Rémond[1], Bilal Baz[1,2], and Georges Aouad[2]

[1] IMT Lille Douai, Univ. Lille, EA 4515 - LGCgE – Laboratoire de Génie Civil et géoEnvironnement, Département Génie Civil & Environnemental, 59000 Lille, France
noura.khalil@imt-lille-douai.fr
[2] Faculty of Engineering, University of Balamand, UOB, Al Koura, Lebanon

Abstract. Printable mortars used in 3D printing of cementitious materials must have rheological behaviour and setting rigorously controlled. In this research, mixes made of two types of cement, ordinary Portland cement (OPC) and Calcium Sulfoaluminate cement (CSA) are adopted to control the printability of a mortar. Extrudability, buildability and comparable compressive strength to that of a traditional mortar are the specifications required by the printable mortar. Different mixes of OPC/CSA cement pastes (ranging between 0 and 10% of CSA) and 2 mixes of OPC/CSA mortars (0 and 7% of CSA) are studied. The cement pastes are studied by isothermal calorimetry and rheometer. Heat of hydration and cement pastes' yield stress increase with the dosage of CSA in the mix. A mortar made out of 7% CSA in previous article [1] is then tested in laboratory and their behaviour is compared to the results of the cement pastes.

Keywords: Sulfoaluminate cement · Heat of hydration · Thixotropy

1 Introduction

One method of additive manufacturing (AM) is 3PD printing (3DP). Objects realized by this technique are automated using CAD software by "joining materials to make objects from 3D model data, usually layer upon layer" [2]. Several types of materials are being so far well covered by different 3DP methods [1, 3, 4]. In the field of construction, 3DP could present some advantages by reducing time, manpower and cost [5, 6].

The development of 3D printing is growing all over the world [7–11]. However, a lack of standards for cementitious materials used in 3DP specifications leads to restricting the construction from being a field of application of this technology [9]. Two basic characteristics are defined for the printability of a mortar: extrudability and buildability. The first deals with the workability of the material that must not be blocked during its extrusion from the nozzle of the printer. The second reflects the stiffness of the extruded mortar to handle the layering without falling down.

Portland cement is the common component of mortars used in previous studies but accelerated differently to adapt their characteristics to the needs of 3DP [10–14].

T. Wangler and R. J. Flatt (Eds.): DC 2018, RILEM Bookseries 19, pp. 53–60, 2019.
https://doi.org/10.1007/978-3-319-99519-9_5

Diverse mixture compositions of 3D printing mortars and different automated printing processes are used in these studies with different nozzle sizes.

Another method for changing the Portland cement reactivity is to mix it with small amounts of CSA. In fact, CSA cement increases the cement hydration resulting in rapid hardening cements. Blends made out of calcium sulfoaluminate cement (CSA) and ordinary Portland cement (OPC) are applied in special applications. However, one inconvenient of this mixture is the expansion resulting from the CSA addition provoking voids formation and lower compressive strength [15, 16].

The main objective of this paper is to characterize an OPC/CSA cement paste from a mortar used in 3DP considered as printable. The following procedure is adopted. First, isothermal calorimetry is used to follow the heat of hydration of different OPC/CSA cement pastes. Then, the acceleration effect of the cement hydration on the rheological characteristics of these binders is followed. The results are used to understand the behavior of the mortars developed in the previous paper [1].

2 Materials and Methods

Ordinary Portland cement (CEM I 52.5) EXTREMAT® CEM I 52.5 N-SR3 CE PM-CP2 NF I (99% Clinker) and a Calcium Sulfo Aluminate Alpenat cement, both from Vicat company have been used in this study. Chemical and mineralogical compositions of both cements are presented in Tables 1, 2, 3 and 4. A new generation polyvalent non chlorated acrylic copolymer superplasicizer (SP)/high water reducer, SIKA VISCOCRETE TEMPO 11 has been added. Carrieres du Boulonnais provided calcareous 0/2 mm crushed sand with 19% particles smaller than 63 μm. These data have been cited in previous article where materials were used in other characterizations [1].

Table 1. Chemical composition of OPC [1]

Composition	Na2O eq.	Fire loss 950 °C	MgO	Al2O3	SO3	Cl-	S2-	Insoluble residue
CEM I	0.46	1.20	0.91	3.65	2.24	0.06	0	0.45

Percentages of OPC and CSA are used to nominate the cement pastes in addition to the label P indicating the Paste as follows P%OPC/%CSA. Therefore, the OPC percentage is listed followed by that of CSA. 2%, 5%, 7%, and 10%, of the total weight of the OPC/CSA cement is being replaced by CSA. These dosages are considered not to affect the reactional mechanisms of the blends since they are lower than 10% [12].

CALMETRIX calorimeter (I-CAL 2000 HPC) is used in order to follow the heat of hydration at 20 °C with a water/cement ratio of 0.5. Materials are kept at 20 °C 24 h before being used. To prepare the paste, manual mixing of 3 min is applied with 40 g of cement and 20 g of water.

Table 2. Mineralogical composition of OPC and main characteristics [1]

Composition	C3S	C2S	C3A	C4AF	Blaine specific surface area (cm^2/g)	Density	Initial setting time (min)
CEM I	62.6	16.5	1.5	13.1	3940	3.19	168

Table 3. Chemical composition of CSA cement [1]

Composition	Na2O eq.	Fire loss 950 °C	SiO2	Al2O3	Fe2O3	CaO	SO3	Cl-
CSA	0.16	3.8	8.16	18.22	7.64	43.60	15.24	0.05

Table 4. Mineralogical composition of CSA cement and main characteristics [1]

Composition	$C4A3\bar{S}$	C2S	C3MS2	C3FT	$C\bar{S}$	Free Lime	Blaine specific surface area (cm^2/g)	Density	Initial setting time (min)
CSA	54.3	29.1	4.5	9.3	0.4	0.2	4500	2.97	42

Rheological measurements are performed with an Anton Paar MCR 102 rheometer in order to assess the evolution of yield stress of cement pastes during the first hour after mixing. The parallel plate geometry has been used with plates of 25 mm, made rough by gluing abrasive paper on their surface. Measurements have been carried out with a gap of 1 mm at a temperature of 20 °C. Cement pastes, made out of 6 g of cement and 3 g of water previously stored at 20 °C, are mixed for 1:40 min, before being placed on the fixed base of the rheometer. After the displacement of the rotating top section to the measuring position, the excess is removed and measures start at 3 min, this moment is considered as the origin of times in the following. Shear stress is measured by applying a constant shear rate of 0.025 s-1 during 30 s, the yield stress corresponding to the peak value obtained. 5 measurements of shear stress are taken during 25 min (each measurement is preceded by a period of rest of 4:30 min).

Table 5 presents the compositions of the two manufactured mortars M100/0 and M93/7 [1]. The same denomination is given to the mortars with a letter M to stand for Mortar instead of P. Label M replaces the P to in order to denominate the Mortars used. Extrudability of the mortars was adjusted by modifying the sand weight.

The fresh properties of mortars have first been studied with a manual device of 3DP at the laboratory scale, using a mastic gun with a conical nozzle to check the extrudability and buildability of the mortar [1].

Table 5. Mixing proportions of M100/0 and M93/7 [1]

Composition	M100/0	M93/7
Sand (g)	850	800
OPC (g)	682.75	627.75
CSA (g)	0	47.25
Water (g)	236.25	236.25
SP (g)	1.76	1.76

3 Results

Heat of Hydration

OPC/CSA mixes present variable heat emission depending on the CSA percentage used as shown in Fig. 1. This evolution over 48 h has been presented in a previous article [1]. In Fig. 1, almost parallel and proportional augmentation of heat flow to the percentage of CSA is shown, except for P98/2 which presents a particular behavior for the 15 first minutes. In a previous article [1], a linear variation was proved between the cumulative heat of hydration and the percentage of sulfoaluminate cement up to 10%. Whatever the CSA dosage, all the cement pastes present similar behaviors, which indicates that the hydration mechanism is not changed for these percentages lower than 10% [12]. Intensity of the peaks during the first hour getting higher when CSA percentage is increasing is confirmed by other studies [12, 16]. Then, controlling the intensities and duration of this peak is done buy varying the CSA percentage.

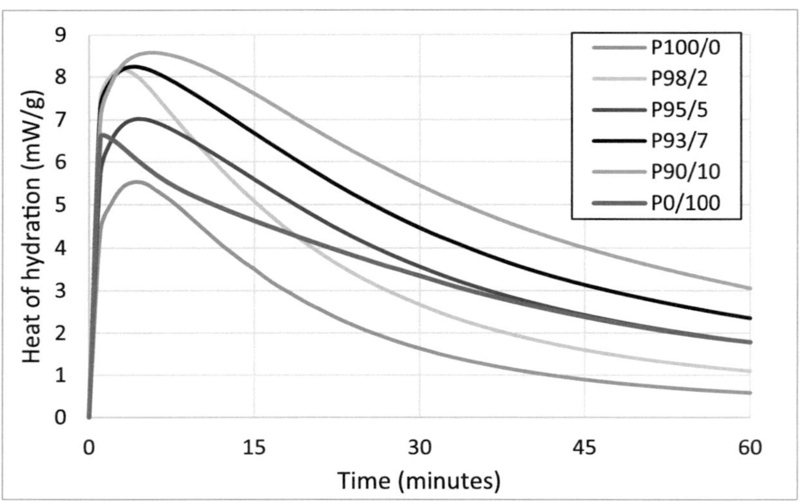

Fig. 1. CSA percentage affecting the increase of the heat emitted by OPC/CSA cement pastes

Yield Stress

The evolution of yield stress is studied along a period of 25 min after positioning the material on the rheometer cell. The materials are left at rest for 270 s before taking the first value and before every new yield stress measurement performed. A constant shear rate of 0.025 s-1 is applied for 30 s for the measurement of yield stress. Figure 2 shows the variation of shear stress and the rest periods for three successive measurements. Yield stress corresponds to the peak value obtained under a shear rate of 0.025 s-1. Two results obtained with 2 cement pastes are presented (P100/0 and P93/7). Results are very reproducible with the pure OPC cement paste (P100/0). When CSA is added to the paste, larger differences are observed between the two replicate measurements. This is probably due to the higher reactivity of the mix which leads to fast variations of the paste behavior. The average of the maximum shear stresses recorded is considered as the yield stresses of each cement paste in the following.

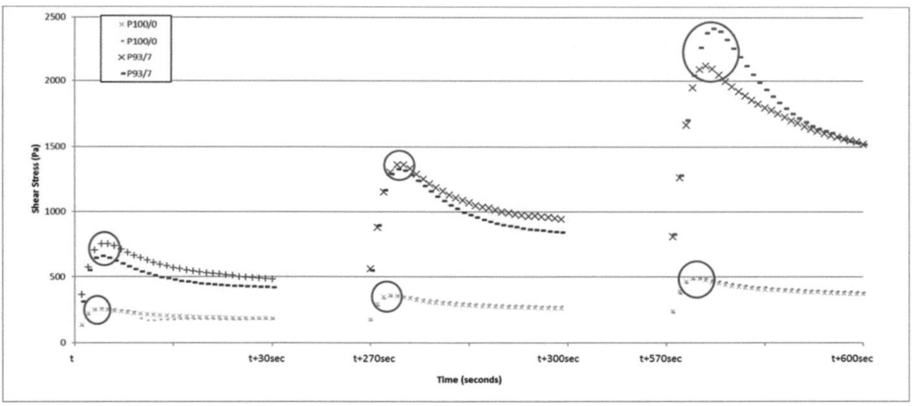

Fig. 2. Shear stress diagrams of P100/0 and P93/7 in function of time at different intervals (note that the time is restricted during the rest period between two consecutive measurements)

Figure 3 presents the variation of yield stress in function of time during the first 1500 s for each cement paste. It is shown to increase when the percentage of CSA increases. The variation seems to increase linearly for small CSA dosages (less than 7%) and not for higher CSA content. Similarly to Roussel et al. [17], we are trying to define the behavior of the cement pastes with the Athix parameter (then to analyze the curves through linear trends). Therefore, in this work the curves can be considered linear during the first 1200 s. The increase of yield stress over the resting time can be calculated according to Roussel's linear model as follow $\tau_0(t) = A_{thix}t + \tau_{0,0}$ [17]. Table 6 presents the Athix and the R2 calculated from the linear equation of Roussel and Fig. 4 the variation of the Athix in function of the CSA content.

The flocculation rate Athix, presented in Table 6 increases when the percentage of CSA added to the cement pastes increases. Blended cement pastes present higher yield stresses than the one made of 100% OPC and 100% CSA.

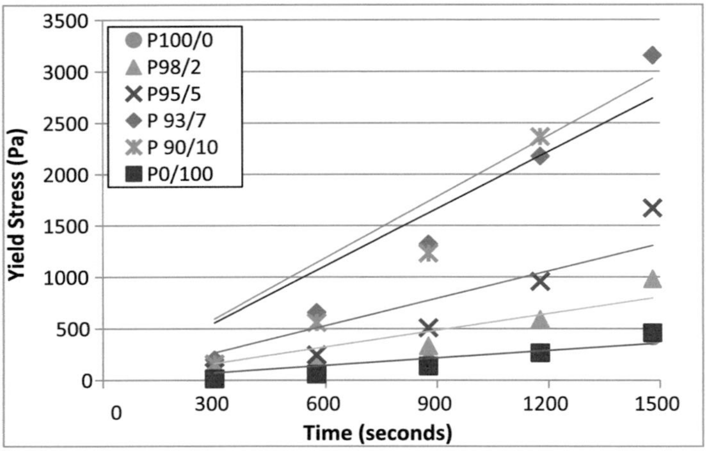

Fig. 3. Yield stress evolution for the 6 cement pastes with the Roussel linear model applied to the first 25 min

Table 6. Flocculation rate Athix and determination factor R^2 of the different cement pastes during the first 25 min

Cement paste	P100/0	P98/2	P95/5	P93/7	P90/10	P0/100
Athix (Pa/s)	0.24	0.54	0.88	1.85	1.98	0.24
R^2	0.88	0.85	0.81	0.9	0.85	0.79

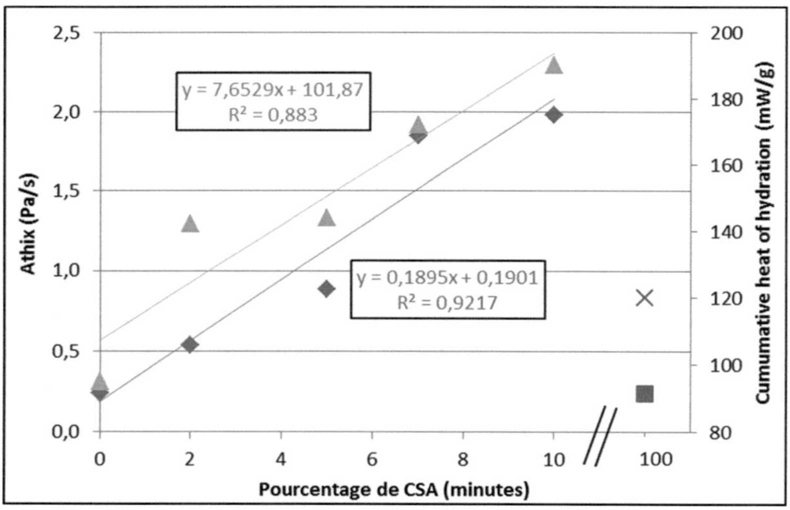

Fig. 4. Variation of Athix and of the cumulative heat flow in function of CSA percentage

Therefore, both the heat of hydration and the flocculation rate depend on the CSA percentage in an OPC/CSA mix. Figure 4 presents the variations of Athix and cumulative heat of hydration as a function of CSA dosage during the first 25 min. Both relationships can be considered linear.

The yield stress of P100/0 and P93/7 could reflect the change of the mortars' rheological behavior. Indeed, these results comply with the printability of the mortars tested with the manual device. The workable mortar M100/0 gains its buildability by replacing 7% of the OPC cement by CSA. The rapid yield stresses evolution of the P93/7 cement paste is resulting from the increase in the heat of hydration. Therefore, the gain in buildability of M93/7 may be interpreted by the change in thixotropy of its cement paste with respect to that of M100/0.

4 Conclusion

OPC/CSA blends have proven to have an effect on both the heat of hydration and rheological behavior of the cement pastes. During the first hour, cement pastes made out of 100% OPC and 100% CSA present lower heat release than OPC/CSA mixes. As well, these blends present higher intensities when the percentage of CSA increases. Shear stress evolution of the OPC/CSA cement pastes during the same period of time is also increasing with the increase of CSA percentage. The rate of the stiffening is increased for the cement pastes containing higher amount of CSA cement and always higher than pure OPC and CSA cement pastes. This is reflected by the better buildability of the mortar containing CSA cement since once printed its shear stress increases rapidly. As well, for mortars, their rheological behavior is influenced by CSA dosage and the intensities of its heat of hydration but not on its reactional mechanism. Therefore, 7% of sulfoaluminate cement with ordinary 93% of Portland cement resulted in giving the required stiffness for the mortar and then being buildable. In addition, it did not reduce the workability of the mortar.

Acknowledgments. This research work has been carried out in the frame of the MATRICE Project, co-funded by the region "Hauts de France" and the European Union with the European Regional Development Fund. The authors also acknowledge the supply of materials from Vicat, Sika and the Carrières du Boulonais.

References

1. Khalil, N., Aouad, G., El Cheikh, K., Rémond, S.: Use of calcium sulfoaluminate cements for setting control of 3D-printing mortars. Constr. Build. Mater. **157**, 382–391 (2017)
2. American Society for Testing and Materials (ASTM). ASTM F2792-10 Standard Terminology for Additive Manufacturing Technologies (2013)
3. Vardhan, G.H., Charan, G.H., Reddy, P.V.S., Kumar, K.S.: 3D printing: the dawn of a new era in manufacturing? Int. J. Recent Innov. Trends Comput. Commun. **2**(8), 2321–2376 (2013)

4. Schwaiger, J., Lueth, T.C., Irlinger, F.: G-code generation for a new printing process based on 3D plastic polymer droplet generation. In: ASME International Mechanical Engineering Congress and Exposition, Volume 2A: Advanced Manufacturing ():V02AT02A006. https://doi.org/10.1115/imece2013-63152
5. Weller, C., Kleer, R., Piller, F.T.: Economic implications of 3D printing: market structure models in light of additive manufacturing revisited. Int. J. Prod. Econ. **164**, 43–56 (2015)
6. Le, T., Webster, J., Buswell, R., Austin, S., Gibb, A., Thorpe, T.: Fabricating construction components using layered manufacturing technology. In: Global Innovation in Construction Conference, Loughborough University, pp. 13–16 (2009)
7. Le, T.T., Austin, S.A., Lim, S., Buswell, R.A., Gibb, A.G.F., Thorpe, T.: Mix design and fresh properties for high-performance printing concrete. Mater. Struct. **45**(8), 1221–1232 (2012)
8. Le, T.T., Austin, S.A., Lim, S., Buswell, S., Law, R., Gibb, A.G.F., Thorpe, A.T.: Hardened properties of high performance printing concrete. Cem. Concr. Res. **42**, 558–566 (2012)
9. Wangler, T., et al.: Digital concrete: opportunities and challenges. RILEM Tech. Lett. **1**, 67–75 (2016)
10. Gosselin, C., Duballet, R., Roux, P., Gaudillière, N., Dirrenberger, J., Morel, P.: Large-scale 3D printing of ultra-high performance concrete - a new processing route for architects and builders. Mater. Des. **100**, 102–109 (2016)
11. Malaeb, Z., Hachem, H., Tourbah, A., Maalouf, T., El Zarwi, N., Hamzeh, F.: 3D concrete printing: machine and mix design. Int. J. Civ. Eng. Technol. **6**(6), 14–22 (2015)
12. Le Saoût, G., Lothenbach, B., Hori, A., Higuchi, T., Winnefeld, F.: Hydration of Portland cement with additions of calcium sulfoaluminates. Cem. Concr. Res. **43**(1), 81–94 (2013)
13. Hargis, C.W., Kirchheim, A.P., Monteiro, P.J.M., Gartner, E.M.: Early age hydration of calcium sulfoaluminate (synthetic ye'elimite, C 4A3S) in the presence of gypsum and varying amounts of calcium hydroxide. Cem. Concr. Res. **48**, 105–115 (2013)
14. Shermanl, N., Beretkal, J., Santoro, L., Valenti, G.L.: Long-term behaviour of hydraulic binders based on calcium sulfoaluminate and calcium sulfosilicate. Cem. Concr. Res. **25**, 113–126 (1995)
15. Chaunsali, P., Mondal, P.: Physico-chemical interaction betwe en mineral admixtures and OPC-calcium sulfoaluminate (CSA) cements and its influence on early-age expansion. Cem. Concr. Res. **80**, 10–20 (2016)
16. Trauchessec, R., Mechling, J.M., Lecomte, A., Roux, A., Le Rolland, B.: Hydration of ordinary Portland cement and calcium sulfoaluminate cement blends. Cem. Concr. Compos. **56**, 106–114 (2015)
17. Roussel, N.: A thixotropy model for fresh fluid concretes: theory, validation and applications. Cem. Concr. Res. **36**, 1797–1806 (2006)

Rheological and Water Transport Properties of Cement Pastes Modified with Diutan Gum and Attapulgite/Palygorskite Nanoclays for 3D Concrete Printing

Siwei Ma and Shiho Kawashima(✉)

Department of Civil Engineering and Engineering Mechanics, Columbia University, 500 West 120th Street, New York, NY 10027, USA
s-kawashima@columbia.edu

Abstract. This paper identifies and addresses two challenges in extrusion-based 3D concrete printing from a materials perspective. The first is the effect of self-weight and the weight of subsequent layers on structural build-up. And the second is the excessive water loss of printed materials due to the absence of formwork. Viscosity modifying admixtures (VMAs) are extensively used in cement-based 3D printing projects to achieve sufficient print quality, shape stability, and printability window. This study aims to evaluate VMAs' effects on the two aforementioned challenges through investigating the evolution of static yield stress under sustained stress at rest and water retention capacity of cement pastes modified with nanoclay and diutan gum.

Keywords: Sustained stress · Yield stress · Water retention · Nanoclay Diutan gum

1 Introduction

Additive manufacturing, generally known as 3D printing, was successfully applied in a large variety of domains, including industrial tooling and medical, in the last two decades. Its application in the construction industry was firstly introduced by Khoshnevis [1, 2] using cement-based materials as the printing material and has since been developed steadily in recent years [3–5].

Unlike most extrusion-based 3D printing processes, in which materials are immediately solidified upon deposition, cement-based filaments are still in the fresh state [5]. For instance, the study [6] estimated that an interlayer time gap would be 19 min for a 108-square-meter house at a printing speed of 60 mm/s, which means the printed cement-based materials are in the fresh state far before the typical initial setting time of 3–5 h. This highlights the importance of rheology, namely the control of shape stability and structural build-up rate of printed cement-based materials. This includes not only the ability to resist its own weight but the weight of subsequent layers on top of it. Some researchers determined that the maximum printing speed allowing the printed structure to stand by itself is $V = \sqrt{3}\,L\,A_{thix}/\rho g h$, where L is printing length, A_{thix} is static yield stress, ρ is the density of printing material, g is gravitational acceleration, and h is

T. Wangler and R. J. Flatt (Eds.): DC 2018, RILEM Bookseries 19, pp. 61–69, 2019.
https://doi.org/10.1007/978-3-319-99519-9_6

printing height [7]. However, A_{thix}, which is determined by a shear rheological approach, usually ignores the effect of self-weight and the weight of subsequent layers, which is referred to as sustained stress in this study. The static yield stress is strongly dependent on the sustained stress applied to the cement paste during rest [8]. As the layer of fresh concrete filament is deposited, its self-weight and weight of the subsequent layer above it would result in a stiffer structure due to the compression of layers. In this study, the effect of sustained stress will be further investigated.

Another issue of cement-based printing materials comes along with the major breakthrough of cement-based 3D printing - the elimination of formwork. The absence of formwork can expand aesthetic freedom and reduce materials, labor and time, as the construction of formwork represents 35–60% of the overall cost of concrete construction [9]. However, for layer-based, additive manufacturing, the freeform components exhibit relatively large exposed surface area, which is susceptible to rapid evaporation of water. It can hinder hydration and induce plastic shrinkage, ultimately impairing the strength and durability of the final printed structure. Thus, water transport behavior and water retention capacity play important roles in cement-based 3D printing.

To achieve the desirable shape stability during printing, viscosity modifying admixtures (VMAs) can be utilized. VMAs can be classified as organic, such as polysaccharides and polyacrylamides, and inorganic, such as clays. Previous work has investigated the effect of sustained stress on static yield stress of plain cement systems without additives [8]. It was found that static yield stress is linearly dependent on the sustained stress during resting. However, the effect of sustained stress on cement systems with VMAs has been seldom reported. To determine the suitability of VMAs for 3D printing applications, their role in water transport abilities during the fresh state must be investigated.

The present work evaluated the effect of nanoclay and diutan gum on the evolution of static yield stress under the influence of sustained stress and water retention capacity.

2 Materials and Methods

2.1 Materials

The cement used is a Class H oil well cement, and its chemical and mineralogical compositions are reported in Table 1. The Blaine fineness is 323 m^2/kg. Distilled water is used in all mixes. The water-cement ratio is 0.34 in all mixes.

Table 1. Chemical composition of Class H oil well Cement.

	SiO_2, %	Al_2O_3, %	Fe_2O_3, %	SO_3, %	CaO, %	MgO, %	C_3S, %	C_2S, %	C_3A, %	C_4AF, %
Compound Amount	22.0	2.8	4.4	2.8	64.4	2.6	62.0	16.3	0	13.4

A highly purified form of the mineral attapulgite, or palygorskite, was the clay chosen for the study. It is rod-like in structure – 1.75 μm in length and 30 nm in diameter [10]. Therefore, the single rod crystal can be referred to as a nanomaterial. However, the rod crystal usually exists in a bundle and aggregate state in nature. To disperse the attapulgite, it was blended with water in a Waring blender for 3 min to produce a suspension. A commercially available diutan gum gel whose solid content is 1% by mass was the gum chosen for this study. Diutan gum is a polysaccharide produced by Sphingomonas bacteria in fermentation [11]. It has high molecular weight and anionic charges.

All cement pastes were prepared using a hand mixer at a speed of 540 rpm. Cement pastes with nanoclay were prepared by adding the cement to the nanoclay suspension, then mixing for 3 min. Cement pastes with diutan gum were prepared by adding cement to water, mixing for 1.5 min, adding diutan gum gel, then mixing for another 1.5 min. Thus, the pastes prepared with the nanoclay and the diutan gum both had a total paste mixing time of 3 min. The material design for sustained stress test is shown in Table 2.

Table 2. Mix compositions of pastes.

Mix	Cement (g)	Distilled water (g)	Nanoclay (g)	Diutan gum gel (g)	Solid volume fraction (%)
Cement	50	17	0	0	48.29
NC	49.82	17	0.15	0	48.29
DG	50	15.75	0	1.25	48.32

2.2 Methods

Rheological Measurement

Rheological measurements were performed in a rotational rheometer to determine static yield stress and interstitial solution viscosity. The temperature was controlled by a circulating water bath maintained at 25 °C.

The static yield stress tests to evaluate the effect of sustained stress during resting were conducted using a vane geometry. All samples were pre-sheared for 600 s at a constant shear rate of 260 1/s to de-flocculate the cement structure until it reached steady-state – this ensured they shared the same shear history. The following step was stress controlled where sustained stresses below the dynamic yield stress were imposed, to ensure no flow, for 1200 s. The cement pastes were subjected to the following applied sustained shear stresses: 0 Pa, 10 Pa, and 20 Pa. Before performing the static yield stress measurement, we initialized the stress to 0 Pa for 5 s. A stress growth measurement was then applied to obtain the static yield stress by applying deformation at a constant shear rate of 0.1 1/s. The shear stress progressively develops to a maximum value and then decays to an equilibrium value. The static yield stress is defined as the peak shear stress value [12].

To better understand the results of water retention (discussed in the next section) the interstitial solution of pastes were tested to measure viscosity using a parallel plate geometry. The cement paste pore solution was extracted using centrifugation. Then the obtained interstitial solution was loaded into the rheometer. The apparent viscosity of the final interstitial solution was tested by a logarithmically increasing shear rate ramp from 1 1/s to 100 1/s, which was applied over 1000 s. The viscosity at 100 1/s was selected as the apparent viscosity of the studied polymer solution.

Water retention Experiments
A modified version of the filter paper method (DIN 18 555-7 [13]) was used to estimate the water retention capacity of the cement paste. A plastic ring was placed on top of the stack of qualitative, creped, fast flow filter papers (Fig. 1).

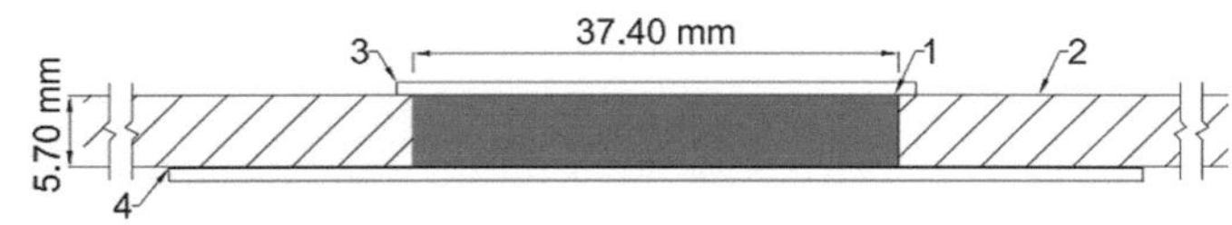

Fig. 1. Arrangement for determining the water retention of freshly-mixed mortars (1. Cement paste, 2. Plastic ring, 3. Cap to prevent evaporation, 4. Filter papers)

To determine the water retention capacity, the cement paste was filled into the ring. After a test time of 20 min, the ring and cement paste were carefully removed from the filter papers. The water retention capacity is calculated from the mass difference of filter papers before and after the test by Eq. 1:

$$\text{Water retention } (\%) = (1 - W_{abs}/W_O) \times 100 \tag{1}$$

where W_{abs} is the absorbed water in filter papers, W_O is the mixing water in cement paste sample.

3 Results

3.1 Sustained Stress Results

As shown in Table 3, sustained stress enhanced the strengthening of the plain paste and paste with diutan gum while the sustained stress effect on paste with nanoclay was inconclusive.

The effect of sustained stress on colloidal systems was discussed by [14]. The researcher proposed that anisotropic microstructures, which are induced by pre-shearing flow, is held by the sustained stress during resting. Hence, jamming at a given sustained stress may induce a change in the microstructure, which was evident by the linear increase in static yield stress with increasing sustained stress levels.

The study [14] further notes that the yield stress is only dependent on the sustained stress that is applied to the material during the liquid-solid transition, while it is insensitive to the sustained stress when the material is in its solid state. In the solid state, the percolated network is formed, and the sustained stress is transmitted only by the solid skeleton: the particles (or aggregates) stick on this percolated network and so

Table 3. Influence of sustained stress on static yield stress of cement pastes with nanoclay and diutan gum.

Cement mixes	Static yield stress (Pa) with different sustained stress (Pa)		
	0[a]	10	20
Plain	163.92[b] [1.52][c]	172.13 [2.82]	180.56 [2.34]
NC	223.36 [3.93]	215.15 [6.61]	235.53 [10.70]
DG	507.66 [22.50]	629.71 [27.73]	708.58 [8.37]

[a]Sustained stress applied during resting, [b]static yield stress, [c]standard error of the measured static yield stress (in Pa).

are insensitive to the applied stress. In contrast, during the liquid-solid transition, at rest cement particles aggregate together to form a percolated network of colloidal interactions. After a few hundred seconds later, the nucleation of CSH strengthens the soft colloidal network into a percolated rigid network [15]. Then the sustained stress may be transmitted through the formed rigid network.

It has been reported that paste with diutan gum shows a lower elastic modulus and retardation of hydration compared with plain cement paste [8]. This marks the steric hindrance effect of diutan gum. Polymer coils fill into the space between cement particles to prevent the formation of a solid skeleton of cement particles. Thus, the cement paste incorporating diutan gum is more sensitive to the sustained stress and so higher increase in static yield stress under applied stress during resting was observed. Meanwhile, sustained stress had a less significant effect on the paste with nanoclay. It may be tied to the enhanced interconnectivity of the fresh state microstructure of cement paste with nanoclay. The study [16] found that nanoclays can decrease critical strain measured by SAOS, which provides evidence that they can provide more contact points within the material and make its structure more interconnected.

3.2 Water Retention Capacity

The water retention capacity of cement pastes with various dosages of diutan gum (0–6.5%) and nanoclay (0–0.5%) were measured and the results were tied to the viscosity of the interstitial pore solution.

In general, diutan gum improved the water retention capacity of cement pastes. Most of the recent studies concluded that the increase of interstitial solution is correlated with the increase of water retention. This is consistent with what we observed in Figs. 2 and 3. Moreover, some studies highlighted the formation of polymer aggregates as the origin of higher interstitial solution viscosity and higher water retention [17, 18].

Attapulgite has a strong ability to absorb water, with an absorption capacity of approx. 200% of its own weight [19]. However, as shown in Fig. 4, water retention was independent of nanoclay dosages. It is likely that the high suction force provided by filter paper exceeded the nanoclay's capacity to hold water. Moreover, the nanoclay did not have any notable effect on the viscosity of the interstitial fluid (Fig. 5), confirming the correlation between the viscosity of the interstitial solution and water retention of the paste system.

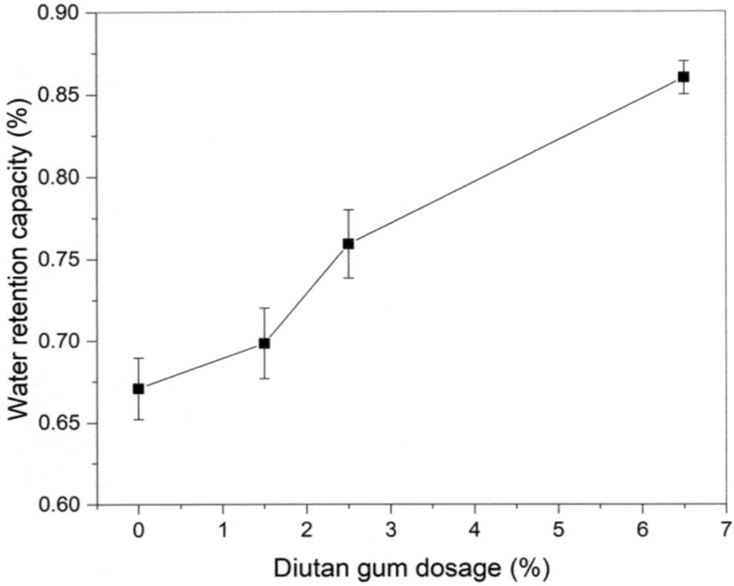

Fig. 2. Effect of diutan gum on water retention capacity.

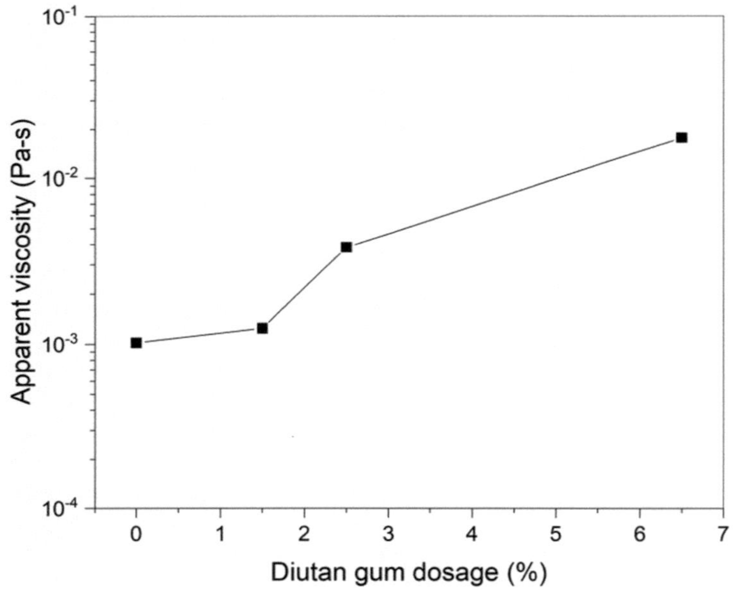

Fig. 3. Effect of diutan gum on apparent viscosity of the interstitial solution.

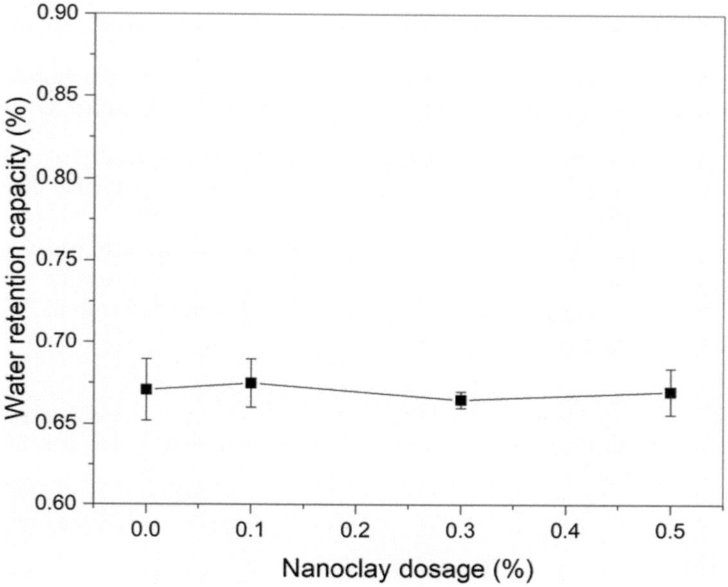

Fig. 4. Effect of nanoclay on water retention capacity

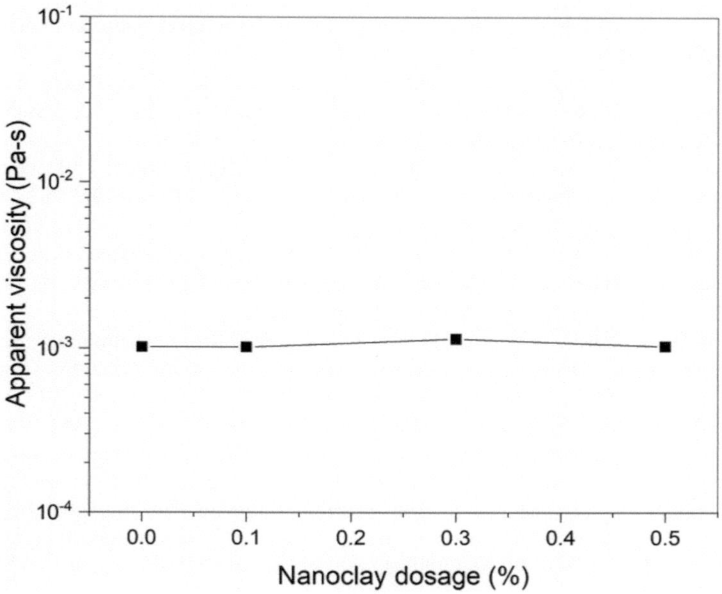

Fig. 5. Effect of nanoclay on apparent viscosity of the interstitial solution.

4 Conclusion

In the current work, the effects of VMAs on structural rebuilding under sustained stress and water retention capacity were studied. The following results were obtained:

1. Sustained stress enhanced the strengthening of the plain cement paste and cement paste with diutan gum while the sustained stress effect on nanoclay cement paste was inconclusive.
2. Diutan gum improved the water retention capacity of cement pastes while the nanoclay did not have any notable effect on water retention capacity.
3. In the studied systems, the water retention was well correlated with the viscosity of the interstitial solution.

Acknowledgments. The authors would like to acknowledge the Thornton Tomasetti Student Innovation Fellowship for financial support, and technical support by the staff of Columbia University's Carleton Laboratory.

References

1. Khoshnevis, B., Dutton, R.: Innovative rapid prototyping process makes large sized, smooth surfaced complex shapes in a wide variety of materials. Mater. Technol. **13**, 53–56 (1998). https://doi.org/10.1080/10667857.1998.11752766
2. Khoshnevis, B.: Automated construction by contour crafting—related robotics and information technologies. Autom. Constr. **13**, 5–19 (2004). https://doi.org/10.1016/j.autcon.2003.08.012
3. Le, T.T., Austin, S.A., Lim, S., Buswell, R.A., Law, R., Gibb, A.G.F., Thorpe, T.: Hardened properties of high-performance printing concrete. Cem. Concr. Res. **42**, 558–566 (2012). https://doi.org/10.1016/j.cemconres.2011.12.003
4. Lim, S., Buswell, R.A., Le, T.T., Austin, S.A., Gibb, A.G.F., Thorpe, T.: Developments in construction-scale additive manufacturing processes. Autom. Constr. **21**, 262–268 (2012). https://doi.org/10.1016/j.autcon.2011.06.010
5. Bos, F., Wolfs, R., Ahmed, Z., Salet, T.: Additive manufacturing of concrete in construction: potentials and challenges of 3D concrete printing. Virtual Phys. Prototyp. **11**, 209–225 (2016). https://doi.org/10.1080/17452759.2016.1209867
6. Kazemian, A., Yuan, X., Cochran, E., Khoshnevis, B.: Cementitious materials for construction-scale 3D printing: laboratory testing of fresh printing mixture. Constr. Build. Mater. **145**, 639–647 (2017). https://doi.org/10.1016/J.CONBUILDMAT.2017.04.015
7. Wangler, T., Lloret, E., Reiter, L., Hack, N., Gramazio, F., Kohler, M., Bernhard, M., Dillenburger, B., Buchli, J., Roussel, N., Flatt, R.: Digital concrete: opportunities and challenges. RILEM Tech. Lett. **1**, 67 (2016). https://doi.org/10.21809/rilemtechlett.2016.16
8. Ma, S., Qian, Y., Kawashima, S.: Experimental and modeling study on the non-linear structural build-up of fresh cement pastes incorporating viscosity modifying admixtures. Cem. Concr. Res. **108**, 1–9 (2018). https://doi.org/10.1016/j.cemconres.2018.02.022
9. Lloret, E., Shahab, A.R., Linus, M., Flatt, R.J., Gramazio, F., Kohler, M., Langenberg, S.: Complex concrete structures: merging existing casting techniques with digital fabrication. CAD Comput. Aided Des. **60**, 40–49 (2015). https://doi.org/10.1016/j.cad.2014.02.011

10. ACTI-GEL® 208 - Active Minerals International, LLC. http://www.acti-gel.com/acti-gel-208#hw. Accessed 21 Apr 2017
11. Plank, J.: Applications of biopolymers and other biotechnological products in building materials. Appl. Microbiol. Biotechnol. **66**, 1–9 (2004). https://doi.org/10.1007/s00253-004-1714-3
12. Liddel, P.V., Boger, D.V.: Yield stress measurements with the vane. J. Nonnewton. Fluid Mech. **63**, 235–261 (1996). https://doi.org/10.1016/0377-0257(95)01421-7
13. Standard DIN 18555-7. Testing of mortars containing mineral binders; determination of water retentivity of freshly mixed mortar by the filter plate method, Dtsch. Inst. Für Normung. https://standards.globalspec.com/std/88279/din-18555-7 (2000). Accessed 19 Feb 2018
14. Ovarlez, G., Chateau, X.: Influence of shear stress applied during flow stoppage and rest period on the mechanical properties of thixotropic suspensions. AIP Conf. Proc. **1027**, 1042–1044 (2008). https://doi.org/10.1103/PhysRevE.77.061403
15. Roussel, N., Ovarlez, G., Garrault, S., Brumaud, C.: The origins of thixotropy of fresh cement pastes. Cem. Concr. Res. **42**, 148–157 (2012). https://doi.org/10.1016/j.cemconres.2011.09.004
16. Qian, Y., Kawashima, S.: Use of creep recovery protocol to measure static yield stress and structural rebuilding of fresh cement pastes. Cem. Concr. Res. **90**, 73–79 (2016). https://doi.org/10.1016/j.cemconres.2016.09.005
17. Poinot, T., Govin, A., Grosseau, P.: Importance of coil-overlapping for the effectiveness of hydroxypropylguars as water retention agent in cement-based mortars. Cem. Concr. Res. **56**, 61–68 (2014). https://doi.org/10.1016/j.cemconres.2013.11.005
18. Marliere, C., Mabrouk, E., Lamblet, M., Coussot, P.: How water retention in porous media with cellulose ethers works. Cem. Concr. Res. **42**, 1501–1512 (2012). https://doi.org/10.1016/j.cemconres.2012.08.010
19. Kawashima, S., Kim, J.H., Corr, D.J., Shah, S.P.: Study of the mechanisms underlying the fresh-state response of cementitious materials modified with nanoclays. Constr. Build. Mater. **36**, 749–757 (2012). https://doi.org/10.1016/j.conbuildmat.2012.06.057

Rheological Control of 3D Printable Cement Paste and Mortars

Scott Z. Jones[1(✉)], Dale P. Bentz[1], Nicos S. Martys[1], William L. George[1], and Austin Thomas[2]

[1] National Institute of Standards and Technology, Gaithersburg, MD, USA
scott.jones@nist.gov
[2] University of New Haven, West Haven, CT, USA

Abstract. Recent advances in concrete construction such as three-dimensional concrete printing (3DCP) have given rise to new requirements on the control of both the hydration and rheology of cementitious materials. To meet these new demands, and to move toward adoption of 3DCP on a commercial scale, in-operando control of hydration and rheology will be required. In this study, two cement paste mixtures containing limestone powder of two different median particle sizes are used to create 3D printed structures with a cement paste printer. Hydration control in the form of acceleration is achieved with the addition of the limestone powder to the cement and rheology control is achieved by using limestone with different median particle sizes. Rheology measurements conducted concurrently with printed structures indicate that yield stress and a measure of thixotropy of the cement paste provide an indicator as to whether a material will produce a multi-filament free-standing structure for a given 3DCP system. Simulations of particles flowing in a pipe are used to study the rheological behavior of paste and mortar. For the case of a mortar, the flow rate of suspended particles (sand) follows the same functional form with driving force as the matrix fluid (cement paste). Shear-induced particle migration increases the density of particles toward the center of the pipe, a result that implies that the aggregates may not be uniformly distributed.

Keywords: Concrete 3D printing · Pipe flow simulations · Rheology control

1 Introduction

Three-dimensional concrete printing (3DCP) is a type of concrete construction where a cement-based material is selectively placed by means of a computer-controlled robotic system. Additive construction by extrusion (ACE) is one approach to 3DCP. Here, a robotic system controls the position and orientation of a nozzle that extrudes layers of cementitious material onto a working surface. Numerous challenges must be addressed before the benefits of this technology can be realized, including how to reliably control rheology and the setting time of the cementitious material [1, 2]. Lim et al. [1] define four characteristics of materials used in ACE applications - pumpability, printability, buildability, and open time. Pumpability is defined to be the ease with which materials are moved through the delivery system, printability is the ease of deposition,

T. Wangler and R. J. Flatt (Eds.): DC 2018, RILEM Bookseries 19, pp. 70–80, 2019.
https://doi.org/10.1007/978-3-319-99519-9_7

buildability describes the resistance of the fresh deposited material to deformation under load, and open time is the period when the aforementioned characteristics are within acceptable limits [1]. In practice, these characteristics are a function of the rheology and hydration of the material and the configuration of the ACE equipment. The interdependence of material and ACE equipment on the quality of the built structure necessitates in-operando measurements of the material to control rheology. This study begins to address this necessity by evaluating two cement pastes in an ACE application. Hydration control is achieved by accelerating the setting time of the material with the addition of fine limestone powder, while rheology control is achieved by utilizing limestone powders with two differing median particle sizes. The two pastes are used to print a free-standing structure with a custom-built cement paste printer. Pumpability and printability are evaluated qualitatively during the print, while buildability is assessed by concurrent measurements of two rheological parameters; (1) yield stress, which is interpreted to be related the vertical build rate; and (2) the thixotropy (structural build-up) of the material [1, 3, 4], which is interpreted to be related to the dimensional stability. Here, the open time of the paste begins when two or more filaments are deposited without collapsing and ends when pumping is no longer. Numerical simulations of the flow of mono-size spheres suspended in Newtonian and non-Newtonian fluids are used to study the feasibility of in-operando measurements and the migration of particles within a pumped and extruded filament.

2 Materials and Methods

Two mixture formulations, A and B, were used in this study. An ASTM C150 [5] Type I/II ordinary portland cement containing approximately 4% limestone (approximately 96% calcite) by mass was used for both mixtures A and B. The Bogue-calculated phase compositions reported by the cement manufacturer are 53.1% C_3S, 14.2% C_2S, 6.5% C_3A and 10.0% C_4AF on a mass basis. The water to powder ratio on a mass basis for both mixtures was 0.28, corresponding to a solids volume fraction of 0.55. 4.0 mL/kg of cement of a polycarboxylate ether (PCE) high-range water-reducing admixture (HRWRA) was used to control the viscosity and yield stress of the paste. Mixture A contains a 50% by mass of powder blend of a 6 μm median particle diameter (D_{50}) limestone and cement, while Mixture B contains a 50%, 25%, and 25% by mass of powder blend of the cement, a 2.2 μm D_{50} limestone, and the 6 μm D_{50} limestone, respectively. The density of paste both mixtures is calculated to be 2016 kg/m^3, assuming a 2% air content.

The pastes were mixed using a two-rotor baking mixer, see [6].[1] The cement and limestone were pre-blended prior to mixing. Approximately 800 g of the blended powder was measured into a plastic container with an internal diameter of approximately 20 cm. During the mixing process, water was added in two stages, with half of

[1] Certain commercial products are identified in this paper to specify the materials used and the procedures employed. In no case does such identification imply endorsement or recommendation by the National Institute of Standards and Technology, nor does it indicate that the products are necessarily the best available for the purpose.

the required water added at the start of mixing. The powder and water were mixed for 30 s on speed 1, at which point the remaining water and PCE HRWRA were added. Mixing continued for 60 s on speed 2, followed by a rest for 90 s. After resting, the paste was again mixed for 60 s on speed 2. After mixing was completed, the paste was either added to the hopper of the cement paste 3D printer or stored in a sealed, insulated container until rheology testing. A polymer-based fused filament fabrication 3D printer was modified to dispense cement paste through a 3 mm nozzle. The printability of mixtures A and B was evaluated by creating a tall, thin structure composed of multiple layers, shown schematically in Fig. 1, with dimensions 4.0 mm (w) × 45.0 mm (l) 76.5 mm (z). The filament (k) cross section was 1.5 mm (h) × 4.0 mm (w). One layer of the structure is composed of two filaments and has a height of 2 h. The geometry is relatively unstable, and a successful structure can only be achieved if the material can withstand the forces exerted during the 3D printing process. To reduce artifacts caused by pumping starts and stops, the structure was built at a nozzle speed of 13 mm/s using a continuous printing pattern, where the z-axis is continuously moving throughout the print. The time required to complete on structures is approximately 3 min.

To relate the printability of mixtures A and B to rheological parameters, rheometery was used with both a strain-controlled and a stress-controlled rheometer calibrated using SRM 2492, following the procedure described in [7]. Yield stress (τ_y) measurements were made using a strain-controlled rheometer with a 25 mm serrated, parallel plate tool and a 1.0 mm gap. A 10 s^{-1} strain rate ($\dot{\gamma}$) was applied for 10 s to homogenize the sample. The sample was allowed to rest for 60 s before a constant 1.0 s^{-1} shear rate at the outer edge of the tool was applied.

Thixotropy of the paste is assessed by the procedure descried in [4]. Here, a 35 mm cross-hatched plate with the gap set to 0.6 mm is used with a stress-controlled rheometer. A shear rate of 100 s^{-1} is applied for 60 s. Immediately after the application of the shear rate, a constant stress of a magnitude approximately 10% of the recorded yield stress is applied. The strain rate as a function of time after the application of stress is recorded. Between each test, the material was stored in a sealed, insulated container.

The dependence of the mortar's rheological properties on the cement paste, shear-induced particle migration, and the feasibility of in-operando measurements are studied using numerical simulations of a dense suspension of particles flowing in a pipe. The flow simulation is based on an algorithm that combines Smooth Particle Hydrodynamics with rigid body motion of spheres. Details of the simulation are described in [8]. Figure 2 shows the inputs to the numerical simulation. Figure 2a and b show a typical configuration of the mortar in the simulation; a pipe filled with mono-size spheres. Mixtures A and B in Fig. 2c demonstrate power-law behavior, e.g., the viscosity, $\mu \sim \dot{\gamma}^{n-1}$ where μ is the viscosity, $\dot{\gamma}$ is the local shear rate, and n is a power law exponent, at the time the first multi-filament structure is created. The dashed lines in Fig. 2c show a typical power-law fluid with $n = 0$ and $n = 1/2$ that may represent the matrix fluid in a mortar simulation. The ratio of the pipe diameter to the diameter of the spheres is approximately 32. This simulation could correspond to particles of 1 mm sand in a pipe of diameter 3 cm or cement particles of about 30 μm diameter in a nozzle of 1 mm diameter.

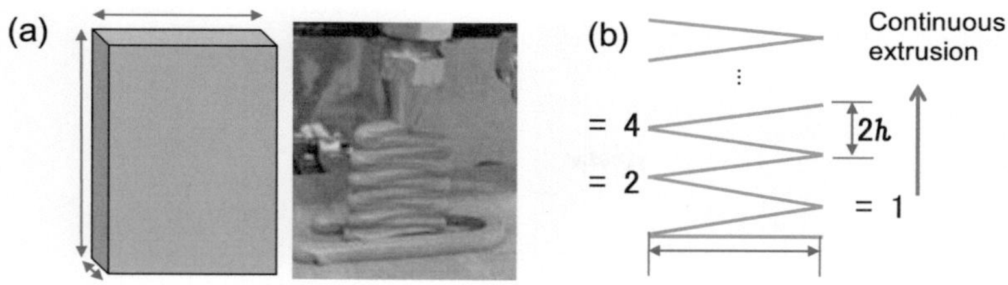

Fig. 1. (a) Schematic of printed structure and (b) schematic of continuous printing process.

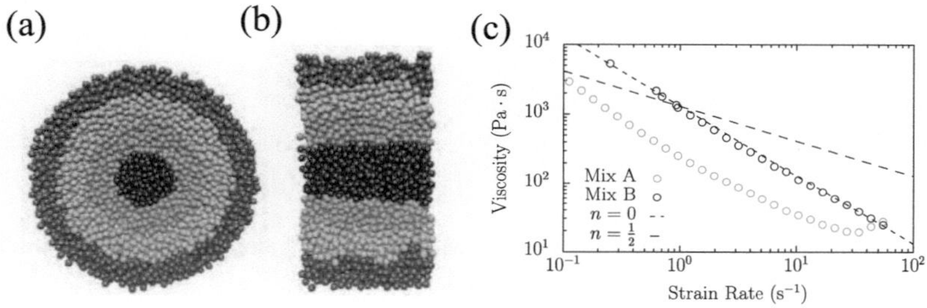

Fig. 2. Axial (a) and side (b) view of suspension in pipe. (c) viscosity of mixtures A and B and power-law fluids with $n = 0$ and $n = 1/2$. The color of spheres is for visualization purposes only.

3 Results and Discussion

3.1 Printing Mixtures A and B

Mixture A is assessed by printing the structures shown in Fig. 3. The open time of this mixture is 20 min, beginning at 50 min after the mix start time. Figure 3a shows that the first free-standing structure contained air voids, indicating poor pumpability. Four layers were deposited before the structure collapsed. At 60 min, Fig. 3b, the number of layers printed increased to six, while voids created by discontinuous filaments were observed. The print at 65 min produced a structure with 9.5 out of 25 layers (38%), indicating an increase in the buildability; however, the collapse of the structure was initiated at a void created by pumping difficulties. The final free-standing structure was printed at 70 min after mixing. Here, seven layers were deposited before the collapse. This structure experienced printability difficulties, as the subsequently deposited layers became shorter in length producing a trapezoidal structure. The presence of visible air voids in the layered structure resulted in collapse before the yield stress of the bottom layer had been exceeded. It is apparent from Fig. 3d, that as hydration progressed, the material became more difficult to pump. This had an adverse impact on printability. The structure begins to narrow as successive filaments are deposited.

Mixture B completed a free-standing structure at 30 min after mixing and was able to print structures until 80 min after mixing, for a 50 min open time. At 30 min, Fig. 4a, 13 layers were printed before collapse of the structure. The mass of these

(a) t = 50 min

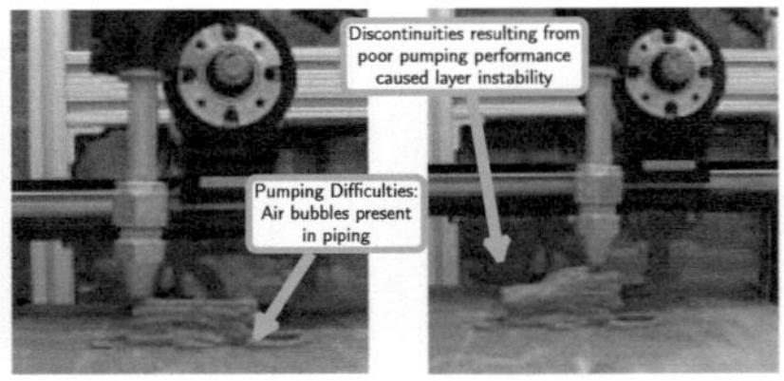

(b) t = 60 min

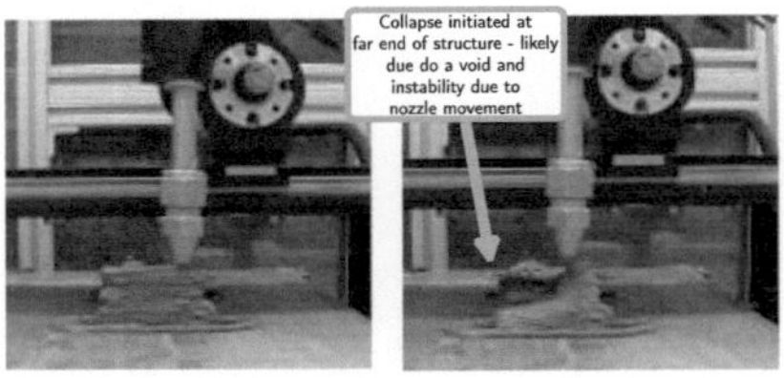

(c) t = 65 min

(d) t = 70 min

Fig. 3. Printing test artifact with mixture A. (a) First free standing structure at 50 min after mixing. (b) At 60 min (c) 65 min. (d) At 70 min

deposited filaments exceeded the yield stress of the first filament, causing the material to flow and the structure to collapse. This failure highlights the critical role that the yield stress of the material plays on the buildability of the printing material. Printing at 40 min after mixing is shown in Fig. 4b. Here, 20 layers (80%) were deposited before collapse of the structure. The movement of the nozzle created a bending moment (see the diagram in Fig. 4b) which caused the bottom layer to collapse. This failure is particularly noteworthy, as it demonstrates the dynamic loading the printed structure experiences during material deposition. Figure 4c and d show a printed structure with 20 layers (80%) and 23 layers (93%), respectively. Both structures fail as a result of the nozzle exerting a bending moment on the structure. As the height of the structure increases, the failure of the structure appears to be related to buckling, as shown if in Fig. 4d and noted in [9]. It is at this stage of printing that the printability of the material has degraded to the point where the edges of the structure cannot be maintained. Collapse of the structure in Fig. 4d toward the top is likely due to this irregular geometry as it becomes the weak point. By 75 min and 80 min, hydration has progressed to the point where pumping, and thus printability, is difficult. In Fig. 4e, 13 layers (52%) are deposited before the 14th layer is not printed. In Fig. 4f, a few layers are printed, but the printability had degraded to the point where the structure does not resemble the intended geometry.

3.2 Rheometry

Yield stress and thixotropy of mixtures A and B are assessed, as a function of time after initial mixing, using rotational rheometry. Figure 5a shows the change in yield stress as a function of time after mixing. Both mixtures A and B exhibited an increase in yield

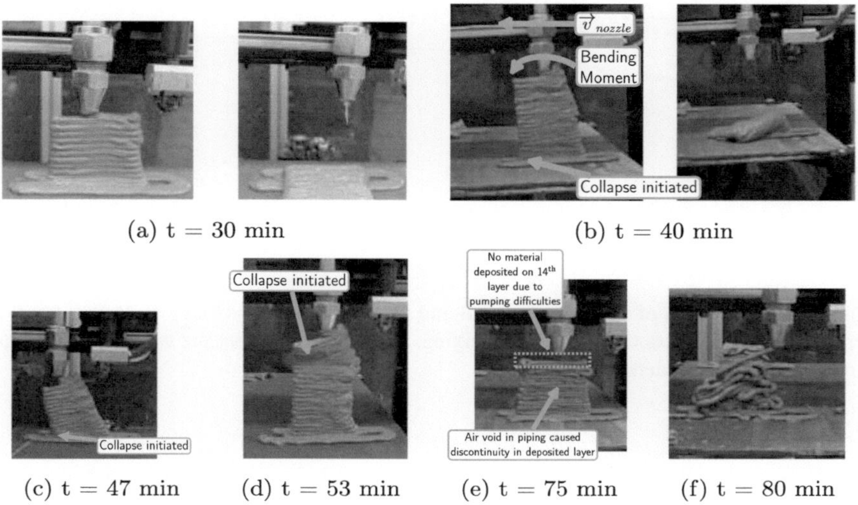

Fig. 4. Printing test artifact with mixture B. (a) First free standing structure at 30 min after mixing. (b) Printed structure at 40 min. (c) Printed structure at 47 min. (d) Printed Structure at 53 min. (e) Printed structure at 75 min. (f) Printed structure at 80 min.

stress which may be modeled assuming an equation of the form of $\tau_y(t) = \alpha e^{\beta t}$, where, *t*, is the time after mixing. The parameters α, β, and their expanded uncertainty at a 95% confidence level, may be estimated by linear least squares regression with the equation in the form $\ln(\tau_y) = \ln(\alpha) + \beta t$. The parameter α represents the yield stress at t = 0 min, or just after mixing, while the parameter β is indicative of the rate at which the yield stress increases. The estimated α parameter for mixture A is (8 ± 2) Pa and for mixture B, α is estimated to be (317 ± 61) Pa. The stress, τ_k, exerted on the bottom filament by the *k* deposited filaments may be calculated by $\tau_k = k\rho_p g_0 h$, where ρ_p is the density of the paste, *h* is the filament height, and g_0 is the standard acceleration due to gravity.

The time after mixing a material requires to obtain a yield stress necessary to build a structure with *k* filaments may be estimated by $t_k = \ln(0.5\tau_k/\alpha)/\beta$, assuming Tresca yield criterion. A multi-filament structure consists of, at least two filaments, i.e., *k* = 1. The stress required to support a single filament (τ_1) is calculated to be 30 Pa. For mixture A, t_1, is estimated to be 11 min. Estimations of t_1 assume yield stress of the deposited material is achieved immediately after deposition. As shown in Fig. 3, the first multi-filament structure with mixture A was achieved at 50 min after mixing. Mixture A demonstrated poor printability and pumpability throughout the open time. Air bubbles created voids in the printed structure which contributed to the shortening filament lengths. For mixture B, the α parameter is about 10 times that of τ_1 indicating a multi-filament structure, is possible immediately after mixing is complete. Neither air bubbles or shortening of filament lengths were not observed in the structures shown in Fig. 4a and b, indicating the printability and pumpability are sufficient to produce a free-standing structure, however, one was not achievable until 30 min after mixing.

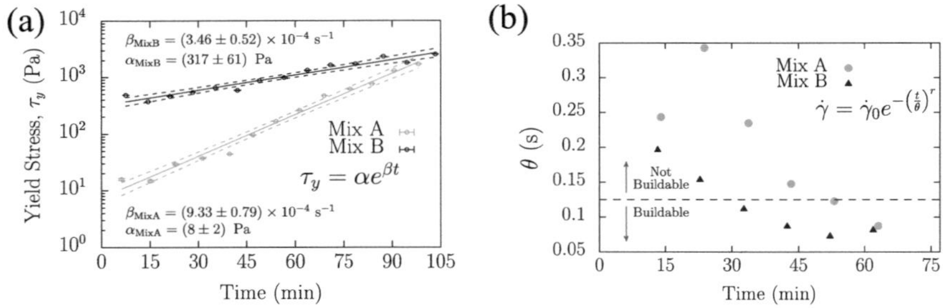

Fig. 5. (a) Yield stress of mixtures A and B and (b) θ parameter assessing structural rebuilding after shearing as a function of time from mixing. Error bars represent the expanded uncertainty estimated at 95% confidence.

The parameter, β, is estimated to be $(9.33 \pm 0.79) \times 10^{-4}\ s^{-1}$ for mixture A. For mixture B, β, is estimated to be $(3.46 \pm 0.52) \times 10^{-4}\ s^{-1}$. Mixture A remains printable and buildable for approximately 20 min, beginning at 50 min after the mix start time with a yield stress estimated to be 125 Pa. In comparison, mixture B remains buildable and printable for 50 min, beginning at 30 min after the mix start time with the yield stress estimated to be 519 Pa. Figure 5a suggests the length of the open time may be partially explained by the β parameter. The open time of mixture A is less than that of mixture B because the yield stress grows rapidly to a point where the material cannot be pumped. This requires the yield stress be above a certain value to support the mass of additional layers. The yield stress values for mixtures A and B at the time of the first free-standing structure differ by 394 Pa. This suggests that the yield stress is not the only rheological parameter governing the onset of buildability. One possible explanation might be the rate at which the material recovers a yield stress after the removal of an applied shearing stress.

Yield stress measurements alone indicate that a multi-filament free-standing structure should be possible before one is achieved. This may be related to the thixotropy of the material. Figure 5b reports the θ parameter from the controlled stress experiments as a function of time from mixing. θ is a measure of the rate at which the strain rate decays as a constant stress is applied to the sample. The strain rate decay is a result of flocs forming in the cement paste to create the microstructure that can withstand an applied stress [4, 5]. The rate at which this structure forms is important to 3D printing applications as it is partially responsible for the buildability and printability of a particular cement paste. As the paste is pumped and extruded, shearing forces break down the microstructure which allows the material to flow. After the paste exits the nozzle, this structure must rebuild to maintain the shape of the filament and support additional deposited material. Once deposited, the rate at which the yield stress increases with time since mixing becomes the governing parameter driving buildability. For mixture A, the first buildable structure occurs at 50 min. This corresponds to $\theta = 0.125$ s in Fig. 5b. Similarly, for Mixture B, the first printable structure occurs at 30 min; again, this corresponds to $\theta = 0.125$ s in Fig. 5b. For this printing system (the paste printer, mixture formulations, and printed structure), $\theta = 0.125$ s indicates a free-standing structure is possible, suggesting that buildability is dependent upon two rheological parameters. The yield stress value governs the number of layers that may be deposited, while the θ parameter indicates the time after mixing when the first buildable structure is possible. θ provides an indication as to the dimensional stability of the structure, a smaller θ implies a faster return to the material's yield stress after the removal of the shear strain. This may have implications for the rate at which filaments may be deposited that can be achieved with a given material as the next layer cannot be deposited before the material has achieved the yield stress.

3.3 Simulations of Flow Properties

Numerical simulations of a dense particle suspension flowing in a pipe are used to extend the rheology of cement paste to mortar. In extrusion-based 3DCP, it is crucial to monitor the rheological properties of the printing material as they change with time. The paste can be treated as a continuum fluid with a non-Newtonian behavior. Its flow is largely controlled by the nozzle, with the possibility of slip effects on the nozzle wall as well as the development of obstructions to flow as the fluid enters the nozzle area. Because of the shear thinning behavior of the paste the highest shear rates are near the nozzle wall. Determining the printability of a mortar introduces additional challenges. The mortar may be thought of as a suspension composed of a matrix fluid (cement paste) with the addition of solid particles (sand). We have found, using scaling arguments, that if the matrix fluid has a power-law behavior, then the mortar's viscosity will have a similar power-law behavior. As a consequence, it can be shown that the typical flow rate of the mortar, V, will scale as $V \sim g^{1/n}$, where g is the driving force. (e.g., body force, pressure gradient). This is important for two reasons: first, from measurements of the matrix fluid alone, one can predict the general flow behavior of the suspension assuming measurements probe the shear rates approximately equal to those found near the fluid/wall interface. Second, one may monitor the rheological behavior of the mortar by evaluating the flow rate as a function of pressure using a few changes in pressure to estimate the power law behavior. To validate this scaling behavior, simulations have been carried out of pipe flow of a dense suspension with a power law matrix fluid where $n = 1/2$. Figure 6a shows the flow velocity profile as a function distance from pipe's center axis, for three different driving forces. The flow rate scales proportionally to g^2 in agreement with scaling predictions. Note that the flow velocity is consistent with the scaling law. We have also tested this scaling law for $n = 1$ (Newtonian fluid) and $n = 3/2$ (shear thickening). For $n = 0$, the velocity profile is flat across the pipe diameter, indicating plug flow [10]. For the shear thinning fluid there appears to be a slip velocity at the pipe wall. However, the simulation strictly obeys a no slip boundary condition at the pipe wall. This apparent slip is actually an artifact of the coarse graining of the velocity field.

The layer-by-layer deposition of material introduces anisotropic characteristics to the printed structure [11]. The anisotropy is a result of the interface between deposited filaments, the strength of which has been shown to be affected by the moisture content at the surface of an extruded filament [12]. Figure 6b shows the initial (blue curve) and near-steady state (green curve) volume fraction of spheres as a function of the radial position for the case of a mono-size sphere suspension at volume fraction 40% with a

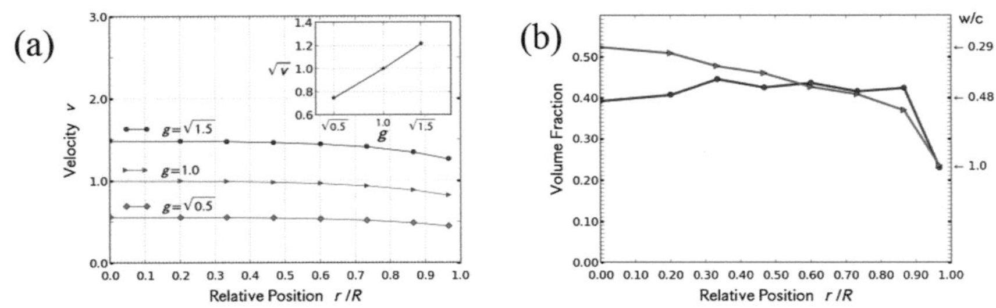

Fig. 6. (a) Velocity profiles for suspension flow in pipe (a). The inset shows how the flow rates scale with driving forces. The radial dependences of volume fraction (b). The solid circles (blue) and triangles (green) represent the initial and late stage volume fractions, respectively.

Newtonian fluid matrix, representing the case of cement in water with a *w/c* of about 0.48. While the initial radial dependence of the local volume fraction is essentially flat, except near the pipe walls due to packing effects [13], the local volume fraction becomes significantly higher in the middle of the pipe as the simulation progresses. Simulation results suggest this shear-induced particle migration rapidly takes place over a flow distance of four to five times the pipe radius. The simulation assumes the spheres are non-interactive. Including the effects of agents that make a suspension flocculate or disperse would have an impact on the final volume fraction profile and will be studied in the future. Particle packing near the wall and shear-induced particle migration due to the locally high shear rates near the pipe wall contribute to the excess water near the surface of a deposited filament. The potential lack of homogeneity of the material may be worth considering for certain cases of printing as the effective *w/c* of this volume fraction profile ranges from 0.29 near the center to 1.0 at the pipe wall. The excess water near the surface of an extruded filament may be a contributing factor which creates a porous microstructure, reduces bond strength, and increases permeability.

4 Conclusion

Success or failure of materials used in extrusion-based 3DCP applications is often described in terms of pumpability, printability, buildability, and open time. Buildability and open time of two pastes used to create multi-filament free-standing structures, fabricated with a cement paste printer, are interpreted in the context of concurrent yield stress and thixotropy measurements. The time after mixing required prior to building a multi-filament structure is estimated from measurements of yield stress and indicates that mixture A will be buildable at 11 min after mixing and mixture B is buildable immediately after mixing. However, this was not observed when mixtures A and B were used to fabricate free-standing structures. Both mixtures were buildable approximately 30 min after the estimated buildability time, a result attributed to the thixotropy of the materials. Results indicate that the length of the open time in this study is regulated by pumpability issues and, therefore, is dependent upon the β

parameter. The slower the change in yield stress, the longer the open time. This assumes the yield stress can support the mass of the additional deposited material. Buildability appears to be linked to the value of the yield stress, i.e., as the yield stress increases, so does the buildability. The rate at which the material gains a yield stress after the application of a shearing stress, is estimated by the θ parameter. Here, $\theta = 0.125$ s appears to correspond with the ability to print a free-standing structure for the specific printer, materials, and printing speed used in this study. Balancing yield stress growth (β) with the number of deposited layers and the rate at which the yield stress is achieved after the removal of a shearing strain (θ) are partially responsible for buildability and printability.

Simulations of paste and mortars have provided insight into the flow of dense suspensions in pipes. It is found that the flow rate of a mortar scales in a fashion consistent with the flow of the matrix fluid. In practice, measuring the change in pressure across a length of a pipe and knowing the flow rate, one could calculate the power law behavior of the mortar suspension. Simulation results also provide insight into the movement of particles within the suspension. Results indicate the particle density increases toward the center of the pipe as a result of the local high shear rates produced between the wall of the pipe and the suspended particles. This phenomenon could have implications for pumpability. While particle migration creates a lubrication layer that facilitates flow in straight pipes, migrating particles could facilitate jamming in the nozzle as particles reorient to move through the contracted diameter. The deposited layers could experience inferior bonding at the paste scale, as the particles have migrated toward the center of the filament during extrusion. In the case of cement paste, migrating cement particles may allow for a locally increased *w/c* at the interfaces between the layers, creating a porous microstructure that may reduce bonding and increase permeability.

References

1. Lim, S., et al.: Developments in construction-scale additive manufacturing processes. Autom. Constr. **21**(1), 262–268 (2012)
2. Wangler, T., et al.: Digital concrete: opportunities and challenges. RILEM Tech. Lett. **1**, 67–75 (2016)
3. Perrot, A., Rangeard, D., Pierre, A.: Structural build-up of cement-based materials used for 3D-printing extrusion techniques. Mater. Struct. **49**, 1213–1220 (2016)
4. Kawashima, S., et al.: Rate of thixotropic rebuilding of cement paste modified with highly purified attapulgite clays. Cem. Concr. Res. **53**, 112–118 (2013)
5. ASTM International, West Conshohocken, PA. ASTM C150/C150M-17, Standard Specification for Portland Cement (2017)
6. Oster 6 speed retractable cord hand mixer with clean start. http://www.oster.com/kitchen-tools/mixers/oster-6-speed-retractable-cord-hand-mixer-with-clean-start/FPSTHM2578.html. Accessed 2018
7. Ferraris, C.F., et al.: Certification of SRM 2492: Bingham paste mixture for rheological measurements. NIST SP 260-174 (2012)

8. Martys, N.S., et al.: A smoothed particle hydrodynamics-based fluid model with a spatially dependent viscosity: application to flow of a suspension with a non-newtonian fluid matrix. Rheol. Acta **49**(10), 1059–1069 (2010)
9. Suiker, A.S.J.: Mechanical performance of wall structures in 3D printing processes: theory, design tools and experiments. Int. J. Mech. Sci. **137**, 145–170 (2018)
10. Byron Bird, R., Armstrong, R.C., Hasseger, O.: Dynamics of Polymeric Liquids, vol. 1, 1st edn. Wiley, Hoboken (1987)
11. Le, T.T., et al.: Hardened properties of high-performance printing concrete. Cem. Concr. Res. **42**, 558–566 (2012)
12. Sanjayan, J.G., et al.: Effect of surface moisture on inter-layer strength of 3D printed concrete. Constr. Build. Mater. **172**, 468–475 (2018)
13. Ridgway, K., Tarbuck, K.J.: Voidage fluctuations in randomly-packed beds of spheres adjacent to a containing wall. Chem. Eng. Sci. **23**, 1147–1155 (1968)

Adapting Smart Dynamic Casting to Thin Folded Geometries

Anna Szabo[1(✉)], Lex Reiter[2], Ena Lloret-Fritschi[1,2], Fabio Gramazio[1], Matthias Kohler[1], and Robert J. Flatt[2]

[1] Gramazio Kohler Research, NCCR Digital Fabrication, ETH Zurich, Zurich, Switzerland
szabo@arch.ethz.ch

[2] Institute for Building Materials, NCCR Digital Fabrication, ETH Zurich, Zurich, Switzerland

Abstract. The first thin folded concrete prototypes produced with Smart Dynamic Casting (SDC) exposed numerous challenges concerning concrete. The SDC process is modelled to explain the increased difficulty to fabricate thin folded members compared to columns. Due to the smaller volume to surface ratio in formworks for thin folded structures the effect of friction is amplified and the process window narrows down. In order to compensate for this, retarded self-compacting mortar mix designs and acceleration strategies are investigated.

Material testing results provide guidelines of how to achieve a uniform hardening rate over the course of an experiment, while preserving sufficient fluidity and dealing with variations in raw materials. For this, offline penetrometer tests are performed to evaluate material properties and online measurements are recorded to follow the strength evolution of the same mix processed with the experimental setup. In addition, the slipping criterion and the deformability of the concrete are tested in a 1:1 scale robotic experiment to evaluate the fabrication feasibility with the adapted mix. This unveils the potential to produce thin folded members for architectural applications.

Keywords: Smart Dynamic Casting · Folded structures · Process window

1 Introduction

At the age of "the third industrial revolution" [1], mass customization offers a solution on the quest for expanding design freedom with digital fabrication technologies by translating computer aided design data to manufacturing information in order to control the fabrication process [2]. In digital fabrication, concrete has attracted attention later than other materials, especially due to the difficulty of controlling the fluid to solid transition [3].

This paper focuses on adapting the material mix formulation and the manufacturing protocol of an existing digital fabrication process, Smart Dynamic Casting (SDC) [4], to thin folded geometries. It offers an approach to assess how precisely material parameters have to be in line with the process boundaries, given the geometry.

T. Wangler and R. J. Flatt (Eds.): DC 2018, RILEM Bookseries 19, pp. 81–93, 2019.
https://doi.org/10.1007/978-3-319-99519-9_8

SDC is a robotic slipforming process that enables to produce non-standard geometries using a single formwork significantly smaller than the structures produced, by shaping concrete in the delicate phase when it changes from a yield stress fluid to a cohesive frictional material [5]. In order to perform successful experiments with SDC, it is essential to control the evolution of concrete yield stress in the early hydration phase and synchronize it with the movement of the formwork [6].

SDC is a gravity driven extrusion process in which the concrete weight in the formwork drives the extrusion, while the friction on the sliding formwork walls opposes it [4]. For the process, the cumulative of the spatially variable shear stress along the formwork surface is responsible for friction [7] (Fig. 1). This depends on many parameters, such as formwork material, material evolution, surface geometry and process rate. The ratio of formwork volume to surface is an important parameter and it will be referred to as hydrodynamic radius in this paper. For geometries with small hydrodynamic radius (thus large surface) friction is affecting the process more in relative terms. This relation is taken into account in the model described in the following section.

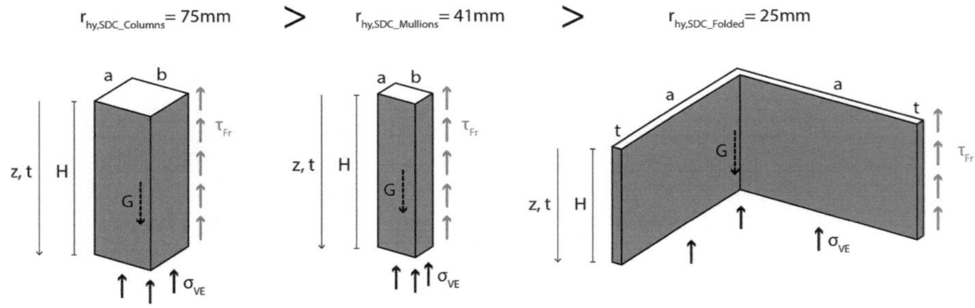

Fig. 1. Force balance for SDC columns, mullions and thin folded structures

Hydrodynamic radius: $r_{hy} = \frac{2 \cdot Volume}{Surface}$
Examples of previous studies using SDC:
Columns: $r_{hy} = \frac{2 \cdot a \cdot b}{2a + 2b}$, with $a = b = 150\,\text{mm} \Rightarrow r_{hy} = 75\,\text{mm}$
Mullions: $r_{hy} = \frac{2 \cdot a \cdot b}{2a + 2b}$, with $a = 100\,\text{mm}, b = 70\,\text{mm} \Rightarrow r_{hy} = 41\,\text{mm}$
Folded structures: $r_{hy} = \frac{2 \cdot 2 \cdot a \cdot t}{4 \cdot a} = t$, with $t = 25\,\text{mm} \Rightarrow r_{hy} = 25\,\text{mm}$

In addition, SDC has three main requirements for the material mix. Firstly, a concrete batch is prepared in a way to provide a long open time – the retarded concrete – that can be activated on demand with an accelerator. Secondly, the retarded material has to be suitable for pumping. Thirdly, the accelerated material has to be self-compacting and fluid enough to fill the formwork, not segregate and harden at a known rate [7].

The geometrical design space of the process is explored by moving a formwork for thin folded structures along spatial trajectories with a 6-axis robotic arm. The exploration follows empirical methodology in which the fabrication feasibility is evaluated through physical experimentation at 1:1 scale. For the first experiments, a rigid formwork system for thin folded structures with 90° fold opening is slipped along trajectories in plane with one of its sides. These experiments enhance the understanding of the difficulties due to the smaller process window and serve as the first step towards exploring the architectural potential of thin, folded concrete members with non-standard geometries enabled by the SDC prefabrication process.

2 Slipforming Process Model

The force balance between the hydrostatic pressure ρgH of concrete in the formwork and the cumulated friction of concrete along the formwork allows for calculating the vertical stress σ_{VE} at the extrusion point:

$$\sigma_{VE} = \rho gH - \frac{2}{r_{hy}} \int_0^z \tau_{Fr}(z)dz \qquad (1)$$

Previously defined failure criteria limit the vertical stress allowed by the slip-forming process to positive values (no cracking caused by tensile stresses) and values lower than twice the yield stress τ_0 at the time of extrusion t_{Extr} (no flow according to Tresca criterion) [7–9]:

$$0 \leq \sigma_{VE}/2 \leq \tau_0(t_{Extr}) \qquad (2)$$

It is assumed that in any point of the formwork, friction $\tau_{Fr}(t)$ is proportional to the concrete yield according to a frictional parameter α_{Fr} and that the yield stress evolves according to a power law scaling with time since concrete placing t:

$$\tau_{Fr}(t) = \alpha_{Fr}\tau_0(t) = \alpha_{Fr}\alpha_C t^{\beta_C} \qquad (3)$$

The force balance (1) can be reformulated with (3), the slipping speed v and integrated over time:

$$\sigma_{VE} = \rho gH - \frac{2}{r_{hy}} \int_0^{t_{Extr}} \tau_{Fr}(t)vdt = \rho gH - \frac{2v\alpha_{Fr}\alpha_C}{r_{hy}(1+\beta_C)} t_{Extr}^{1+\beta_C} \qquad (4)$$

Using the failure criteria (2), the range of the relative balance of weight and yield stress can be described with the global parameter representing the effect of friction $\xi = \frac{H\alpha_{Fr}}{r_{hy}(1+\beta_C)}$, involving only a frictional parameter, the hydrodynamic radius, the power law exponent of yield stress evolution and the formwork height. These define the process window:

$$\frac{H\alpha_{Fr}}{r_{hy}(1+\beta_C)} \leq \frac{\rho gH}{2\tau_0(t_{Extr})} \leq 1 + \frac{H\alpha_{Fr}}{r_{hy}(1+\beta_C)} \qquad (5)$$

Therefore, the material that leaves the formwork has to have a minimum yield stress to prevent flow-out and has to cause as small friction as possible (yield stress below an upper limit) to avoid crack formation.
Examples:

$$r_{hy} \to 0 \Rightarrow \infty \leq \frac{\rho g H}{2\tau_0(t_{Extr})} \leq \infty \Rightarrow \text{Process impossible}$$

$$r_{hy} \to \infty \Rightarrow 0 \leq \frac{\rho g H}{2\tau_0(t_{Extr})} \leq 1 \Rightarrow \text{Only sufficient strength required}$$

When the factor ξ is larger than 1, requirements related to material control become substantially harsher, as the compressive strength at the time of extrusion has to be lower than the hydrostatic pressure, while it is required to evolve at the rate of vertical building after extrusion. In addition, as this factor increases the upper and lower relative limits of hydrostatic pressure to compressive strength ratio are approaching. Thus allowing for a smaller variation of yield stress at the time of extrusion.

Taking relation (5) the effect of a change of power law exponent of yield stress or hydrodynamic radius on friction and the process window can be quantified as:

$$R_\beta = \frac{\xi_2}{\xi_1} = \frac{(1+\beta_1)}{(1+\beta_2)}; R_{r_{hy}} = \frac{\xi_2}{\xi_1} = \frac{r_1}{r_2}$$

It is evident from the hydrodynamic radius calculation that folded structures are more difficult to extrude than previously studied column geometries. An illustration of this is that our initial experiments on thin folded structures used a material mix that had been successfully used for robust production of facade mullions in a demonstration building (SDC NEST Mix), but that proved inadequate in this different situation. Thus, the first prototypes produced with the rigid formwork system for thin folded structures showed various failures such as crack formation, complete failure at low heights and bad surface quality.

In order to successful produce thin folded elements, a wider process window is required in which the yield stress of the material stays in the slipping range for a longer time. Thus, the formulation of the retarded base mix needs to be adjusted to the new hardening requirements and the acceleration strategy has to be redefined. These modifications are discussed in the next sections.

3 Materials and Methods

The mix design for this application (SDC FS_2) is derived from a previous SDC composition (SDC NEST Mix) however it provides a wider process window. Modifications to the previous composition are based on a set of offline, online and robotic tests as detailed below, involving penetration resistance tests, slump tests and slip-forming experiments.

The concrete is a high performance self-compacting mortar comprising 0–4 mm siliceous sand aggregates, a CEM I 52.5R Portland cement and silica fume. Water, sucrose and superplasticizer (BASF MasterGlenium ACE 30) are added to the mixing water as admixtures unless otherwise stated. The concrete is mixed using forced action mixers for 7 min. This composition is referred to as retarded concrete. A set accelerator (SikaRapid C-100) is added prior casting at a time upon water addition equivalent to the conditions in the slipforming process. The concrete is mixed again using a forced action mixer or the mixing reactor for 3 min. This composition is referred to as activated concrete. A summary of previous and modified composition is given in Table 1.

Table 1. Mix formulations used for online, offline and robotic tests

Component (kg/m^3)	SDC NEST Mix	SDC FS_1 Mix	SDC FS_2 Mix
Sand	1368	1368	1368
Cement	617	623	615
Silica fume	39	33	32
Water	249.3	256.1	259.0
Superplasticizer	1.18[a]	1.28[b]	1.55[b]
Sucrose	0.75	0.69	0.68
Accelerator	28.60	8.99–14.31[c]	8.86–14.11[c]

[a]Delayed addition (15 min after mixing water)
[b]Addition directly to the mixing water
[c]Increasing amount over time

3.1 Initial Robotic Experiments

The robotic experiments aim to produce thin folded structures by adapting the experimental setup of SDC [4] to this different geometrical system with smaller process window. The setup consists of several components that need to be synchronized during the production: the material, the digital design and fabrication tool and the 6-axis robot with the attached custom end-effector in form of a rigid formwork system for thin folded structures (Figs. 2 and 3). The interaction of these components allows for producing a number of physical prototypes at 1:1 architectural scale.

The initial robotic experiments aim to validate the use of the SDC NEST mix with different formwork geometry and mixing reactor. A large batch of retarded concrete is prepared and incrementally dosed with a progressive cavity pump into the mixing reactor. In there it is mixed with the accelerator forwarded in small increments by a peristaltic pump. The mixing reactor was separately developed and consists of a funnel providing inlets for concrete and accelerator at different heights, dispenses by overflow and is stirred by a pin mixer type shaft. The activated concrete flowing out is filled into a rigid formwork for thin folded structures attached to an ABB IRB4600 robotic arm performing the slipping along digitally defined trajectories. The formwork encloses a volume of 2.5 cm thickness, 50–50 cm length and 30 cm height.

Fig. 2. Fabrication setup

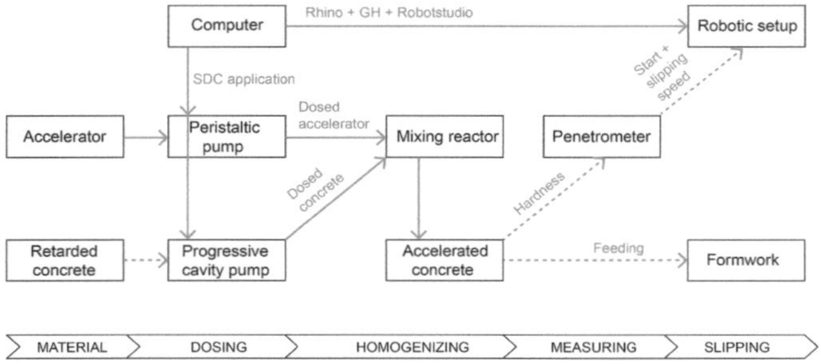

Fig. 3. Schematic representation of the SDC fabrication setup for thin folded structures

During the robotic experiments the material is evaluated by monitoring the surface quality of the prototype, the occurrence of failure types: local flow-out, crack formation or buckling. In addition, day-to-day variability of environmental conditions (mostly temperature) and concrete composition is taken into account by sampling two online penetrometer measurements (described in Sect. 3.3) right before the formwork filling is started. These measurements are used to calibrate the actual time needed to gain enough yield strength for slipping and to adjust the process speed. Based on the results the constant slipping speed is set for a value between 12 and 20 mm/min for the experiments in the initial experimental phase. The pumping rate follows the slipping speed, thus it is constant for both the retarded concrete and the accelerator over time with the SDC NEST mix.

3.2 Offline Tests

The offline tests are a set of experiments performed in laboratory environment on smaller quantities of material without the experimental setup. These material tests aim to modify the previous composition (SDC NEST Mix) to meet the more strict process requirements of SDC folded structures. The material constraints are manifold. The experiment described here concerns the effect of the accelerator dosage and the time of its addition on structural buildup.

First, as a reference, the SDC NEST Mix is tested with unvaried accelerator amount over the expected duration of retardation. Then to measure the influence of accelerator addition, another retarded concrete batch is prepared with the modified composition (SDC FS_1 mix) and is subsequently activated with increasing amount of accelerator (2.5–4% by binder weight) in volumes of 2.5 L and at times of 1, 2, 3, 4 and 5 h upon water addition. The accelerator content is chosen so that the concentration used at 5 h is equivalent to the upper limit indicated by the manufacturer. The activated batches are poured into containers 10 min after accelerator addition for yield stress measurements performed with a penetrometer. The penetrometer is mounted on a triaxial, computer controlled unit to measure at different positions according to a pattern. Its needle ends in a cylindrical head (d = 19 mm, h = 4 mm) that is immersed into specimens with a constant speed of 1 mm/s, while recording the force resisting penetration. The first peak in the recorded data is related to the yield stress [9]. 14 measuring points are defined, and measurements are performed at intervals of 5 min, starting from 30 min after acceleration. A target penetration resistance of 6–7 N is considered ideal for slipping.

3.3 Online Tests

The online penetrometer tests aim to model the material processing during a robotic experiment by dosing the retarded concrete and accelerator with digitally controlled pumps and mixing them together in a funnel (Fig. 3). With these tests, the material formulation of the most successful offline test (SDC FS_1) is adapted to compensate for the variations in the raw materials and the conditions of acceleration.

The accelerator dosage follows the dynamic acceleration strategy, it is increased every 30 min (2.5–4% by binder weight) from the first until the fifth hour after water addition. The activated concrete flowing out is collected every 10 min for single yield stress measurements with the penetrometer. These measurements are performed 40 min upon outflow (subsequent casting), to measure the yield stress of a material shortly before the formwork would be slipped off.

Parallel to the online tests offline tests are also conducted to monitor the impact of the experimental equipment (mainly the use of the mixing reactor) on activation by relating it to the results of the smaller scale laboratory tests. For the offline penetrometer measurements three batches from the retarded concrete with the volumes of 2.5 L are accelerated at times of 1, 3 and 5 h upon water addition. The accelerator dosage for these tests also show increasing tendency over time, it varies according to the dynamic acceleration strategy of the online test.

3.4 Robotic Experiment as Material Evaluation

The aim of this robotic experiment is to evaluate the material formulation with dynamic acceleration strategy chosen based on the online tests (SDC FS_2) by manufacturing a thin folded prototype with the SDC process. The robotic fabrication setup and the material monitoring method (described in Sect. 3.1) remain unchanged.

This experiment aims to produce the least problematic geometry possible. The robot moves the rigid formwork for thin folded geometries along a straight vertical slipping trajectory with a constant speed of 7.5 mm/min. The pumping rate of retarded concrete over time is constant, however the amount of accelerator is increased (ranging from 2.5 to 4% of binder content) over the duration of the experiment every 30 min to provide uniform yield stress evolution and wider process window for the material leaving the formwork along its whole cross section.

4 Experimental Results

4.1 Initial Robotic Experiments

7 prototypes are produced with the SDC fabrication setup for thin folded geometries using the SDC NEST Mix formulation (Fig. 4).

Fig. 4. The initial thin folded concrete prototypes using the SDC NEST Mix

The start of the slipping for all of these robotic experiments is determined with the help of two online measurements, measured 20 and 30 min after flow out from the funnel and subsequent casting. In the case of the initial experiments these measurements show values ranging from 16–17 N. These values are above the ideal slipping range of 6–7 N thus the slipping starts with a delay resulting in a further decreased process window. Detecting the slipping range in time with high accuracy is proven to be challenging due to the fast yield stress evolution of the SDC NEST Mix, the efficiency of the mixing reactor and the varying environmental conditions. The slipping speed ranging from 14 to 20 mm/min for the different experiments is synchronized with the material hardening. Failures such as excessive crack formation, bad surface quality and low final height can be observed on the prototypes (Fig. 4).

4.2 Offline Tests

The maximum value of the first peak of each penetrometer measurement is displayed in Figs. 5 and 6 as penetration resistance as a function of the time after concrete casting. The measurements on the activated samples at given times are fitted with power law. The yield stress evolution is faster for the SDC NEST Mix and shows significant differences depending on the time of accelerator addition. It is slower and more uniform for the SDC FS_1 Mix. By increasing the accelerator dosage over time, this undesirable effect is compensated with the composition SDC FS_1. Additionally, SDC FS_1 reaches the target value for slipping more consistently than SDC NEST Mix. The exponent of the yield stress evolution is rather constant for SDC_NEST, while it increases for SDC FS_1.

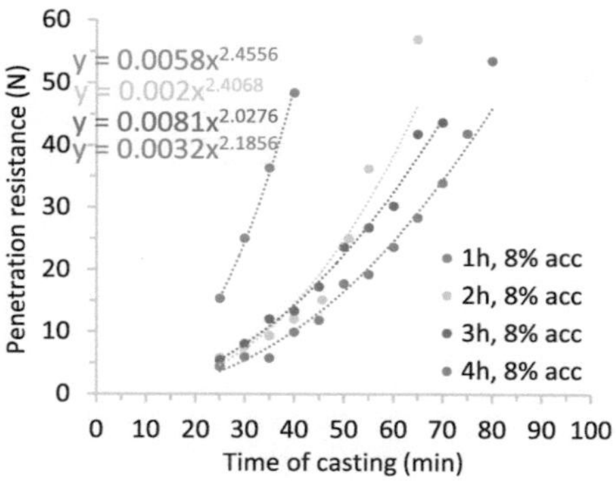

Fig. 5. Penetration resistance of the SDC NEST Mix over time for batches of retarded concrete accelerated 1, 2, 3 and 4 h after water addition with unvaried accelerator amount

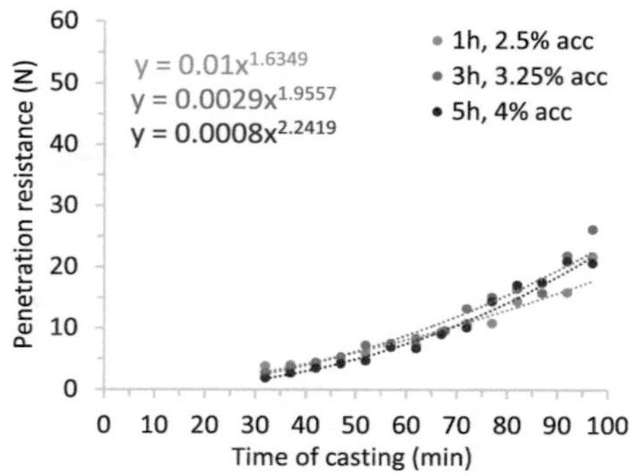

Fig. 6. Penetration resistance (N) of the SDC FS_1 Mix over time for batches of retarded concrete accelerated 1, 3 and 5 h after water addition with increasing accelerator amount

Mix design iterations not described here led to the composition SDC_FS_1 and additional experiments verified that the concrete is stable and self-compacting during placing and that at least 5 h of retardation of the retarded mix are provided.

4.3 Online Tests

As a next step, the adjusted material composition from the offline tests (SDC_FS_1) is characterized with online penetrometer measurements on the SDC process equipment for thin folded structures to adjust it further.

The results of the online test with the final composition (SDC FS_2 Mix) are shown in Fig. 7 as penetration resistance over time relative to the time of acceleration. The data shows a slight yield stress increase as the time of acceleration increases. Offline tests are performed to complement the online tests using the same composition (Fig. 8). Opposing tendencies are observed. In fact in the offline test slower yield stress increase for batches accelerated later is observed. The increase of yield stress evolution exponent is in agreement with the offline tests.

4.4 Robotic Experiment as Material Evaluation

The formulation of the SDC FS_2 mix for robotic experiments with the SDC setup for thin folded geometries is validated by a thin folded prototype of 140 cm (Fig. 9).

The start of the slipping is determined with two online measurements at 20 and 30 min after casting similarly to the initial robotic experiments. However, as this material stays longer in the ideal range for slipping than the initial SDC NEST mix it is easier to detect the start of slipping. The first online measurement showed 3.9 N penetration resistance while the second 9.8 N (slightly beyond the ideal slipping range). The slipping is started right after the second measurement with the speed of 7.5 mm/min. The formwork was approximately 80–85% full and slightly faster

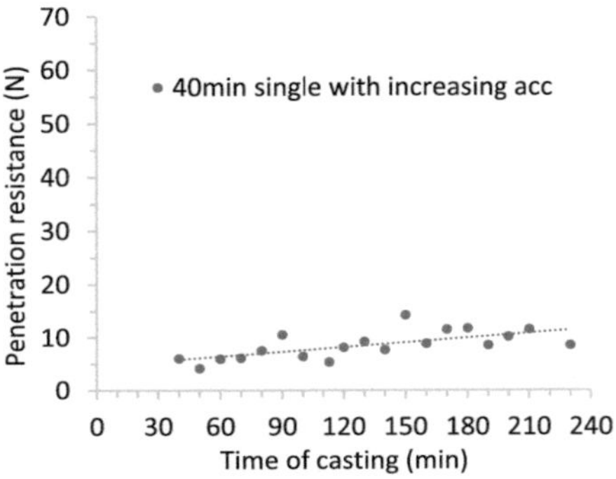

Fig. 7. Penetration resistance of the SDC FS_2 Mix accelerated continuously in the mixing funnel with increasing accelerator amount

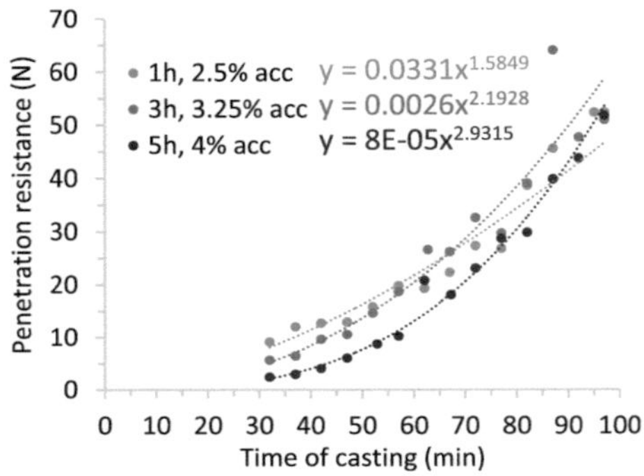

Fig. 8. Penetration resistance of the SDC FS_2 Mix batches accelerated 1, 3 and 5 h after water addition with increasing accelerator amount

material evolution is estimated compared to the online measurements performed prior to the robotic experiment. The workability of the concrete is sufficient during the process. Minor defects can be observed due to additional formwork layers for capillary oiling however the overall surface quality is satisfactory.

5 Discussion of Results from Online, Offline and Robotic Experiments

In the offline tests, the SDC NEST Mix shows non-uniform yield stress evolution over the timeframe of the experiment and a fast yield stress increase, leading to a narrow time window in which the yield stress is in a certain range considered necessary for slipping. These time dependencies make the control of the experiment more difficult requiring constantly adjusted slipping speed over time thus leading to a high risk of either fracturing or flow out [9]. The extensive crack formation can be observed on the prototypes of the first robotic experiments (Fig. 4).

The adjustments of the SDC NEST Mix involves reducing the accelerator dosage to 4% of the binder weight, as specified by the manufacturer to mitigate risks associated with the addition of nitrates [11]. Additionally reducing the silica fume allows for adding less superplasticizer [12] and simplifying the mixing procedure to direct superplasticizer addition, all without negatively affecting the uniform slipping rate [13]. The variability of sand deliveries further required some optimization of the superplasticizer dosage between FS_1 and FS_2 [14]. These modifications were necessary for other considerations than structuration rate but they nevertheless impacted this important property, making material optimization and adjustment more delicate [15–17]. This observation highlights the importance of measuring and controlling yield stress evolution especially for slipforming in applications with small hydrodynamic radii.

Fig. 9. The first thin folded concrete prototype produced with the SDC FS_2 Mix

By mastering the yield stress evolution despite different times of accelerator addition and dosages and by overall decreasing the structuration rate (SDC FS_1 and FS_2) the slipping time frame can be kept constant with respect to the process. Consequently, the surface quality is more consistent and has fewer defects. In this way, the mix improved the predictability of the yield stress increase by the dynamic acceleration strategy. Additionally, as the exponent of the yield stress evolution is higher towards the end of the process in offline measurements (Fig. 8), the contribution of friction is expected to decrease during the process even when the acceleration happens with the funnel. Finally, the fluidity of the retarded concrete could be improved (allowing for better pumping) by aiming for the same final workability and adding less accelerator for activation (maximum 4% for the SDC FS_1 Mix instead of the 8% of the initial SDC NEST Mix).

However, the offline tests can only provide an approximate material composition for the robotic experiments due to the differences in equipment for acceleration and dosing. Thus, online tests with the mixing reactor are needed to obtain material data that can be directly translated to the robotic experiments. The faster structuration at later stages in the online experiment could be a consequence of accumulation of material with longer residence time in the mixing reactor. Iterative and empirically evaluated modifications in the acceleration dosage reduced such effects until SDC FS_2 mix met all requirements for moving on to a robotic experiment.

The success of the robotic experiment depended on the synchronization of the robot movement and the hardening rate of the material. The right timing for the start of the slipping was provided by the online measurements and the constant slipping speed was possible by the virtue of the uniform strength gain. However, for more complex slipping trajectories inline measurements could be beneficial or even necessary to provide information about the material properties along the exit of the formwork. Recent offline ultrasound measurements hold the promise to be integrated into the formwork and take this role [10]. Additionally, slim and especially cantilevering geometries will raise further the requirements for the fast yield stress evolution outside the formwork to avoid buckling.

6 Conclusion

The difficulty to produce thin folded concrete prototypes with the material previously developed for column geometries can be explained with the smaller process window available for slipping. In the formwork for thin folded geometries by its smaller volume to surface ratio (hydrodynamic radius) the friction is affecting the process to a greater extent. However, the adjustment of the material mix and the dynamic acceleration strategy led to slower, uniform yield stress evolution thus increasing the time frame in which the process window conditions are met. Thus, the process became more robust and met the requirements of thin folded geometries. Finally, the thin folded concrete prototype produced serves as the first step towards exploring the full potential of SDC to manufacture a broader range of non-standard thin members for architectural applications.

Acknowledgements. The research is pursued in the interdisciplinary framework of the National Competence Centre of Research (NCCR) Digital Fabrication funded by the SNSF at ETH Zürich.

The authors thank Heinz Richner, Andi Reusser, Michael Lyrenmann for technical assistance, Tom Mundy, Lukas Sigrist, Shaun Wu and Federico Giacomarra for their help during the experiments. The authors are also grateful for the constructive discussions with Nicolas Roussel and Timothy Wangler.

References

1. The third industrial revolution. The Economist (2012)
2. Gramazio, F., Kohler, M., Willmann, J.: The Robotic Touch: How Robots Change Architecture, pp. 286–301. Park Books, Zürich (2014)
3. Wangler, T., Lloret, E., Reiter, L., Hack, N., Gramazio, F., Kohler, M., Bernhard, M., Dillenburger, B., Buchli, J., Roussel, N., Flatt, R.: Digital concrete: opportunities and challenges. RILEM Tech. Lett. **1**, 67–75 (2016)
4. Lloret, E., Shahab, A.R., Linus, M., Flatt, R.J., Gramazio, F., Kohler, M., Langenberg, S.: Complex concrete structures: merging existing casting techniques with digital fabrication. Comput. Aided Des. **60**, 40–49 (2014)
5. Lloret, E., Mettler, L.K., Shahab, A.R., Gramazio, F., Kohler, M., Flatt, R.J.: Smart dynamic casting: a robotic fabrication system for complex structures. In: Proceedings of 1st Concrete Innovation Conference, Oslo, Norway (2014)
6. Reiter, L., Palacios, M., Wangler, T., Flatt, R. J.: Putting concrete to sleep and waking it up with chemical admixtures. In: Proceedings 11th Canmet ACI International Conference Superplast Chem Admix Concr, ACI, Ottawa. Ottawa, ON, Canada, pp. 145–154 (2015)
7. Lloret Fritschi, E., Reiter, L., Wangler, T., Gramazio, F., Kohler, M., Flatt, R.J.: Smart dynamic casting: slipforming with flexible formwork—inline measurement and control. In: Proceedings of 2nd Concrete Innovation Conference, Tromso, Norway (2017)
8. Reiter, L., Wangler, T., Roussel, N., Flatt, R.J.: The role of early age structural build-up in digital fabrication with concrete. Cem. Concr. Res. (2018). https://doi.org/10.1016/j.cemconres.2018.05.011
9. Mettler, L.K., Wittel, F.K., Flatt, R.J., Herrmann, H.J.: Evolution of strength and failure of SCC during early hydration. Cem. Concr. Res. **89**, 288–296 (2016)

10. Sayers, C.M., Dahlin, A.: Propagation of ultrasound through hydrating cement pastes at early times. Adv. Cem. Based Mater. **1**, 12–21 (1993)
11. Myrdal, R.: Accelerating admixtures for concrete. SINTEF REPORT Building and Infrastructure Concrete and Innovation Center, Project No: 3D006020, pp. 1–35 (2007)
12. Hommer, H.: Effectiveness of polycarboxylate superplasticizers in ultra-high strength concrete: the importance of PCE compatibility with silica fume. J. Eur. Ceram. Soc. **29**, 1847–1853 (2009)
13. Flatt, R.J., Schober I.: Superplasticizers. Understanding the Rheology of Concrete. Roussel, N. (ed.) Woodhead publishing, pp. 144–208 (2012)
14. Puertas, F., Santos, H., Palacios, M., Martinez-Ramirez, S.: Polycarboxylate superplasticizer admixtures: effect on hydration, microstructure and rheological behavior in cement pastes. Adv. Cem. Res. **17**, 77–89 (2005)
15. Cheung, J., Jeknavorian, A., Roberts, L., Silva, D.: Impact of admixtures on the hydration kinetics of Portland cement. Cem. Concr. Res. **41**, 1289–1309 (2011)
16. Marchon, D., Flatt, R.J.: Impact of chemical admixtures on cement hydration (Chap. 12). In: Aïtcin, P.-C., Flatt, R.J. (eds.) Science and Technology of Concrete Admixtures, pp. 279–304. Woodhead Publishing, Sawston (2016)
17. Marchon, D., Mantellato, S., Bessaies, H., Kawashima, S., Serina, B.: Hydration and rheology control of concrete by admixtures for digital fabrication. Cem. Concr. Res. (2018). https://doi.org/10.1016/j.cemconres.2018.05.014

Enhancing Printable Concrete Thixotropy by High Shear Mixing

Aileen Vandenberg[1](✉), Hela Bessaies-Bey[2], Kay Wille[1], and Nicolas Roussel[2]

[1] University of Connecticut, Storrs, CT 06269, USA
aileen.c.vandenberg@gmail.com
[2] IFSTTAR, 77420 Marne-La-Vallée, France

Abstract. Our results show that the storage elastic modulus as a function of time increases at a higher rate for cement paste mixed at higher vesus lower mixing intensity. Hence, higher mixing appears to be enhancing thixotropy. Using calorimetry analysis we find that higher mixing decreases the setting time and enhances the peak of the heat flow. By analyzing the nanoparticles present in the suspending fluid of the cement paste, we show, in accordance with literature, that an appropriate combination of mixing energy and superplasticizer dosage promotes hydration by scratching hydrates from the surface of cement particles, stabilizing them in the suspending fluid and hence generating additional nucleation surfaces. These results open the door for the design of printing heads including high-shear micro mixers allowing for a faster liquid-to-solid transition of the printable material.

Keywords: Thixotropy · Nanoparticles · Mixing · Cement · Calorimetry DLS

1 Introduction

With the advent of concrete digital fabrication paving new ways of construction [1], controlling cement paste thixotropy is crucial to ensuring the desired final state of the material is achieved. Each digitally fabricated concrete layer has to evolve from a fluid suspension during mixing and pumping to a cohesive material after deposition that maintains a strong interface with, and mechanical resistance to, the next deposited layer [2]. Consequently, thixotropy is very important for concrete digital fabrication as it enhances the structuration of cement paste.

From a practical view, thixotropy is mainly affected by CSH nucleation [2]. Thus, a very thixotropic concrete can be created by increasing the rate of nucleation of hydration products in the mixture. One way to increase this rate is by increasing the energy input of mixing [3]. Henceforth, in this work we focus on how mixing could enhance thixotropy.

T. Wangler and R. J. Flatt (Eds.): DC 2018, RILEM Bookseries 19, pp. 94–101, 2019.
https://doi.org/10.1007/978-3-319-99519-9_9

2 Methods and Materials

2.1 Materials

The cement used in this study was CEM I 52.5R, equivalent to ASTM Type I, Ordinary Portland cement of specific gravity 3.15. Its mineralogical composition was obtained through differential thermal and thermogravimetric analysis (DTA-TG). Particle size distribution was measured by laser diffraction using a Malvern Mastersizer 5 instrument (Malvern Instruments Ltd., Malvern, UK). Results are reported in Table 1. In this study, a commercial polycarboxylate ether-type (PCE) high range water reducer was used in liquid form containing 29.5% of dry polymer.

Table 1. Mineralogical composition and size of the cement powder used in this study.

Material	C_3S	C_2S	C_3A	C_4AF	SiO_2	D_{v10}	D_{v50}	D_{v90}
Cement	64%	14%	2.5%	14%	21%	1.49 μm	8.11 μm	28.75 μm

2.2 Mixing Protocol

A delayed mixing protocol was chosen to reduce the possibility that the aluminates would co-precipitate with the studied admixtures [4]. While complete prevention of co-precipitation is never the case in fresh cement paste, this protocol was repeatable [5]. Cement pastes were prepared using a Turbo test Rayneri VMI mixer with four mixing speeds chosen – 840 rpm, 1400 rpm, 2100 rpm, and 2800 rpm. The diameters of the mixing blade and beaker were 3 cm and 8 cm, respectively. The mixing protocol was a three stage process. In the first stage, 200 g of cement was added to 90% of the water and then mixed for 90 s. In the second stage, the cement paste was left to rest for 20 min, while in the third stage the paste was continuously mixed for 120 s, where at the 60 s mark the remaining water and the PCE was added to the mixing paste. The water-to-cement (w/c) ratio was 0.3 and the PCE dosage was 0.4% solid content of PCE by weight of cement (bwoc). The amount of water in the PCE solution was considered in the w/c calculation. After the last stage of mixing, the paste was left to rest for another 15 min in order for the polymer to reach equilibrium in the system. The preparation protocol was the same for the quartz powder system.

For the DLS measurements, the paste was centrifuged at 1000 times the gravitational acceleration ($g = 9.81\ m\ s^{-2}$) for 5 min. The suspending liquid was then extracted and filtered with a polyvinylidene fluoride (PVDF) 0.45 μm membrane (Millipore).

2.3 Rheological Measurements

Dynamic rheological experiments were conducted using a Bohlin C-VOR shear rheometer equipped with a Vane geometry. The Vane tool diameter, outer cup diameter and depth were 25 mm, 50 mm and 60 mm, respectively. After the mixing process the paste was loaded into the geometry and pre-sheared at 150 s^{-1} for 150 s before the start of the time sweep dynamic test. In order to measure the evolution of the kinetics of the rigid interactions network, an oscillating strain of amplitude lower than the rigid critical strain [6] was applied. This ensures that the system is purely elastic and the measured stress is in phase with the applied oscillating strain [2]. Thus, the oscillating strain of amplitude was set to 0.03% at a frequency of 1 Hz. To capture the elastic storage modulus fully the time of the test was set to 2 h.

2.4 Isothermal Calorimetry

The thermal power and heat of hydration of the cement pastes were monitored for the first 48 h using a TAM Air microcalorimeter (Thermometrics) at a constant temperature of 20 °C. Samples were mixed according to the mixing procedure and then 5 g of paste was placed inside the calorimeter cell. All samples were balanced with a reference cell having the same heat capacity.

2.5 Dynamic Light Scattering Measurements

Dynamic Light Scattering (DLS) is generally used to measure the size distribution by intensity of particles in a liquid medium. In this work, the measure scattered intensity was further used to estimate the concentration of particles remaining in the interstitial pore fluid after mixing [7, 8]. This was done by first recording the scattered intensity. Then the area underneath the intensity curve from 30 to 1000 nm was computed. This area, referred hereafter as the derived count rate (DCR) of the system, was then used to represent the concentration of nanoparticles in the interstitial pore fluid. Measurements were performed using a Zetasizer nano S from Malvern Instruments, operating at an incident light wavelength of 633 nm, scattered light detection angle of 173°, and constant temperature of 25 °C. Averages of three measurements per specimen were taken.

3 Results

3.1 Rheological Measurements

We plot in Fig. 1 the elastic storage modulus versus time for cement paste mixed at 840 rpm versus 2800 rpm. Our results show that the storage elastic modulus as a function of time increases at a higher rate for the cement paste mixed at 2800 rpm versus 840 rpm. In Fig. 2 we plot the ratio of the elastic storage modulus of the two speeds. We can see it is linear and that the higher speed is approximately four times higher after 2 h.

3.2 Isothermal Calorimetry

The heat evolution during the hydration of cement paste mixed at four different mixing speeds is presented in Fig. 3. It is seen that the paste mixed at 2800 rpm reaches its maximum peak almost 4 h before the paste mixed at 840 rpm. The maximum peak of this higher mixed paste reaches almost 1 mW g^{-1} higher in heat flow than the lower mixed paste. In general, our results show that increasing the mixing speed shifts the

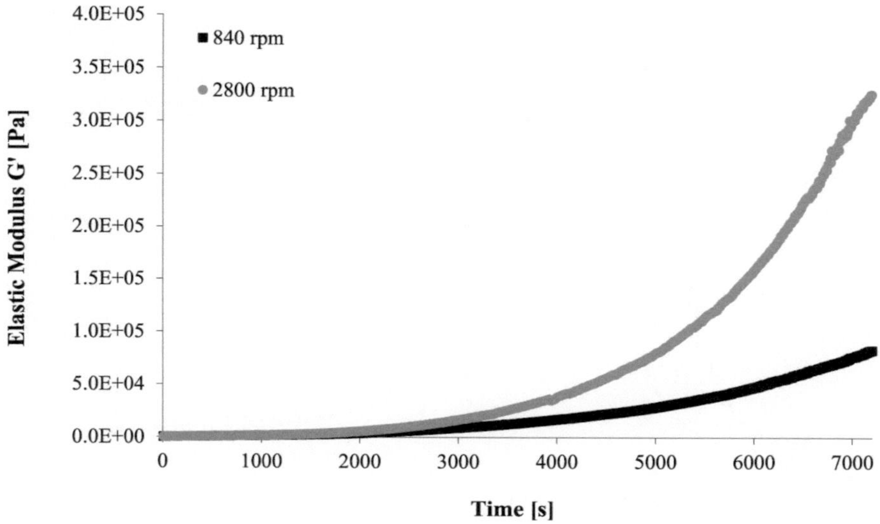

Fig. 1. Effect of mixing intensity on the elastic storage modulus of cement paste samples prepared with w/c ratio of 0.30 and 0.4% of PCE bwoc.

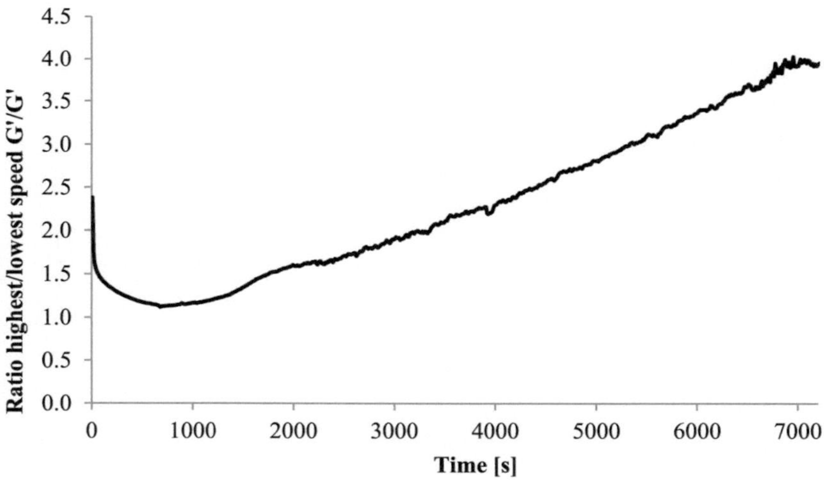

Fig. 2. The ratio between the elastic storage modulus of the 2800 rpm versus 840 rpm.

heat flow curve left, increases the slope and the maximum peak of the heat flow, but does not affect the overall shape of the curve.

We plot in Fig. 4a the rate of the heat flow during the acceleration period and in Fig. 4b the time of the maximum peak of heat flow during the acceleration period as a function of mixing speed. There appears to be an almost linear relation for both of these figures with higher mixing showing higher heat flow rates and shorter times to reach the maximum peak.

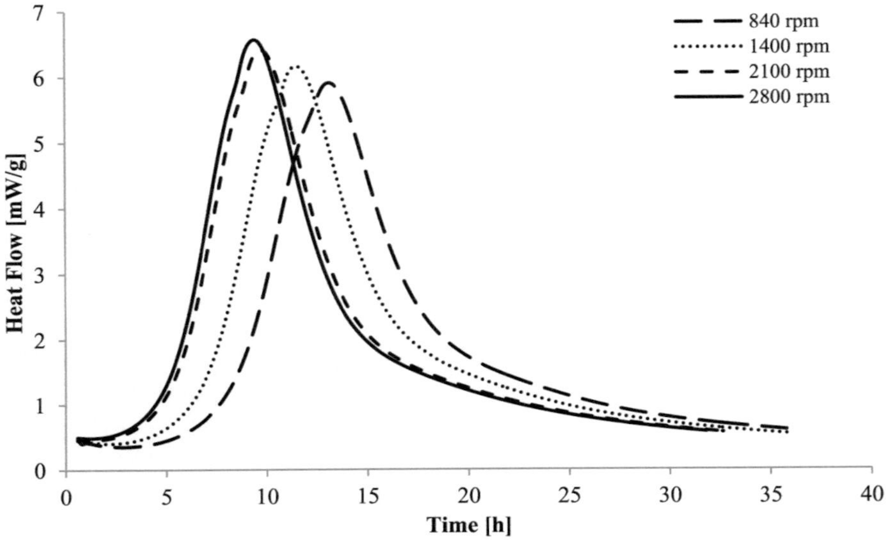

Fig. 3. The heat flow of cement paste (w/c 0.3, PCE dosage 0.4%) mixed at different speeds.

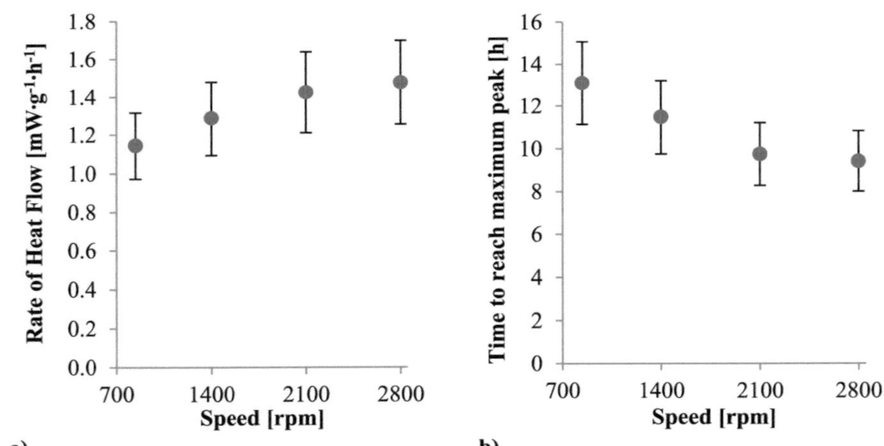

Fig. 4. (a) The slope of the heat flow curve during the acceleration period and (b) the time to reach the maximum peak as a function of mixing speed.

3.3 Dynamic Light Scattering Measurements

We show in Fig. 5 the results of the DLS measurements. We observe that higher mixing enhances the concentration of nanoparticles in a linear fashion. The DCR for the highest speed is half an order higher than the lowest speed.

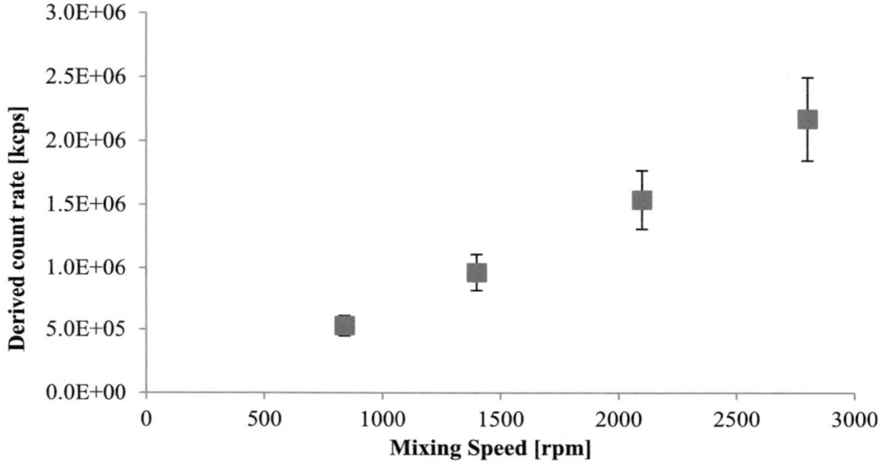

Fig. 5. Cement powder in water (W/C 0.50) with 0.8% PCE dosage (by weight of cement) and increasing mixing speed.

4 Discussion

The main thing we have shown is that higher mixing enhances the thixotropy, the kinetics of hydration, and the concentration of nanoparticles in the suspending fluid of cement paste. These phenomena have been observed before.

Han and Ferron [9] found that the mixing rate enhanced both the elastic storage modulus and the rate of heat evolution. Juilland et al. [3] were able to demonstrate that two kinetics of cement hydration were affected by the mixing rate – the induction period and the acceleration period. An increase in the mixing rate led to a shortening of the induction period by decreasing the diffuse electrical double layer, which in turn increased the dissolution rate, and hence, the transportation rate of the ions that detached from the cement surface into the bulk pore fluid. The increase in the height of the main heat evolution peak was due to the mechanical action of the mixing, resulting in an increase in the detachment of C-S-H from the surface and their density.

Caruso et al. [8] highlighted the formation of nanoparticles in cement pore solution containing superplasticizers. They asserted that these nano-size particles are either nano-C-S-H, nano-ettringite, nano-AFm, or interamolecular complexes of polymer and cations. However, they could not draw any further conclusions on whether one type of nanoparticle was more present than another and did not consider the consequences of the variations in mixing speed on the generation of these nanoparticles.

Consequently, we suggest that an appropriate combination of mixing energy and super-plasticizer dosage promotes hydration by scratching hydrates from the surface of cement particles, stabilizing them in the suspending fluid and hence generating additional nucleation surfaces.

The physical origins of thixotropy originate mainly from the rigid interactions network of C-S-H nucleation [2]. Thus, with higher mixing power the number of C-S-H bonds created by hydration increases and hence why thixotropy is enhanced and the setting time decreases.

5 Conclusion

This study aimed to understand how mixing can enhance the thixotropy of a cement paste. Our results show that the storage elastic modulus as a function of time increases at a higher rate for the cement paste mixed at higher versus lower mixing intensity (i.e. approximately four times higher after 2 h). Using calorimetry analysis we found that higher mixing decreases the setting time and enhances the peak of the heat flow. By analyzing the nanoparticles present in the suspending fluid of the cement paste we show, in accordance with literature, that an appropriate combination of mixing energy and super-plasticizer dosage promotes hydration by scratching hydrates from the surface of cement particles, stabilizing them in the suspending fluid, and hence, generating additional nucleation surfaces.

These results open the door for low cost cement acceleration through just the mixing as well as the design of printing heads including high-shear micro mixers allowing for a faster liquid-to-solid transition of the printable material.

References

1. Lloret, E., Shahab, A.R., Linus, M., Flatt, R.J., Gramazio, F., Kohler, M., Langenberg, S.: Complex concrete structures: merging existing casting techniques with digital fabrication. Comput.-Aided Des. **60**, 40–49 (2015)
2. Roussel, N., Ovarlez, G., Garrault, S., Brumaud, C.: The origins of thixotropy of fresh cement pastes. Cem. Concr. Res. **42**, 148–157 (2012)
3. Juilland, P., Kumar, A., Gallucci, E., Flatt, R.J., Scrivener, K.L.: Effect of mixing on the early hydration of alite and OPC systems. Cem. Concr. Res. **42**, 1175–1188 (2012)
4. Flatt, R.J., Houst, Y.F.: A simplified view on chemical effects perturbing the action of superplasticizers. Cem. Concr. Res. **31**, 1169–1176 (2001)
5. Hot, J., Bessaies-Bey, H., Brumaud, C., Duc, M., Castella, C., Roussel, N.: Adsorbing polymers and viscosity of cement pastes. Cem. Concr. Res. **63**, 12–19 (2014)
6. Nachbaur, L., Mutin, J.C., Nonat, A., Choplin, L.: Dynamic mode rheology of cement and tricalcium silicate pastes from mixing to setting. Cem. Concr. Res. **31**, 183–192 (2001)
7. Bessaies-Bey, H., Baumann, R., Schmitz, M., Radler, M., Roussel, N.: Organic admixtures and cement particles: competitive adsorption and its macroscopic rheological consequences. Cem. Concr. Res. **80**, 1–9 (2016)

8. Caruso, F., Mantellato, S., Palacios, M., Flatt, R.J.: ICP-OES method for the characterization of cement pore solutions and their modification by polycarboxylate-based superplasticizers. Cem. Concr. Res. **91**, 52–60 (2017)
9. Han, D., Ferron, R.D.: Effect of mixing method on microstructure and rheology of cement paste. Constr. Build. Mater. **93**, 278–288 (2015)

Discrete Element Simulations of Rheological Response of Cementitious Binders as Applied to 3D Printing

Pu Yang, Sooraj Kumar A. O. Nair, and Narayanan Neithalath(✉)

Arizona State University, Tempe, AZ, USA
Narayanan.Neithalath@asu.edu

Abstract. This paper aims to model the extrusion-based 3D printing process of a plain ordinary Portland cement (OPC) paste using the discrete element method (DEM), and outlines the methodology adopted to evaluate the linkage between particle scale processes and extrusion process. A mini slump test is used to define the rheological model to be used in DEM, and extract the relevant parameters. They are then implemented in a scaled-down extrusion printing model to determine the influence of particle-scale effects on extrusion force. The DEM model is able to capture the differences in extrusion load-displacement responses similar to the experiments. Refinements to the model based on extracted parameters are also discussed.

1 Introduction

3D printing (also referred to as additive manufacturing or digital fabrication) of cement-based materials has been receiving widespread attention in the recent past [1–4]. 3D printing is gaining prominence in civil construction because of its potential to enable labor and energy reductions, speed up the construction process, and reduce construction-related risks [5, 6]. This technology also offers extensive shape optimization where architectural shapes for multi-functional applications can be built economically [7, 8]. It is well accepted that, along with advances in robotics and automation processes, it is very important to develop cementitious materials that are amenable to additive manufacturing, in order to create robust structures using this technique. The complexities and variabilities associated with cement-based materials pose real challenges in the layered extrusion process which is the most common method for 3D printing of cement-based materials. From a binder rheology standpoint, it is imperative that the material should flow cohesively though the extruder, but at the same time, be able to "set" fast enough once extruded so that it can maintain its shape and carry the weight of the overlaid layers. This requires control of particle shapes and sizes capable of retaining paste cohesion to enable flow as well as yield stress development, and adequate chemical reactions. Selection of the source materials and their combinations become critical in such as case, as has been shown in [9]. Moreover, the processing techniques also influence the rheology – yield stress is a function of the compacting pressure that the paste is subjected to while being extruded, the geometry of the extruder, the surface roughness of the container walls etc. The efficiency of

T. Wangler and R. J. Flatt (Eds.): DC 2018, RILEM Bookseries 19, pp. 102–112, 2019.
https://doi.org/10.1007/978-3-319-99519-9_10

printing, thus is a function of a number of materials and processing related parameters, knowledge of which is essential in successful digital fabrication of cement-based binders.

As explained earlier, characterization of paste flow is important in ensuring rheological control during printing. The interaction between the rheological characteristics and processing parameters are better studied through a combination of experimental and simulation tools. For fresh pastes and concrete, discrete element method (DEM)-based simulations are appropriate to provide insights into the particle scale processes occurring during extrusion-based printing, and to relate them to the macro-scale response of the entire system. In this paper, we model the extrusion process of a plain ordinary Portland cement (OPC) paste using DEM, and outline the methodology adopted to evaluate the linkage between particle scale processes and extrusion process.

2 Discrete Element Method (DEM) and the Model Used

Discrete element method (DEM) is widely used to simulate the macro-mechanical response of granular materials by computing the individual motion and interactions of a large number of discrete elements with certain microscale properties [10]. Several commercial and open-source DEM solvers are available to model particulate systems including fresh concrete. A number of recent publications deal with DEM simulations of flow of self-consolidating concrete (SCC), spraying of fiber reinforced mortars, and blending of grains in a concrete mixer [11–15]. This paper uses a Particle Flow Code program, PFC2D, developed by ITASCA Consultants to simulate the extrusion and printing response of cementitious systems. For computational expediency, only 2D simulations are carried out. For all the DEM simulations, Newton's law o is used to determine the motion of each particle and the contact law (local constitutive model describing the inter-particle force-displacement relationship) is adopted to calculate the contact force between the particles. Thus, the simulation results (or the macro-scale behavior) is highly influenced by the contact laws (or the chosen micro-scale properties). Hence, it is critical to choose appropriate contact laws to ensure that the macroscale behavior of the material is appropriately captured.

2.1 Burger's Model

To simulate the rheological properties of fresh cementitious pastes, Burger's model [16] is used in this study. As shown in Fig. 1, Burger's model contains a Kelvin model acting in series with a Maxwell model in both normal and shear directions. The Kelvin model is constituted of a linear spring and a dashpot in parallel, while the Maxwell model is a combination of a linear spring and a dashpot in series. It also should be noted that the Burger's model acts over a vanishingly small area and can only transmit force. A contact with Boolean value of 0 or 1 determines whether the normal contact can sustain tensile force or not, and a slider with friction coefficient limits the shear force according to Coulomb law.

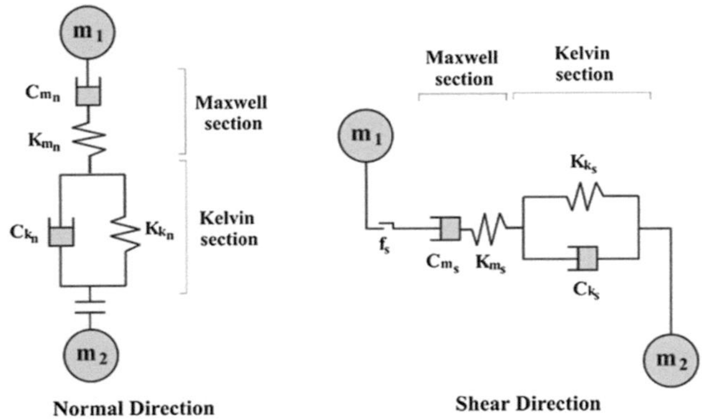

Fig. 1. Burger's model for particle contact in normal and shear direction [17].

In both the normal and shear direction, the deformation of the Burger's model u is the sum of the deformations of Kelvin section u_k and Maxwell section u_m:

$$u = u_k + u_m \tag{1}$$

The deformation of Maxwell section is a combination of the dashpot u_{mc} and spring u_{mk}:

$$u_m = u_{mk} + u_{mc} \tag{2}$$

Equation 1 can be written as:

$$u = u_k + u_{mk} + u_{mc} \tag{3}$$

The first and second derivatives of Eq. 3 are:

$$\begin{aligned} \dot{u} &= \dot{u}_k + \dot{u}_{mk} + \dot{u}_{mc} \\ \ddot{u} &= \ddot{u}_k + \ddot{u}_{mk} + \ddot{u}_{mc} \end{aligned} \tag{4}$$

The force-displacement equation of the Burger's model in each direction is then given by a second-order differential equation [17]:

$$f + \left[\frac{C_k}{K_k} + C_m\left(\frac{1}{K_k} + \frac{1}{K_m}\right)\right]\dot{f} + \frac{C_k C_m}{K_k K_m}\ddot{f} = \pm C_m \dot{u} \pm \frac{C_k C_m}{K_k}\ddot{u} \tag{5}$$

Where K_k and C_k represents the stiffness and viscosity of Kelvin section and K_m and C_m represent the stiffness and viscosity of Maxwell section. The symbol $\pm$ correspond to the cases of normal and shear direction, respectively. A second subscript is used to distinguish the Kelvin and Maxwell parameters in normal and shear direction (as shown in Fig. 1), where n denotes the normal direction and s denotes the shear direction.

A central difference approximation of the finite difference scheme for the time derivative is then used to update both the normal force $\vec{F}_n$ and shear force $\vec{F}_s$. Shearing (sliding) begins at the contact point where $\vec{F}_n$ and $\vec{F}_s$ satisfies the Mohr-Coulomb rupture criterion:

$$\left\|\vec{F}_s\right\| - f_s\left\|\vec{F}_n\right\| \leq 0 \tag{6}$$

Here, f_s is the inter-particle frictional coefficient.

2.2 Calibration of Burger's Model Using Mini Slump Test

In order to calibrate the Burger's model for plain cement pastes, a mini slump test was used. The mini slump cone has a top diameter of 19 mm, a bottom diameter of 38 mm, and a height of 57 mm. The simulated mini slump cone had top and bottom diameters of 5 mm and 10 mm, and a height of 15 mm, to maintain equivalence with the actual mini slump cone. The smaller volume of the mini slump cone is intended to reduce the number of particles required for simulation, and thus speed up the computations.

The particle size distribution of OPC is shown in Fig. 2. The OPC has a very wide PSD, with particles ranging from sub-micron size to ~ 100 μm. Simulating all the sizes in OPC will result in a very large number of particles, especially the small particles, thus necessitating heavy computational effort. In this study, therefore, the cement is considered to have particles in the 5 μm to 20 μm range (which accounts for 60% of the total volume of particles). In order to further enhance computational efficiency, the simulated particles are assumed to be 10 times larger than the actual sizes. DEM is quite insensitive to the size of particles (within limits) since it describes the inter-particle contacts by the contact law. The contact law describes the particle interactions, and is independent of spherical particle size. Thus, the particles are divided into two bins - 50 μm to 100 μm, and 100 μm to 20 μm size ranges, with volume fractions of 0.435 and 0.565 respectively, corresponding to those in the actual experiments. The final packing containing around 3,000 particles. The density of each particle is also reduced by 100 times so that each particle remains the same mass as a real cement particle. This is to ensure that the gravity acts in the real and simulated systems similarly, since slump of cement paste is a gravity-driven phenomenon.

Facet walls are used to create a trapezoidal geometry as shown in Fig. 3(a). Circular particles of the size distribution as mentioned above are used to create a dense packing inside the geometry. During the particle packing stage, a linear spring contact law is assigned to all the interparticle and particle – wall contacts. An extremely low value of frictional coefficient (0.01) is used to minimize the frictional force. Gravity is activated to help maximize the packing density.

After the generating the target virtual specimen, Burger's model is activated for all the inter-particle contacts. The top and sidewalls are assigned a zero frictional coefficient and the frictional coefficient of the bottom wall (the surface on which the slump cone is placed) is set to 0.70, which is same as that for the polycarbonate material used

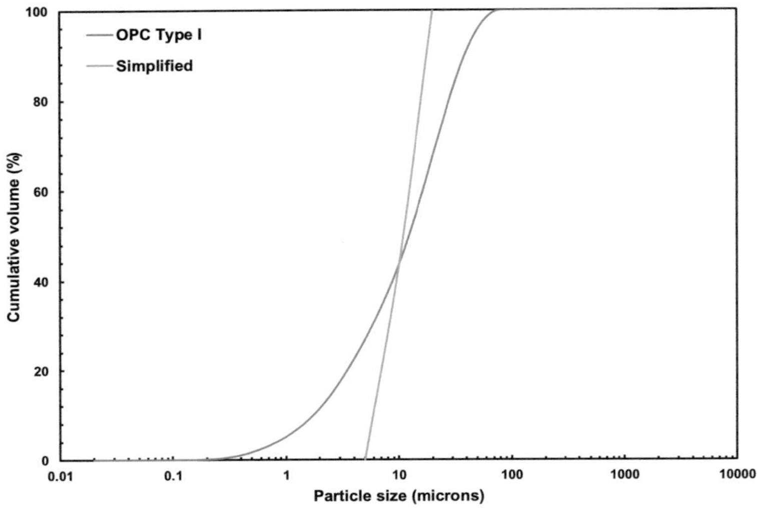

Fig. 2. Particle size distribution (PSD) of OPC. The simulation considers a part of this PSD only to reduce computational effort.

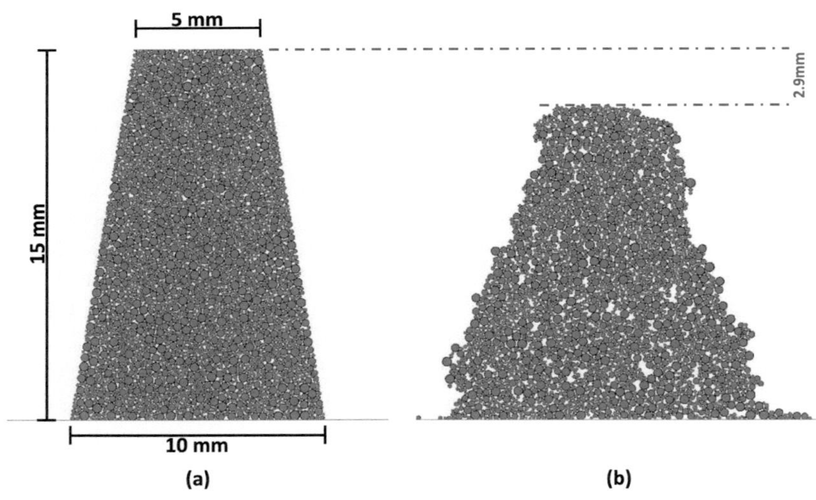

Fig. 3. Simulation of slump test on plain OPC paste: (a) before slump test and (b) after slump test.

in the experiments. The system is then solved until it reaches a balance state (where the total unbalanced force of the system is smaller than a threshold value; the default value of 1e−5 is used here). Figure 3(a) shows the simulated specimen inside the mini slump cone. The top and side walls are then removed from the system and the drop in height is recorded. The simulation is allowed to run until there is no significant difference in drop of the sample height. A user-defined function is used to find the position of the highest particle in the sample, and this information is used to calculate the height drop of the numerical sample. With a dual 16-core system having two Intel® Xeon® E5-2623 CPUs (clock speed 3.0 GHz) and 64 GB RAM, the computational time for a single slump test is about 12 h.

Several simulations using Burger's law (see Fig. 1) were carried out, in an effort to match the relative slump values (ratio of the drop in height to the specimen height) between the experiments and simulations, by varying the parameters of the Burger's model. While some publications have used the same parameters for the normal and shear components of the model, our parameteric study converged better with the shear components being equal to 0.1 times the value of the normal components. Thus, K_{k_n}, C_{k_n}, K_{m_n}, and C_{m_n} were assigned values of 1E06, while K_{k_s}, C_{k_s}, K_{m_s}, and C_{m_s} were assigned values of 1E05. In some other studies of similar systems, lower values (of the order of 1E03) were used for these parameters, but such low stiffness values resulted in overlapping spheres, which is not a characteristic of hard sphere systems. The inter-particle friction coefficient f_s was assigned a value of 0.05. Note that the Burger's law parameters were assigned based on the assumed particle packing in the model. A packing corresponding to the actual porosity of a cement paste with a water-to-cement ratio (w/c) of 0.40 will have a porosity of 0.56, which leads to instability in the slump prediction. Thus, as a first approximation we use a dense packing, which is what other studies on slump prediction have used [11, 13]. The calibrated values are used in DEM models of extrusion-based 3D printing as shown in the following section.

3 Extrusion Experiments and DEM Modeling

3.1 Extrusion-Based 3D Printing and Extrusion Rheology

A customized BCN3D printer with a Cartesian configuration (Fig. 4(a)), outfitted with a 100 cm^3 syringe ram extruder was used to print the pastes under controlled pressure and flow rate. The ram extruder is a rigid, high-density polyethylene (HDPE) barrel of length (L_b) 100 mm and diameter (D_b) 35 mm. The nozzle is either tapered as shown in Fig. 4(b) with entry and exit diameters of 10 mm and 4 mm respectively, or having a constant diameter of 10 mm or 4 mm respectively. In addition to the nozzles, a plain orifice opening of 10 mm was also considered. Details of cement pastes used for 3D printing, and their rheological characteristics can be found in [9]. In this paper, the results are restricted to those for a 0.40 w/c plain cement paste.

Extrusion rheology of the cement paste was carried out using all the four geometries mentioned above. The syringe extruder of the desired opening geometry, filled with the paste, was placed in a specially fabricated holder to enable the top of the plunger to be pushed by the upper platen of a servo-controlled MTS load frame (4.45 kN maximum load). The ram was moved at a velocity of 25 mm/min, which was similar to the printing speed. The force-ram displacement relationships for all the four geometries were recorded. Figure 5 shows the force-ram displacement relationships for OPC pastes extruded through the four geometries (shown in the inset).

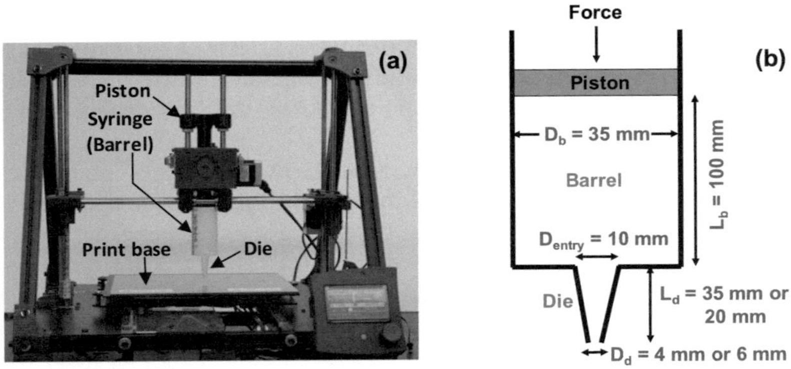

Fig. 4. (a) The syringe extrusion printer used in this study, and (b) geometry of the syringe extruder.

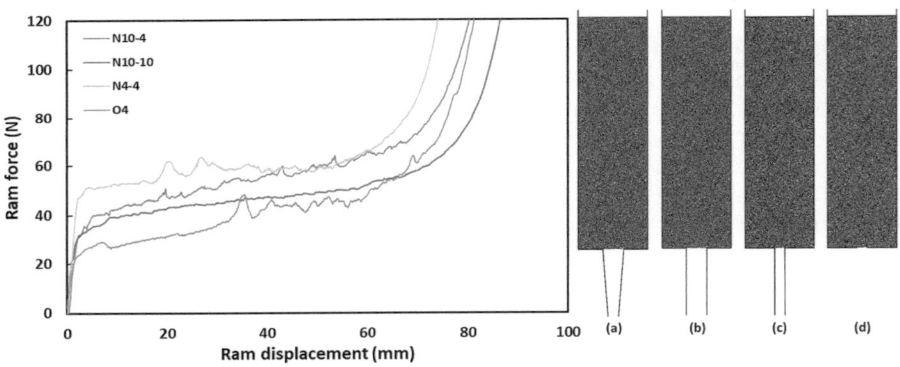

Fig. 5. Experimental force-ram displacement relationship from extrusion rheology for the OPC pastes.

In Fig. 5, O and N indicates the geometries with either the orifice (case d in the figure), or the nozzles (cases a to c) respectively, and the numbers next to them indicate the entry and exit diameters. The experimental behavior could be divided into three stages. The force-displacement response shows an initial increase in force under instantaneous consolidation (Stage 1) which ends with the shaping force, which is the force required to push the paste through the die, followed by a force plateau where the shaping force has been attained and extrusion happens without much increase in force (Stage 2). After a certain ram displacement (consolidation), when the ram approaches the end of the barrel, a rapid increase in extrusion force results because of the dead zone that is created (Stage 3). The paste extruded through the orifice shows the lowest extrusion force since the friction along the nozzle (die) wall is absent in this case. For a large diameter entry and exit (10 mm; N10-10), the force is lower than the case with the same entry diameter and a smaller exit diameter (4 mm; N10-4). While the force to shape the paste into the nozzle is roughly the same here (notice the end of the initial rising portion that is broadly similar for both the cases), the tapering nozzle results in an increased force requirement to extrude the paste. The friction in the constricting die walls also could contribute to this enhancement in force. The highest force is observed for the system with the smallest nozzle entry and exit diameter (N4-4). The difference in extrusion force between the orifice alone configuration (O4) and the N4-4 configuration can be attributed almost entirely to die wall friction.

3.2 DEM Modeling of Extrusion Rheology and Printing

In this section, a 2D DEM model of extrusion is developed based on the calibrated Burger's model. While the extrusion nozzle in the experiment had a 30 mm diameter and 100 mm height, the numerical extrusion process is carried out on a scaled model 10 mm wide (since it is in 2D) and 35 mm tall, preserving the dimensional ratio of the experiment. Hence, orifice or nozzle diameters of 4 mm in the experiments is scaled down to 1.5 mm, and 10 mm to 3 mm. The particle size distribution and numerical particle packing procedure described in the section on slump simulation are used in the extrusion simulation also. During the contact assignment stage of sample preparation, the frictional coefficient of the barrel and nozzle walls and the top cap (ram) is set to a value of 0.1, and the frictional coefficient of the bottom wall, which is the surface on to which the paste is extruded, is set to 0.7. The frictional coefficient of the top and side walls (0.10) has been arbitrarily chosen for the simulations reported in this paper, but a parametric study on the influence of wall friction is underway. The prepared dense-packed numerical samples have 10,000 particles. A constant ram speed of 20 mm/s is assigned to the top cap (roughly 60 times faster than in the real extrusion case, to ensure efficient completion of the simulation process). In cases where printing is coupled with extrusion, a printing speed of 100 mm/s (again 60 times faster than the real experiments) is assigned to the syringe and nozzle walls as well as the top cap to simulate the printing process. The simulation is continued until the top cap (ram) reaches a displacement of 30 mm. Using the computer configuration mentioned earlier, a single printing simulation takes about 24 h.

A comparison between Figs. 5 and 6 show that the general trends in the force-displacement curves as a function of extruder geometry (orifice/nozzle, diameters of entry and exit) are captured by the DEM simulation. The initial rising portion, which is the result of early consolidation (Stage 1), and the region of relatively constant force (Stage 2) are captured in the simulations also. The rapid increase in extrusion force due to the ram approaching the end of the barrel and the consolidated paste requiring a larger force to be extruded (Stage 3) is not always predicted by the DEM simulation for the pastes and geometries studied. The simulated extrusion forces are lower as compared to those in the experiments, likely due to a combination of factors including the coarser particles in the simulation volume, lower friction forces assumed etc. However, it is plausible that some of these effects are offset due to the higher particle packing in the extrusion simulations.

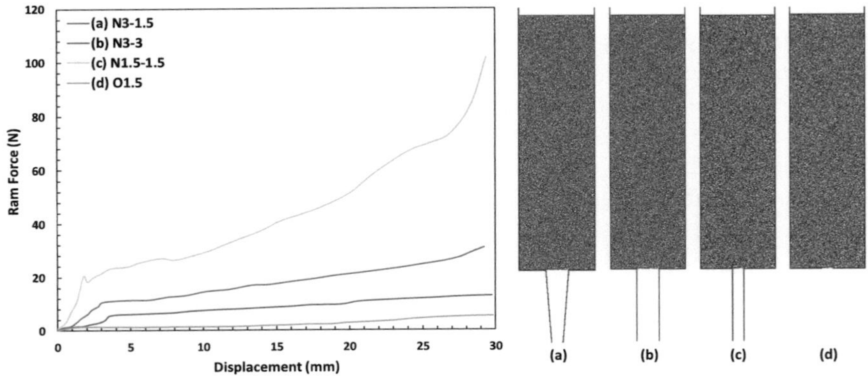

Fig. 6. Simulated force-ram displacement relationship from extrusion modeling of the OPC pastes.

It is interesting to note from the simulations that the extrusion through an orifice (the O4 case) requires very little force, while the experiment shows a significant shaping force requirement. This could be attributed to the very low friction coefficient value used for the barrel wall (0.1 in this case). While the same value is considered for the other cases also, the constriction in sizes of the die entry results in a non-insignificant shaping force. This emphasizes the need to consider the effects of friction in the barrel and die walls. Accounting for the proper friction coefficient also could result in the simulations capturing the rapid increase in force towards the end of the extrusion process.

There are other significant effects in addition to wall friction that need to be accounted for in the simulations, which are being considered, but not explained in this paper. The ratio of the die entry and exit diameters to the maximum particle size in the system will influence the extrusion forces. For example, a die entry/exit of 4 mm diameter in the actual experiment results in a ratio of 40 (considering maximum particle size to be 100 μm), while for the corresponding simulation, this ratio is 7.5 (1.5 mm die entry/exit, and 100 μm maximum size of the simulated particle). Furthermore, a dense packing of multi-size spheres is used in the simulation (with a packing density of 0.85), while for the actual experiments, the packing density of particles is 0.44.

Figure 7 shows images at different times in the DEM simulation of a two-layer printing process. These simulations are carried out with the intention of understanding the effects of overburden pressure on the print layers, and thus to better calibrate the interparticle coefficients, which was accomplished using a slump test. Since yield stress of the printed layer is what will enable printing of stable overburden layers, it is anticipated that the Burger's law parameters will be more effectively calibrated using experimental yield stress values (even though one can argue that the slump test is an indirect measure of yield stress). Since yield stress is a function of particle sizes in the system and the interparticle effects, the influence of considering larger particles for the simulation can be minimized by choosing the appropriate Burger's law parameters.

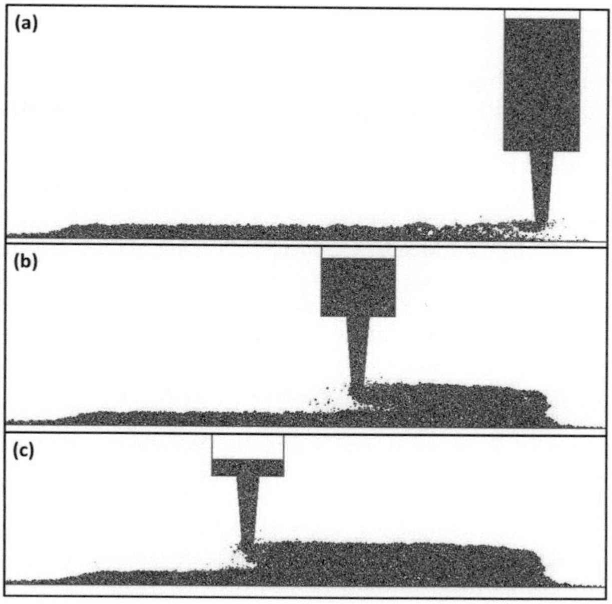

Fig. 7. Snapshots in time of DEM simulation of two-layer 3D printing.

4 Conclusions

This paper has discussed the development of a DEM model for extrusion-based 3D printing of cement-based materials using a standard printable OPC paste an example. Burger's model was used as the particle-scale model to represent the particle interactions in the normal and the shear directions. A virtual mini slump test was performed on a system with maximum packing density (since high porosities as those in fresh cement pastes cannot be simulated adequately without particles losing contact with each other, thus rendering the model parameters invalid) to calibrate the normal and shear parameters of the Burger's model. As a first approximation, these parameters were assumed, and the best fit set of parameters that matched the experimental data was chosen. Virtual extrusion experiments were carried out on the cement paste with four different geometries of the extrusion cell. Three of the geometries had a barrel and a nozzle, and one had a barrel and an orifice. The simulations were able to match the experimental trends adequately, even though the actual values were different because of the assumptions in the parameter estimates. Other significant effects such as wall friction needs to be incorporated into the model. The framework established can be used to link particle-scale and processing effects in the 3D printing of cement based materials.

Acknowledgments. The authors sincerely acknowledge support from National Science Foundation (CMMI: 1353170) towards the conduct of this study.

References

1. Panda, B., Tan, M.J.: Experimental study on mix proportion and fresh properties of fly ash based geopolymer for 3D concrete printing. Ceram. Int. **44**(9), 10258–10265 (2018)
2. Ma, G., Li, Z., Wang, L.: Printable properties of cementitious material containing copper tailings for extrusion based 3D printing. Constr. Build. Mater. **162**, 613–627 (2018)
3. Hambach, M., Volkmer, D.: Properties of 3D-printed fiber-reinforced Portland cement paste. Cem. Concr. Compos. **79**, 62–70 (2017)
4. Asprone, D., Auricchio, F., Menna, C., Mercuri, V.: 3D printing of reinforced concrete elements: technology and design approach. Constr. Build. Mater. **165**, 218–231 (2018)
5. Chi, H.-L., Wang, X., Jiao, Y.: BIM-enabled structural design: impacts and future developments in structural modelling, analysis and optimisation processes. Arch. Comput. Methods Eng. **22**, 135–151 (2015)
6. Buswell, R.A., Soar, R.C., Gibb, A.G., Thorpe, A.: Freeform construction: mega-scale rapid manufacturing for construction. Autom. Constr. **16**, 224–231 (2007)
7. Gosselin, C., Duballet, R., Roux, P., Gaudillière, N., Dirrenberger, J., Morel, P.: Large-scale 3D printing of ultra-high performance concrete–a new processing route for architects and builders. Mater. Des. **100**, 102–109 (2016)
8. Lim, S., Buswell, R.A., Valentine, P.J., Piker, D., Austin, S.A., De Kestelier, X.: Modelling curved-layered printing paths for fabricating large-scale construction components. Addit. Manuf. **12**, 216–230 (2016)
9. Nair, S.A.O., Alghamdi, H., Arora, A., Mehdipour, I., Sant, G., Neithalath, N.: Microstructural packing, rheology, and ram extrusion characteristics of 3D-printable cementitious binders. Cem. Concr. Res. (Submitted)
10. Cundall, P.A., Strack, O.D.: A discrete numerical model for granular assemblies. Geotechnique **29**, 47–65 (1979)
11. Ness, C., Ooi, J.Y., Sun, J., Marigo, M., McGuire, P., Xu, H., Stitt, H.: Linking particle properties to paste extrusion flow characteristics using discrete element simulations, ArXiv Prepr. 160806188. (2016)
12. Mechtcherine, V., Gram, A., Krenzer, K., Schwabe, J.-H., Bellmann, C., Shyshko, S.: Simulation of fresh concrete flow using discrete element method (DEM): theory and applications. Mater. Struct. **47**, 615–630 (2014)
13. Mechtcherine, V., Shyshko, S.: Simulating the behaviour of fresh concrete with the distinct element method-deriving model parameters related to the yield stress. Cem. Concr. Compos. **55**, 81–90 (2015)
14. Tan, Y., Cao, G., Zhang, H., Wang, J., Deng, R., Xiao, X., Wu, B.: Study on the thixotropy of the fresh concrete using DEM. Procedia Eng. **102**, 1944–1950 (2015)
15. Remond, S., Pizette, P.: A DEM hard-core soft-shell model for the simulation of concrete flow. Cem. Concr. Res. **58**, 169–178 (2014)
16. Zelelew, H.M.: Simulation of the permanent deformation of asphalt concrete mixtures using discrete element method (DEM). Washington State University (2008)
17. Cundall, P.A.: PFC user manual. Itasca Consulting Group, Inc., Mineapolis (2004)

Mechanics and Structure

Three-Dimensional Printing Multifunctional Engineered Cementitious Composites (ECC) for Structural Elements

Yi Bao, Mingfeng Xu, Daniel Soltan, Tian Xia, Albert Shih, Herek L. Clack, and Victor C. Li(✉)

University of Michigan, Ann Arbor, MI 48109, USA
vcli@umich.edu

Abstract. Three-dimensional printing (3DP) has great potential to facilitate fabrication of structures with smart functions. This research aims to develop an effective and efficient method to fabricate multifunctional structural elements using Engineered Cementitious Composites (ECC) through 3DP. To this end, ECC slabs measuring 304.8 mm by 76.2 mm by 12.7 mm (length by width by thickness) are prepared for experimental testing. Titanium dioxide nanoparticles are incorporated in the slabs to deliver photocatalytic functionality for chemical reduction of gaseous air pollutants. Two schemes for incorporating titanium dioxide nanoparticles into the ECC slabs are investigated and compared. 3DP is employed to fabricate the slabs and compared with the conventional cast-in-mold fabrication method. The photocatalytic functionality of different slabs is evaluated through nitrogen oxides abatement testing under ultraviolet light. The concentration of nitrogen oxides is measured in real time. After the nitrogen oxides abatement testing, all slabs are tested to failure under four-point bending to evaluate their flexural properties. The results show that 3DP is promising to fabricate multifunctional ECC structural elements with improved efficiency.

Keywords: Engineered Cementitious Composite · Functionally-graded composite · Multifunctional element · Photocatalysis · Three-dimensional printing

1 Introduction

Compared with conventional concrete construction methods, three-dimensional (3D) printing has great potential to improve the construction process by reducing the construction time, labor, cost, hazards, waste, and impacts on environment; hence, 3D printing (3DP) is attracting increasing interest in the construction industry. As an additive manufacturing approach, 3DP facilitates a diverse range of architectural and structural designs [1]. For elements with complex geometries, 3DP can greatly reduce the manufacturing cost associated with the complex geometries. For instance, non-monolithic structures [2–4] have been 3D printed to save materials and reduce self-weight. This gives architectural and structural designers freedom to make each design unique and architecturally appealing, while simultaneously optimizing structural performance and material use. In addition, without significantly increasing the

T. Wangler and R. J. Flatt (Eds.): DC 2018, RILEM Bookseries 19, pp. 115–128, 2019.
https://doi.org/10.1007/978-3-319-99519-9_11

manufacturing cost, functionally-graded structures can be 3D printed in layers to deliver multi-functions [2, 3], such as thermal-insulation and self-cleaning functions.

Since the mid-1990s when the first attempt was made at using cement-based materials for freeform fabrication of architectural scale structures [5], major technical developments have been made in manufacturing approaches in printing material, process, design, and printer. Various types of high-performance cement-based mortar have been prepared to deliver adequate printability [1, 6]. Large-scale 3D printers with automatic robotic arms or cable-suspended robots have been developed to provide greater capabilities in 3DP [1, 7, 8]. In recent years, architectural scale 3DP has been implemented to build a children's castle in the US [9], a 131 m^2 hotel in the Philippines [9], ten 200 m^2 houses, a 400 m^2 two-story building and a five-story apartment building in China [10, 11], and a 250 m^2 office building in Dubai [12]. In addition, an 8-meter long bridge was assembled using 3D printed segments over a river in the Netherlands [1]. These projects demonstrate the great potential and feasibility of architectural scale 3DP with concrete materials.

Concrete is brittle in tension and requires reinforcement to prevent catastrophic collapse of structures. In conventional construction, concrete is cast in a formwork where steel reinforcement is positioned prior to concrete placement, forming composite structural elements with a degree of tensile ductility. However, the placement of steel reinforcement is incompatible with the 3DP process, antithetical to the speed, ease, and design freedom promised by the freeform, bottom-up 3DP paradigm. There are two approaches to reinforcing cement-based materials. The first approach is to place reinforcement to form a composite structure. Attempts include: (1) placing fiber mesh between printed layers [1], which interferes with the printing process; (2) installing reinforcing bars during printing [3], which limits the shape of the printed structure and/or direction of reinforcement due to difficulty in bending reinforcing bars, as well as compromising the efficiency of 3DP; (3) placing reinforcement in a 3D printed formwork that is filled with concrete afterward [1]; and (4) placing steel wires in printed layers [13], which only provides reinforcement along the print path. The second approach is to impart tensile load-carrying capacity intrinsically in a composite material. Fiber reinforced concrete have been proposed as a 3DP material [2, 14, 15]. However, the tension-softening fiber reinforced concrete is not expected to provide sufficient structural resiliency in large-scale structures.

Engineered Cementitious Composite (ECC), a high-performance fiber reinforced cementitious composite, is a promising candidate for self-reinforced 3DP concrete with tensile strain-hardening property [16]. ECC has the unique characteristics of high tensile ductility and tight crack width [17]. A typical ECC has tensile ductility of over 3% (several hundred times higher than that of traditional concrete), and average crack width below 60 μm even when highly strained beyond the elastic limit [16]. The tensile ductility of ECC enhances structural safety [18–20] and has been utilized in coupling beams of tall building cores for earthquake resistance and in highway bridge deck to resist fatigue loads [21, 22]. The high tensile ductility of ECC may be exploited in 3DP to eliminate or significantly reduce steel reinforcement. With randomly oriented short fibers at a moderate volume fraction (typically less than 2%), ECC has been extruded into structural elements [23]. Apart from extrusion, the rheology of ECC has been tailored for various methods of processing, including self-consolidating casting [24], shotcreting [25], and 3DP [16]. In addition to the unique mechanical properties, ECC can be tailored to possess smart functionalities, such as self-cleaning [26] and air-purifying [27] that can be highly valuable to architectural elements.

The primary objective of this research is to develop an effective and efficient method to fabricate self-reinforced, multifunctional structural elements using multi-functional ECC through 3DP. Titanium dioxide (TiO_2) nanoparticles are incorporated to deliver photocatalytic functionality, such as air purification. Two different schemes for incorporating TiO_2 nanoparticles into ECC slabs are investigated. 3DP is used to fabricate the slabs and compared with the conventional casting method. The air-purifying functionality of TiO_2-doped ECC slabs is evaluated through nitrogen oxides (NO_x) abatement test. The mechanical properties of TiO_2-doped ECC are evaluated through compressive, tensile and flexural tests.

2 Experimental Program

2.1 Materials

In this study, two ECC mixtures developed in previous studies [16] were tailored as the 3D printing materials. The mix proportions and contents of admixtures were slightly modified to obtain more favorable extrudability and buildability for 3DP. The mix proportions are shown in Table 1. The physical properties of the polyvinyl alcohol (PVA) fibers are listed in Table 2. The volume fraction of the PVA fiber was set at 2%. Titanium dioxide nanoparticles (nano-TiO_2) with an average particle size of 25 nm were used to prepare photocatalytic ECC (PC-ECC) [26].

Table 1. Mix design (unit in kg/m^3).

Designation	Type I Portland cement	Calcium aluminate cement	Fly ash	Silica sand	Silica fume	Flour silica	Water	ANC	VMA	HRWR	PVA fiber	TiO_2
ECC	792	55	253	495	110	55	400	6	6	6	26	0
PC-ECC	500	30	1112	620	0	0	420	6	6	15.5	35	85

Table 2. Manufacturer specified properties of the PVA fibers.

Length (mm)	Diameter (μm)	Elongation (%)	Density (kg/m^3)	Young' modulus (GPa)	Tensile strength (MPa)
8	39	6	1300	42.8	1600

Type I Portland cement, calcium aluminate cement, and ASTM Class F fly ash were used as the binder. The chemical and physical properties of two types of cement and fly ash are listed in Table 3. The strength activity index of the fly ash was 91% at 7 days and 97% at 28 days. Finely grounded silica sand with an average grain size of 75 μm was adopted. An attapulgite nanoclay (ANC) was used to impart thixotropic characteristic into the fresh ECC mixtures [27]. A commercial hydroxypropyl methylcellulose viscosity modifying agent (VMA) and a polycarboxylate-based high-range water reducer (HRWR) were used to improve the rheological properties, such as the viscosity and flowability of fresh ECC mixtures.

Table 3. Chemical composition and physical properties of cements and fly ash.

Chemicals	Type I Portland cement	Calcium aluminate cement	Fly ash
SiO_2 (%)	19.6	$\leq$6.0	39.8
Al_2O_3 (%)	4.8	50.0–53.0	20.0
Fe_2O_3 (%)	2.9	$\leq$3.0	9.7
CaO (%)	63.5	$\leq$40.0	18.9
MgO (%)	2.2	$\leq$1.5	3.7
SO_3 (%)	2.6	$\leq$0.4	1.95
Na_2O equivalent (%)*	0.57	–	2.0
Loss of ignition (%)	2.6	–	0.9
Density (g/cm^3)	3.15	3.00–3.10	2.71

*Na_2O equivalent = $Na_2O + 0.658K_2O$

2.2 Mix Processing

The materials were mixed using a 45-L Hobart mixer. Because the mixing process, including the sequence and time of ingredient addition, affects the fresh properties of the cementitious paste, the procedure was conducted according to explicit schedules.

For the mixing of the ECC mixture, dry ingredients (excluding ANC, VMA, and MS) were mixed first for 5 min (min). A portion of the water content was added at time zero (t = 0), followed by the HRWR at t = 30 s and the MS content at t = 1:00 min. Remaining portion of water was mixed with the ANC for the ANC exfoliation process [27]. The portion of water separated out for this process was about 0.7 L to 1.0 L, depending on the batch size, sufficient to cover the blender blades and for rinsing of the blender jar to ensure nearly all of the exfoliated ANC content was added to the mix. The ANC was exfoliated in water (blended) for at least 3 min. before being added to the mix. The ANC and associated water were added to the mix at t = 6 min. Half the VMA content was added at t = 8 min, followed by the addition of fibers in small handfuls between t = 9 min and t = 12 min, after which the remaining VMA was added. Mixing continued until t = 28 min to ensure even dispersion and activation of the viscosity modifying ingredients.

For the mixing of the PC-ECC mixture, dry ingredients were mixed first for 5 min. Then, HRWR was introduced to water, and the HRWR solution was added and mixed for 5 min. Finally, fibers were added and mixed for 5 min.

2.3 Test Specimens

Design of Specimens. In this study, two methods to fabricate the slab specimen measuring 304.8 mm by 76.2 mm by 12.7 mm (length by width by thickness) are compared, and two methods to incorporate TiO_2 nanoparticles in ECC specimens are investigated, as listed in Table 4. The slab specimens are used to represent structural elements, and tested to investigate the air-purifying functionality and flexural properties. For each designation, three specimens were prepared and tested.

The first TiO_2 incorporating method (specimens S1 and S2) was to disperse the TiO_2 nanoparticles uniformly in the PC-ECC mixture when the mixture was prepared in a mixer. The TiO_2 content (mass replacement of binder) was set at 5%. The second method (specimen S3) was to print a PC-ECC layer on the surface of the ECC layer, forming a functionally-graded composite. The thickness of the PC-ECC layer was 40% that of the total thickness of the slab specimen. In the second method, the amount of TiO_2 used was 40% that for the first method.

Compressive and tensile properties of the ECC and PC-ECC mixtures were respectively evaluated by testing cubic specimens that measured 50 mm by 50 mm by 50 mm and dogbone-shaped specimens. The dogbone specimen, as shown in Fig. 3(a), measured 380 mm in length and 13 mm in thickness. Within the 80 mm gauge length, the width of the specimen was 30 mm.

Fabrication Methods. As indicated in Table 4, two methods were used for the specimen fabrication, which are casting and printing. In the casting method, material was placed at the center of the specimen, and tapped to let material flow to the two ends by hand. The mold for casting was placed on a vibration table for material consolidation. After casting, the specimens were air cured at a relative humidity of 50% ± 10% (mean value ± one standard deviation) at room temperature (22 °C ± 2 °C) until testing at 28 days.

For the printing of specimens, a caulk gun was used for extruding the materials, to simulate the extrusion process of 3D printing. This method was applied in our previous study on development of ECC for 3D printing [16]. Figure 1(a) and (b) show the caulk gun and extruding process, respectively. Material was manually extruded from the nozzle on the tube at a horizontal speed of about 5 mm/s to 10 mm/s. The caulk gun was used to test the extrudability and buildability of the materials. The test is useful to

Table 4. Information of specimens.

Designation	Fabrication method	Structural material	Illustration (unit in mm)*
S1	Cast in mold	PC-ECC	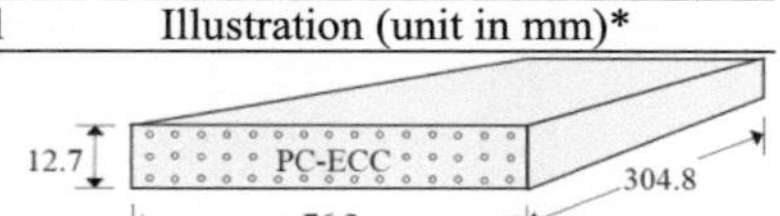
S2	Cast in mold	ECC, PC-ECC	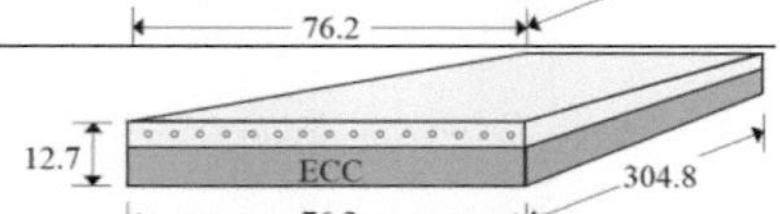
S3	Print in place	ECC, PC-ECC	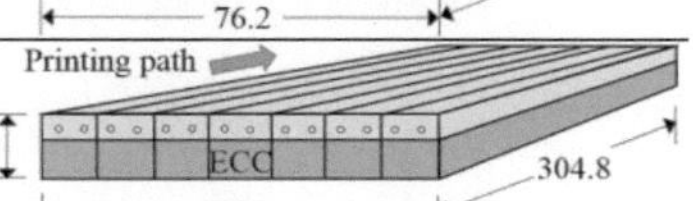

ensure the adequate printability before the material is used in a powered concrete pump. After printing, the specimens were air cured at a relative humidity of 50% ± 10% (mean value ± one standard deviation) at room temperature until testing at 28 days.

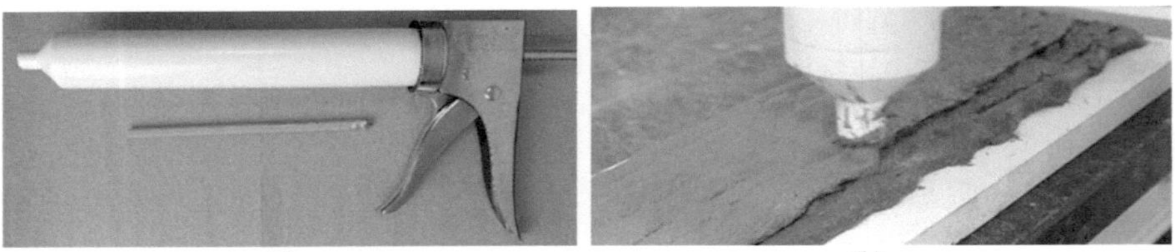

Fig. 1. Extruding method: (a) caulk gun with an inner diameter of 8 mm and (b) extruding material using the caulk gun. The caulk gun is mechanically actuated for extrusion of ECC filaments.

2.4 Test Methods

Nitrogen Oxides Abatement Test. The photocatalytic characteristics of the specimens were evaluated through NO_x abatement test. Figure 2 illustrates the test setup. Mixed nitrogen oxide (NO) and nitrogen (N_2) pressurized gas and pressurized air in two containers are respectively supplied to a mixing chamber. Valves and flow meters are used to control the volume ratio of the two flows from the two containers. The mixed gas flows through a reactor that has two ultraviolet (UV) lamps with a power of 30 W. The specimen is placed under the UV lamps in the reactor. An electrical fan is used to improve the uniform distribution of NO in the reactor. The outlet gas flows through a NO_x analyzer (Eco Physics CLD 64) to measure the concentration of NO_x in the outlet gas. The NO_x analyzer has a sensitivity of 0.001 ppm.

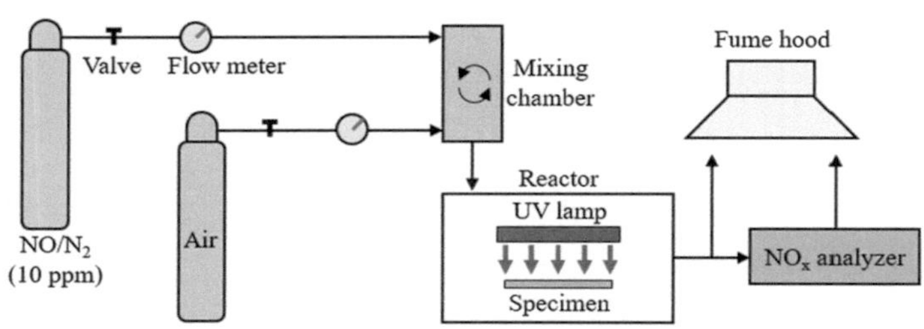

Fig. 2. Nitrogen oxides abatement in the reactor was measured.

Mechanical Properties Test. The compressive strength and tensile properties of the ECC and PC-ECC were evaluated through compressive and uniaxial tensile tests, respectively. The loading direction was along the longitudinal direction of the printed filaments of the specimens. The compressive strength was evaluated by testing cubic specimens, as recommended by ASTM C109 [28], using a loading machine with a 10 N resolution at a loading rate of 0.14 MPa/s. The tensile properties were evaluated by testing dogbone specimens, as shown in Fig. 3(a), according to the JSCE recommendations [29]. Tensile test was conducted under displacement control and the displacement rate was kept at 0.5 mm/min. The applied load was measured using an embedded load cell with a load capacity of 10 kN. The elongation of the gauge length was measured using two external linear variable displacement transducers (LVDTs) at both sides of the specimen. The results from the two LVDTs were averaged.

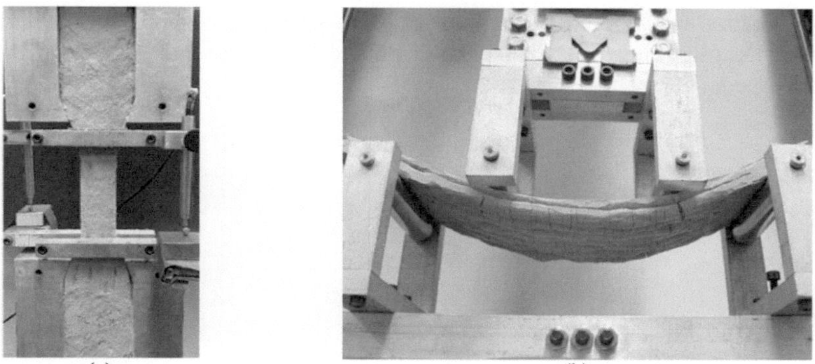

(a) (b)

Fig. 3. Test setup: (a) uniaxial tensile test and (b) four-point bending test. Superior tensile ductility and multiple cracking characteristics are demonstrated.

The flexural behaviors of the slab specimens (Table 4) were evaluated through four-point bending test, as shown in Fig. 3(b), in accordance with ASTM C1609 [28]. The span length was 228.6 mm. The distance between the two loading lines was 101.6 mm, so the distance between the support and loading line was 63.5 mm. Flexural test was conducted under displacement control and the displacement rate was kept constant at 0.5 mm/min. The applied total load and displacement were respectively measured using an embedded load cell with a load capacity of 10 kN and an embedded high-precision displacement transducer. The flexural strength was calculated using Eq. (1), in accordance with the recommendation in JSCE.

$$f_u = \frac{6M_u}{bh^2} = \frac{3P_u(63.5)}{(76.2)(12.7)^2} = 0.0155P_u \quad (1)$$

where, f_u represents the flexural strength (unit in MPa), P_u represents the peak load (unit in N), b and h denote the width (76.2 mm) and depth (12.7 mm) of the slab, respectively.

3 Experimental Results and Discussions

3.1 Printability

The printability of the printing materials was evaluated through experimentation using the caulk gun. The extrudability was evaluated by extruding the materials through the nozzle with an inner diameter of 8 mm, as shown in Fig. 1(b). The buildability was evaluated by depositing ten layers of printed materials within 10 min, as shown in Fig. 4. Each layer was about 5 mm ± 1 mm (mean value ± one standard deviation) in height and 10 mm ± 1 mm in width. These tests showed that the developed ECC mixtures have acceptable extrudability and buildability for fabricating structural elements. However, the use of caulk gun for manual operation has less control for consistent extruding rate and the distance between the extruder and the printed layers. The extruding rate influences the geometry of the extruded filaments. Figure 4 shows that the cross section of the filaments is uneven. The distance between the extruder and the printed layers might influence the shape of printed layers and interlayer bond between adjacent layers.

Fig. 4. Depositing 10 layers of printed filaments within 10 min indicates adequate buildability.

3.2 Nitrogen Oxides Abatement

Figure 5(a) and (b) show the NO_x abatement test results. Figure 5(a) compares two sets of data from different S3 specimens (Table 4). The results from different specimens agree well with each other. When UV light is turned on at 50 min, NO_x concentration rapidly decreased due to photocatalytic reactions [26]. In the presence of TiO_2, which is a photocatalyst, hydroxyl radicals are produced under UV exposure and convert NO_x into nitrate ion (NO_3^-). Figure 5(a) shows that the NO_x concentration is reduced from 1 ppm to 0.2 ppm (80% reduction) within 3 min, and the concentration stabilizes at about 0.04 ppm (96% reduction) within 10 min. After UV light is turned off at 80 min, the concentration gradually increases until reaching 1 ppm at about 210 min.

Figure 5(b) compares the measurement results from the specimens S1, S2 and S3. Different specimens demonstrate comparable efficiency in NO_x abatement, in terms of the rate of NO_x concentration decrease. However, the TiO_2 content of S2 and S3 is only 40% that of S1, indicating that the second method for incorporating TiO_2 have higher efficiency in using TiO_2.

3.3 Mechanical Properties

Compressive and Tensile Properties. Table 5 lists the experimental results of compressive and tensile properties of the ECC and PC-ECC. The average result of three specimens and their coefficient of variation (COV) are reported. The tensile strength represents the tensile stress at the peak load in the uniaxial tensile test, and the ultimate strain represents the tensile strain when the load drops to 90% of the peak load. The printed cubic specimens demonstrated comparable compressive strength with the specimens cast in mold, which is in agreement with the previous study [16]. This is a result of two competing mechanisms. On one hand, the printed specimens have more interlayer voids. On the other hand, the extruding process tends to densify the microstructure of the printed layers, reducing air voids.

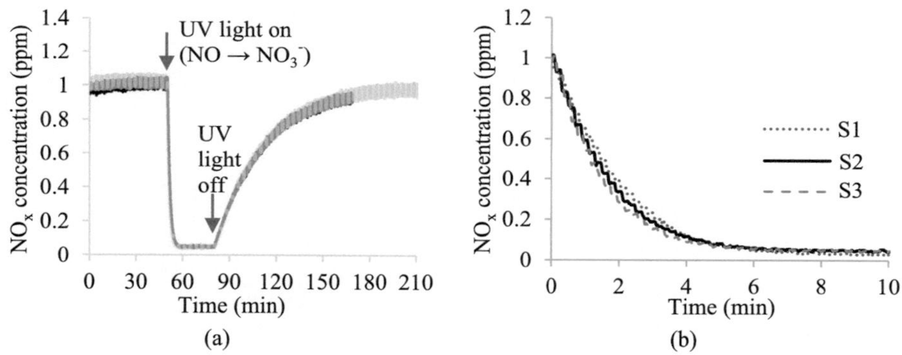

Fig. 5. Nitrogen oxides abatement test results at 28 days: (a) two S3 data sets showing rapid reduction in NOx concentration when UV light is on, and gradual rise when UV light is off, and (b) S1, S2 and S3 show comparable abatement efficiency; S3 shows slightly higher reaction rate.

Table 5. Experimental results of compressive and tensile properties.

Designation	Compressive strength		Ultimate tensile strength		Ultimate strain	
	Average (MPa)	COV* (%)	Average (MPa)	COV (%)	Average (%)	COV (%)
Cast ECC	25.9	4.1	5.2	5.0	2.2	11.5
Printed ECC	24.2	8.3	4.7	6.6	2.4	13.9
Cast PC-ECC	42.8	5.5	5.6	6.2	3.6	9.6
Printed PC-ECC	39.2	7.5	5.5	5.1	3.1	12.0

*Coefficient of variation (COV) = Standard deviation/Average × 100%.

Figure 6(a) and (b) show the tensile stress-strain curves of the ECC and PC-ECC mixtures at 28 days. Both mixtures demonstrated strain-hardening behaviors and significant tensile ductility. Figure 7(a) and (b) show the crack patterns in the cast and printed PC-ECC dogbone specimens, respectively. Multiple cracking is observed in both types of specimens.

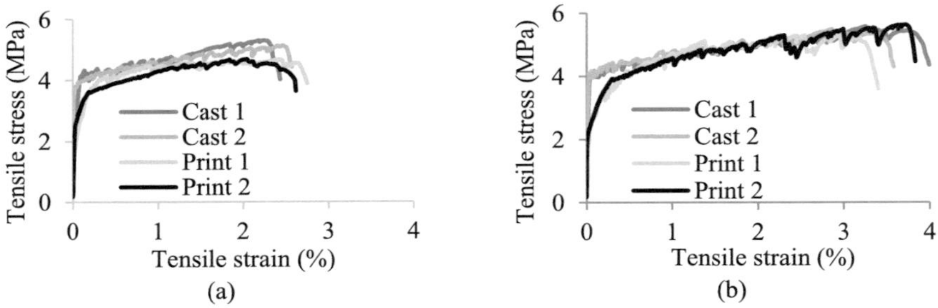

Fig. 6. Tensile stress-strain curves at 28 days: (a) ECC and (b) PC-ECC. Superior tensile ductility is demonstrated by each of the tensile stress-strain curves.

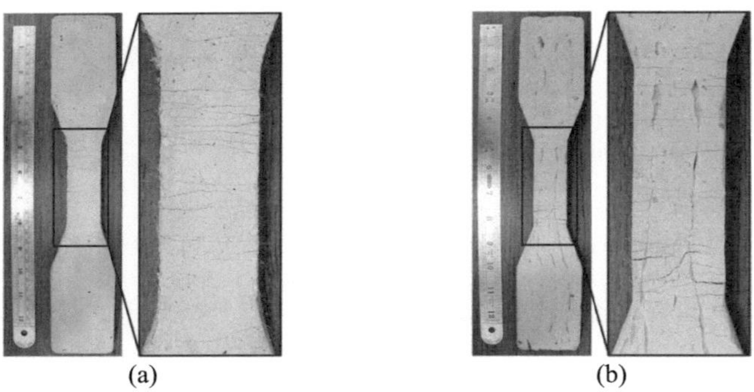

Fig. 7. Multiple cracking is exhibited in PC-ECC: (a) cast specimen and (b) printed specimen.

For the ECC mixture, the printed specimens had slightly lower tensile strengths than the specimens cast in mold. Compared with the cast specimens, the tensile strength was reduced from 5.2 MPa to 4.7 MPa (by 9.6%), while the ultimate strain was increased from 2.2% to 2.4% (by 9.1%). This is a result of the above two competing mechanisms. For the PC-ECC mixture, the printed specimens had a comparable tensile strength with the cast specimens, while the ultimate strain was reduced from 3.6% to 3.1% (by 13.9%). For both the ECC and PC-ECC, the first cracking strength of the printed specimens is about 50% that of the cast specimens. This is likely due to the uneven cross sectional area within the gauge length of the dogbone specimens due to the manual operation using the caulk gun. The uneven cross section is indicated in Fig. 4. The crack was initiated at the weakest section along the gauge length.

Flexural Properties. Table 6 lists the test results of the flexural properties of the specimens. The ultimate deflection represents the average deflection of the two loading lines when the load drops to 90% of the peak load. Energy dissipation represents the dissipated energy up to the moment when the load drops to 90% of the peak load.

Table 6. Experimental results of flexural properties.

Designation	Flexural strength		Ultimate deflection		Energy dissipation	
	Average (MPa)	COV (%)	Average (mm)	COV (%)	Average (J)	COV (%)
S1	11.3	4.0	19.4	10.3	5.9	8.0
S2	12.1	5.1	22.0	4.6	7.5	5.2
S3	10.8	4.9	21.3	5.9	6.5	3.3

Figure 8(a) to (c) plot the load-deflection curves of the tested specimens at 28 days. Each curve is composed of three portions: (1) Linear elastic portion, where the load approximately linearly increases with deflection; (2) Strain-hardening portion, where the load increases with deflection with a decreasing slope until the peak due to presence of cracks and (3) Descending portion, where the load decreases with deflection due to development of cracks.

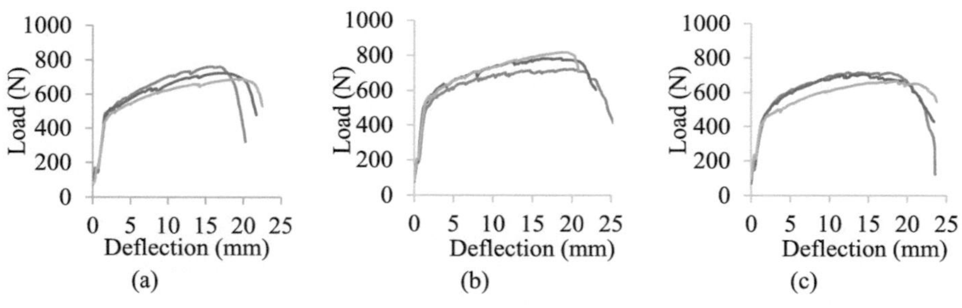

Fig. 8. Flexural test results at 28 days: (a) S1, (b) S2 and (c) S3. Superior tensile ductility is demonstrated by each of the load-deflection curves.

Figure 9(a) to (c) compare the flexural properties of S1 to S3. Figure 9(a) shows that the scheme for incorporating TiO_2 through functionally-graded composite led to higher flexural strengths, likely due to the higher tensile strength and ultimate tensile strain of the PC-ECC mixture than those of the ECC mixture (Table 5). For each scheme, the printed slabs had slightly lower flexural strength than the cast slabs, likely due to the relatively low interlayer bond between filaments in the printed specimens.

Figure 10(a) and (b) show the crack patterns in the functionally-graded composite specimens S2 and S3, respectively. Densely distributed microcracks are observed at the surface of the slab specimens under flexure.

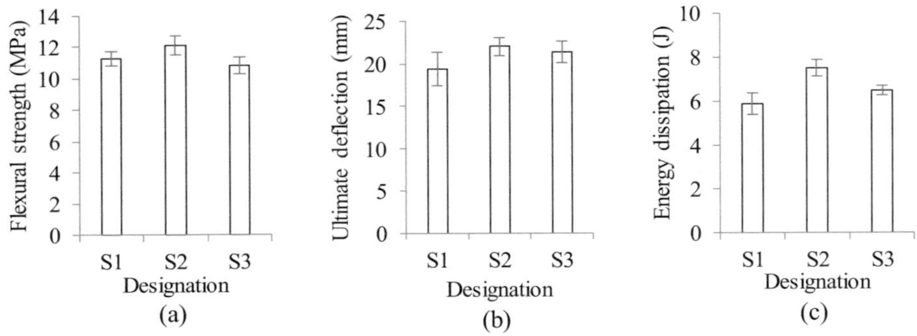

Fig. 9. Effects of TiO_2 incorporation schemes and specimen fabrication methods on flexural properties at 28 days: (a) flexural strength, (b) ultimate deflection, and (c) energy dissipation. The error bars represent the standard deviations of the flexural properties.

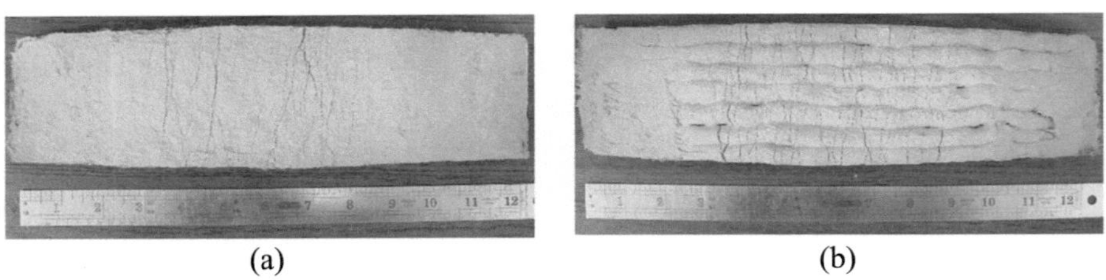

Fig. 10. Crack patterns in functionally-graded composite specimens tested under flexure: (a) S2 and (b) S3. Densely distributed microcracks are observed at the surface of the slab specimens.

4 Conclusions

Based on the above investigation, the following conclusions can be drawn:

The developed ECC and PC-ECC mixtures have acceptable extrudability and buildability for fabricating structural elements. The materials were extruded through the nozzle with an inner diameter of 8 mm, and ten layers of materials were deposited within 10 min. Further research is needed to understand the effects of the extruding rate and the distance between the extruder and the printed layers on the printability.

All of the specimens demonstrated high efficiency in the NO_x abatement testing. With presence of the multifunctional slabs under UV light, NO_x concentration in the reactor was reduced by 80% within 3 min. The slabs that incorporated TiO_2 nanoparticles in different schemes had comparable rates of NO_x abatement based on their rates of decreasing concentration. The slabs with coating and the slabs with functionally-graded structure had high efficiency in using TiO_2.

For both the ECC and PC-ECC mixtures, the printed specimens had comparable compressive strengths with the cast specimens. For the ECC specimens, the tensile strength of the printed specimens was 9.6% lower than that of the cast specimens, while the ultimate strain of the printed specimens was 9.1% higher than that of the cast specimens. For the PC-ECC mixture, the printed specimens had comparable tensile strength with the cast specimens, while the ultimate strain was reduced by 13.9%. For both the ECC and PC-ECC, the first cracking strength of the printed specimens is about 50% that of the cast specimens.

The scheme for incorporating TiO_2 through functionally-graded composite led to higher flexural strengths, likely due to the higher tensile strength and ultimate tensile strain of the PC-ECC mixture than those of the ECC mixture. For each scheme, the printed slabs had slightly lower flexural strength than the cast slabs, likely due to the relatively low interlayer bond between filaments in the printed specimens.

Acknowledgements. This research is funded by the University of Michigan MCubed cross-disciplinary research funding program for innovative, multi-disciplinary collaborative research.

References

1. Bos, F., Wolfs, R., Ahmed, Z., Salet, T.: Additive manufacturing of concrete in construction: potentials and challenges of 3D concrete printing. Virtual Phys. Prototyp. **11**(3), 209–225 (2016)
2. Lim, S., Buswell, R.A., Le, T.T., Austin, S.A., Gibb, A., Thorpe, T.: Developments in construction-scale additive manufacturing processes. Automat. Constr. **21**, 262–268 (2012)
3. Gosselin, C., Duballet, R., Roux, P., Gaudillière, N., Dirrenberger, J., Morel, Ph: Large-scale 3D printing of ultra-high performance concrete – a new processing route for architects and builders. Mater. Des. **100**, 102–109 (2016)
4. Tay, Y.W.D., Panda, B., Paul, S.C., Mohamed, N., Tan, M.J., Leong, K.F.: 3D printing trends in building and construction industry: a review. Virtual Phys. Prototyp. 1–16(2017)
5. Pegna, J.: Exploratory investigation of solid freeform construction. Autom. Constr. **5**(5), 427–437 (1997)
6. Hambach, M., Volkmer, D.: Properties of 3D-printed fiber-reinforced Portland cement paste. Cem. Concr. Compos. **79**, 62–70 (2017)
7. Wu, P., Wang, J., Wang, X.: A critical review of the use of 3D printing in the construction industry. Autom. Constr. **68**, 21–31 (2016)
8. Barnett, E., Gosselin, C.: Large-scale 3D printing with a cable-suspended robot. Addit. Manuf. **7**, 27–44 (2015)
9. Rudenko, A.: 3D Concrete House Printer (2017). http://www.totalkustom.com
10. Starr, M.: Dubai unveils world's first 3D-printed office building (2016). http://www.cnet.com/news/dubai-unveils-worlds-first-3d-printed-office-building
11. Starr, M.: World's first 3D-printed apartment building constructed in China (2016). https://www.cnet.com/news/worlds-first-3d-printed-apartment-building-constructed-in-china
12. Scott, C.: (2016). https://3dprint.com/138664/huashang-tengda-3d-print-house
13. Salet, T.: 3D Concrete printing – a journey with destination unknown. Invited presentation at NSF Workshop on Additive Manufacturing for Civil Infrastructure Design and Construction, Arlington, 13–14 July 2017
14. Le, T.T., Austin, S.A., Lim, S., Buswell, R.A., Gibb, A., Thorpe, T.: Mix design and fresh properties for high-performance printing concrete. Mater. Struct. **45**, 1221–1232 (2012)
15. Kazemian, A., Yuan, X., Cochran, E., Khoshnevis, B.: Cementitious materials for construction-scale 3D printing: laboratory testing of fresh printing mixture. Const. Build. Mater. **145**, 639–647 (2017)

16. Soltan, D.G., Li, V.C.: A self-reinforced cementitious composite for building-scale 3D printing. Cem. Concr. Compos. **90**, 1–13 (2018)
17. Wang, S.: Micromechanics Based Matrix Design for Engineered Cementitious Composites. University of Michigan, Ann Arbor (2005)
18. Fischer, G., Li, V.C.: Deformation behavior of fiber-reinforced polymer reinforced engineered cementitious composite (ECC) flexural members under reversed cyclic loading conditions. ACI Struct. J. **100**(1), 25–35 (2003)
19. Fischer, G., Li, V.C.: Advanced composite materials in flexural members for auto-adaptive structural response modification. In: 1st FIB Congress Concrete Structures, pp. 147–156 (2002)
20. Li, X., Bao, Y., Xue, N., Chen, G.: Bond strength of steel bars embedded in high-performance fiber-reinforced cementitious composite before and after exposure to elevated temperatures. Fire Safety J. **92**, 98–106 (2017)
21. Li, M., Ranade, R., Kan, L., Li, V.C.: On improving the infrastructure service life using ECC to mitigate rebar corrosion. In: 2nd International Symposium on Service Life Design for Infrastructure, vol. 1, pp. 1–8 (2010)
22. Li, X., Wang, J., Bao, Y., Chen, G.: Cyclic behavior of damaged reinforced concrete columns repaired with high-performance fiber-reinforced cementitious composite. Eng. Struct. **136**, 26–35 (2017)
23. Stang, H., Li, V.C.: Extrusion of ECC-material. In: Proceedings of High Performance Fiber Reinforced Cement Composites 3 (HPFRCC 3), pp. 203–212 (1999)
24. Kong, H.J., Bike, S., Li, V.C.: Constitutive rheological control to develop a self-consolidating engineered cementitious composite reinforced with hydrophilic poly(vinyl alcohol) fibers. Cem. Concr. Compos. **25**(3), 333–341 (2003)
25. Kim, Y.Y., Kong, H.J., Li, V.C.: Design of an engineered cementitious composite (ECC) suitable for wet-mix shotcreting. ACI Mater. J. **100**(6), 511–518 (2004)
26. Zhao, A., Yang, J., Yang, E.: Self-cleaning engineered cementitious composites. Cem. Concr. Compos. **64**, 74–83 (2015)
27. Kawashima, S., Chaouche, M., Corr, D.J., Shah, S.P.: Rate of thixotropic rebuilding of cement pastes modified with highly purified attapulgite clays. Cem. Concr. Res. **53**, 112–118 (2013)
28. ASTM Standard Volume C, ASTM International, West Conshohocken, PA
29. Japanese Society of Civil Engineers (JSCE). Recommendations for Design and Construction of High Performance Fiber Reinforced Cement Composites with Multiple Fine Cracks (HPFRCC), Japan (2008)

Large Scale Testing of Digitally Fabricated Concrete (DFC) Elements

Freek Bos[1(✉)], Rob Wolfs[1], Zeeshan Ahmed[1], and Theo Salet[1,2]

[1] Eindhoven University of Technology, Eindhoven, The Netherlands
f.p.bos@tue.nl
[2] Witteveen+Bos Consulting Engineers, Deventer, The Netherlands

Abstract. Case study projects based on Digitally Fabricated Concrete (DFC) are presented in an increasing pace around the globe. Generally, though, it is not reported what structural requirements (if any) these structures meet and how compliance to these requirements was established. Published material research is often not connected to the presented case studies, and even when it is, it is not necessarily obvious their small scale results can be applied to full scale structures as some scale effects should be anticipated. Caution is required as DFC related material tests are still under development and scale effects in DFC have hardly been studied. Therefore, it is recommendable to perform large scale testing, in the range of 1:5 to 1:1, if DFC is applied to actual use structures. This paper presents such testing for two projects, a pavilion in Denmark (not realized) and a bridge in the Netherlands (realized). In both cases, elements printed with the 3D Concrete Printing facility of the Eindhoven University of Technology were intended for actual load bearing performance. The conservative designs past the test requirements, but nevertheless some important findings with regard to element manufacturing and structural behaviour were experienced. It is concluded that large scale testing remains advisable for DFC structures as long as not all relevant aspects of the technology are quantitatively understood, at least when new concepts are being applied.

Keywords: 3D Concrete Printing (3DCP) · Experimental testing
Scale effect · Additive manufacturing

1 Introduction

Case study projects based on Digitally Fabricated Concrete (DFC) are presented in an increasing pace around the globe, as shown by regular updates on websites and databases such as www.3ders.org and www.am4ae.com. Generally, though, it is not reported what structural requirements (if any) these structures meet and how compliance to these requirements was established.

Meanwhile, material research into structural properties of DFC and its dependencies on process specific characteristics such as anisotropy, is also enjoying rapid growth as evidenced by recent publications [1–3]. However, published material research is often not connected to the presented case studies, and even when it is, it is not necessarily obvious their small scale results can be applied to full scale structures as

T. Wangler and R. J. Flatt (Eds.): DC 2018, RILEM Bookseries 19, pp. 129–147, 2019.
https://doi.org/10.1007/978-3-319-99519-9_12

some scale effects should be anticipated. Ahmed et al. [4] have already reported initial findings of the effect of scaling, but they focused on design and construction aspects, rather than structural performance.

The development of DFC has entered into a hazardous phase. Without a proper understanding of structural properties and scale effects, the worldwide enthusiasm for the technology could result in structures that have not been thoroughly proven. If that were to result in a collapse, that could endanger the (speed of the) development of the technology as a whole, in addition to the effects on an individual project. As long as these uncertainties exist, large scale testing of structural elements is an essential tool that should be applied to prove their validity and avoid structural failures. In Europe, the Eurocode 0 (in particular Appendix D) [5] provides the legal framework for this approach, designated as 'design by testing'.

2 TU/e 3D Concrete Printing and Scale Effects

For two reasons, caution is required when applying results of small scale material tests on full scale DFC structures. First, these tests themselves are still under development. In particular, the range of process settings in which test results are valid, is largely unknown (e.g. environment temperature, pump pressures, interlayer interval time, etc.). It is thus also unclear which settings need to be recorded. Current studies focus on the effects of material composition, directional dependency of structural properties (anisotropy), and interlayer interval time. However, many more parameters and settings could have an influence on the (consistency of) structural properties, such as temperature, post-mixing and friction in the wet phase transport system, print speed, nozzle height, and so on.

Second, the effect of scale has hardly been investigated. It is already known that the strength of concrete is scale dependent due to the strain energy release dependency of developing micro cracks. Shrinkage and creep are also well known phenomena to occur in concrete that are in themselves not scale dependent, but rather have a more pronounced effect on larger structures. Furthermore, it should be noted that printable mortars often contain few large aggregates and a high cement content, which contributes to relatively high shrinkage and creep.

In DFC, however, more effects could influence the (distribution of) local material properties, and thus also the structural response of elements or structures (although not all of them are necessarily negative). Some of these include:

- effects of DFC system operation, e.g. increasing system temperatures in use, (partial) flow obstruction by hardening particles, variance in nozzle-to-object distance in high object prints,
- effects of object geometry, e.g. pressure caused by self-weight of an object, interlayer interval time,
- effects of design, e.g. interactions with other materials and joints,
- statistical effects, e.g. due to a varying quality of print material, or scatter in interlayer strength.

A systematic inquiry into these potential scale effects falls outside the scope of this study, but two projects are being discussed that have been designed, printed, and tested around the same time in 2017 (June–September):

- a pavilion for the municipality of Nyborg, Denmark (not realized),
- a bicycle bridge in Gemert, the Netherlands (realized).

Both projects were printed with the 3D Concrete Printing (3DCP) facility of the Eindhoven University of Technology (TU/e) and a custom designed print mortar described by Bos et al. [6]. The print mortar has been named Weber 3D 115-1. It was used for the printing of both projects. Initial structural material properties were experimentally determined by Doomen [7] and Slager [8], and are listed in Table 1. Directional dependency and interface interval time were taken as research variables. Characteristic strengths were calculated by Witteveen+Bos consulting engineers from the average results and sample size, and are given in Table 2.

Table 1. Experimentally determined structural properties of Weber 3D 115-1 print mortar. The tensile strength in w-direction is valid for an interlayer interval time of up to 2 h approximately. For the directional dependency, a relative orientation of axis u, v, w is used [6], indicating the direction parallel to the print in the horizontal plane, perpendicular to the print direction in the horizontal plane, and vertically perpendicular to the print direction (or parallel to the robot arm), respectively.

Property	Dir.	Age	Symbol	Value
Density		28 days	ρ	2,000 kg/m^3
Modulus of elasticity		28 days	E	19,000 MPa
Average compressive strength	u	28 days	$f_{cm,u}$	23.2 MPa
	v	28 days	$f_{cm,v}$	21.5 MPa
	w	28 days	$f_{cm,w}$	21.0 MPa
Average tensile strength	u	28 days	$f_{tm,u}$	1.9 MPa
	v	28 days	$f_{tm,v}$	1.6 MPa
	w	28 days	$f_{tm,w}$	1.3 MPa
Creep factor		56 days	φ	3.0
Shrinkage		56 days	ε	1.5

It should be noted that at this point experimental sample sizes are still relatively small. As a result, the characteristic strengths are quite low compared to the average experimentally determined values.

Even though some variable dependencies have been established, experiences with printing sessions have learned that several more exist which have not yet been studied quantitatively, such as temperature (ambient, system), print speed, and the length of the hose between the pump and the print nozzle. This is an important reason not to be satisfied with material tests only, but to use large scale tests to validate assumptions about structural behaviour.

Table 2. Characteristic strength values calculated by Witteveen+Bos consulting engineers, based on average experimental values and sample size. Separate values were determined only for the u- and w-directions as they are the most distinctly different and most relevant for the structural response.

Property	Dir.	Age	Symbol	Value
Characteristic compressive strength	u	28 days	$f_{ck,u}$	13.6 MPa
	w	28 days	$f_{ck,w}$	10.1 MPa
Characteristic tensile strength	u	28 days	$f_{tk,u}$	1.48 MPa
	w	28 days	$f_{tk,w}$	0.85 MPa

3 Nyborg Studio

3.1 Project and Design

A studio-type structure was planned in the town of Nyborg, Denmark, as part of a larger construction project for student housing. The Sketch Design by architect I. Moltke, shown in Fig. 1, featured 3 closed, curved wall elements that were to be connected with straight timber frames containing the windows and doors. The curved walls would be manufactured through off-site 3D Concrete Printing, with the TU/e facility, as this was expected to allow better process and quality control than on-site printing.

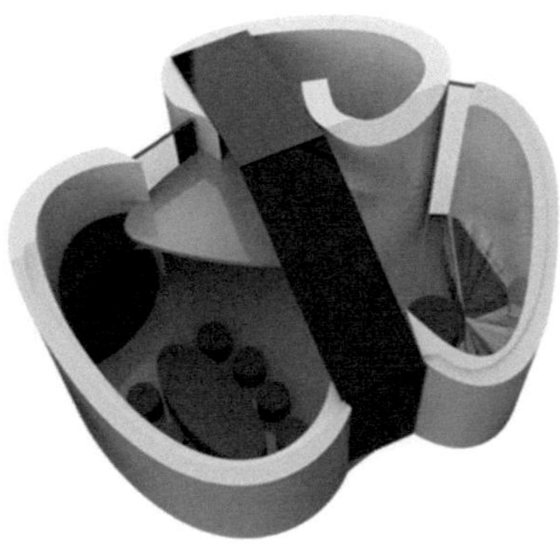

Fig. 1. Sketch design of Nyborg Pavilion by I. Moltke.

The principle of the wall design was elaborated by the architect, structural engineer Witteveen+Bos and the TU/e research team. Probably the most determining factor was the stringent insulation requirements. Danish building codes require an R-value of R = 10 m^2K/W, which ruled out an internal zig-zag concrete pattern or solid cast concrete, as is seen in similar projects [9–11]. Rather, the wall section design consists of a printed inner and outer face with a thickness t = 60 mm, corresponding to 1 filament. The faces, following a sine curve pattern for architectural effect and to increase object stability during printing, are separated by a 305 mm wide cavity that is filled with sprayable PUR insulation foam. The faces are connected by GFRP wall ties (Refus 1 Wall Tie, 365 mm/14"), spaced 285 mm apart horizontally (1 period of the sine) and 300 mm vertically (30 print layers; effectively corresponding to 11.7 pcs/m^2), to transmit axial loads from one face to another.

3.2 Test Program en Specimens

Subsequently, a test phase with large scale elements was determined and conducted. The structural principle of the wall is simple: the inner and outer faces are joined by the wall ties with hinged connections that distribute the acting horizontal loads (Fig. 2). Thus, the faces may be loaded in flexural tension under wind load or collision. Since the faces are not joined with shear connections, the faces are slender when loaded in compression and may be prone to buckling and subsequent flexural tension failure. Considering the structural design and expected load cases, three tests were considered appropriate: a compression/buckling test, a bending test, and an impact test.

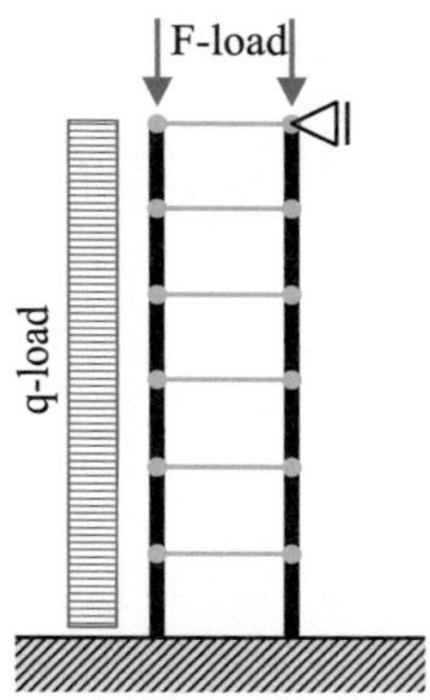

Fig. 2. Structural principle of wall design (vertical section).

The final pavilion geometry was yet unknown when the test program started. Because of the intended due date, the design and test phases partly developed in parallel. The test specimens, nevertheless, had to be designed so that they would yield useful data. As it was clear the three wall elements would each be different and testing complete mock-ups of them would be difficult anyway, it was decided to test approximately 750 mm wide straight segments of the wall section design. An important advantage is that results would be relatively easy to understand, calculate, and compare. Furthermore, in terms of structural response, testing straight segments was a conservative approach as the overall wall curvature would have a positive effect on the structural resistance.

The test specimens were obtained from printed 'rectangular' sections of 2.5 m height (Figs. 3 and 4) (the maximum print height of the TU/e gantry facility is 2.5–2.8 m, depending on the print table design and the print head equipment that is being used). The short sides of the 'rectangles' were to be removed by sawing them of, to obtain a typical wall segment. The use of automatically embedded cable reinforcement [12, 13] was considered for the final design of the pavilion. However, it was not included in the test specimens as the direction of the cables would not influence their response in the test but would complicate the printing and specimen preparation process.

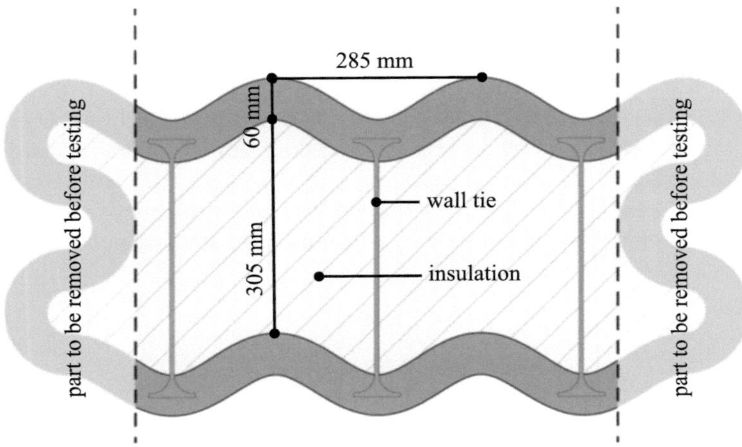

Fig. 3. Horizontal section of test element.

3.3 Printing

The specimens were printed on June 6th and 7th, 2017. To achieve the filament section of 10 × 60 mm^2 (instead of the standard 10 × 40 mm^2 section), a customized nozzle was made that also enabled the application of cable reinforcement later in the project. This nozzle features a back-printing opening to allow proper placement of the cable in the filament, rather than the previously applied down-printing opening at the bottom of a nozzle (see further discussion in Sect. 4). The print speed was reduced by 20% to 80 mm/s to account for the larger filament section. The print time for one object was approximately 2 h.

The strength and stiffness development of the applied print mortar Weber 3D 115-1 was insufficient to print to a height of 2.5 m, without collapse of the dormant mortar. Therefore, an accelerator (weber.ad snel) was dissolved in the water that was used to mix with the mortar. This reduced the setting time of the concrete from 2–3 h to 15–30 min, which was expected not to negatively influence the filament interlayer bond, but enough to allow stable printing.

During and after printing, the object was sprayed with a curing agent (BMP Curing Compound AC) to avoid dehydration of the object. Due to its size, the usually applied method of wrapping in foil was unpractical. Two objects were printed to their intended

Fig. 4. Test element during printing

height of 2.5 m. The printing of the third specimen had to be cancelled at 1.33 m height due to unexpected inconsistencies in the filament. After initial curing, the insulation PUR was sprayed into the cavity in a number of stages over several days. The sides were removed on June 28th, 2017, with a hand-held circular concrete saw operated by an external professional.

3.4 Observations Before Testing

In pre-test observations of the specimens, several issues were noted. First of all, the layer thickness t was 70 mm over almost the entire height (except approximately the final 10 cm), thus exceeding the designed filament width of 60 mm. This difference is attributed to the pressure of subsequent concrete layers that are being added while the concrete is still in dormant state.

Furthermore, one week before testing and before the short sides were removed by sawing, vertical cracks were observed in the short sides of each of the three specimens. These had not been there before, and must probably be attributed to shrinkage of the PUR insulation caused by gas exchange. Test samples from a simultaneously running MSc project [14] show this effect very clearly (Fig. 5). The 1.3 m element suffered 1 vertical crack on the right-side. The 2.5 m elements both suffered a vertical crack on each short side. Not all cracks extend over the full height, but over a very significant part nevertheless. However, since all cracks occurred in areas that were removed before testing, they did not influence the results.

After the short sides were removed, it was also discovered that the test elements featured several horizontal cracks through the layer interfaces. It is suspected these are caused by peak stresses around the wall ties that occur during curing of the concrete, in combination with a low interface adhesion caused by the backward printing nozzle (further discussed in Sect. 4).

Fig. 5. Shrinkage of PUR foam causing fracture in printed concrete (from [14]). The long edges have broken and moved inward, which seems to indicate a force pulling them in that direction. This could be explained by PUR shrinkage due to gas exchange.

3.5 Vertical Flexural Test

The vertical flexural test was performed on July 4th, 2017, as a 3-point bending test on 'standing' elements, with the load acting horizontally (Fig. 6a). The support span was 224 cm. The load is applied in the middle, and distributed over 60 cm height and the full width of the element through steel profiles. Due to the sine-shape of the element, the load is transferred into the concrete over the tops of the sine curves. The forces are measured at the loading jack and the 2 support points. By deducting one from the other, the reaction friction on the ground plate can be calculated.

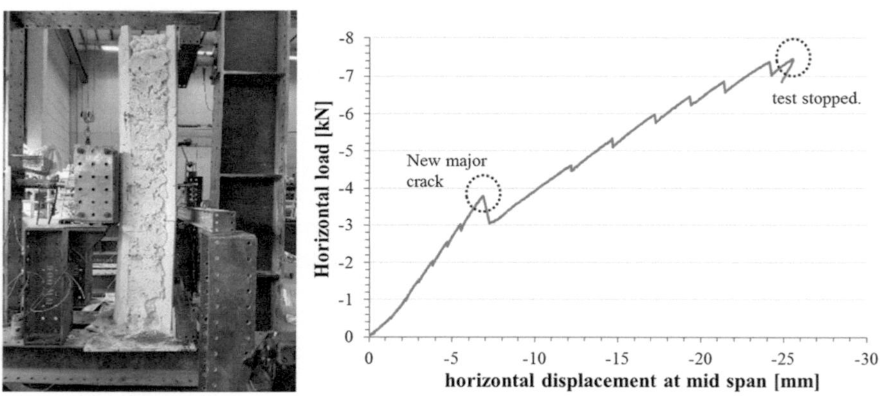

a b

Fig. 6. a, b Vertical flexural test set-up, and resulting load-displacement graph.

Specimen 2.5-1 measured 249.5 cm in height, the width of the outer (loaded) face was 66.5 cm, while the inner (supported) face was 71.0 cm wide (each face included two full sine waves with a period of 28.5 cm). It was subsequently subjected to the bending test.

The design moment corresponded with a load of F = 1.13 kN. The load cell displacement was set at 1 mm/min. To assess the development of damage in the element, the test was stopped when the following load levels were reached: 0.1 kN, 0.5 kN, 1.0 kN, 1.5 kN, 2.0 kN, 2.5 kN, 3.0 kN, 7.5 kN. Due to the nature of the applied load jacks, the load is released slightly at those points, which can be recognized in the load-displacement graph of Fig. 6b. Up to a load of F = 3.8 kN, linear load-displacement behaviour was observed. Slowly, the existing cracks open up when the load increases. From 2.0 kN load onwards, this becomes observable with the bare eye. The largest crack is then 0.4 mm wide. At 3.8 kN load, a new crack occurs in the back side, and the element stiffness is reduced. A new crack occurs at 4.5 kN and from 5.5 kN, the top part seems to be starting to shear of the bottom parts. The load-displacement behaviour remains practically linear though. The test was stopped at 7.5 kN (more than 6 times the design load), and the load was released from the jacks. The element recuperated practically all of its deformations.

Specimen 2.5-2 was measured and tested in a similar manner, but stopped at a load of F = 4 kN, to save it for compression testing. The behaviour was comparable to that of specimen 2.5-1.

Both elements were well capable of carrying the design load without collapse, in spite of the pre-existing horizontal cracks. The design, therefore, was considered safe for these load conditions. It was expected that an improvement in the print nozzle could help to avoid the horizontal cracks in the final manufacturing (see also Sect. 4).

3.6 Compression Test

The compression test was performed once on each face of test element 2.5-2. In the first test, the outer face was loaded, the inner face was loaded in the second test. This element had previously been tested in 3-point bending, with the inner face corresponding with the flexural tension side.

Figures 7a and b show the respective test setups. The load is applied by a hydraulic jack that pushes on a 150 mm steel profile that is placed over the top of the loaded face. This way, the other face, with wall ties and PUR, acts as stabilizing horizontal (buckling) support. The design load could not be established since the design was not yet finalized yet, but was conservatively estimated at 10 kN. The Euler buckling load of each sheet individually was well above 200 kN, however, excentricities could significantly reduce this figure, and buckling could thus not be ruled out completely on forehand.

The load cell displacement was set at 2 mm/min at the start. The loading speed was reduced to 1 mm/min after a load of 2 kN was reached. Note that this displacement is measured over the whole system, including the board interlayer between steel beam and test element, etc. The total deformation of the concrete itself is much smaller. The vertical compressive deformation of the loaded sheets was measured over a 1.0 m length at the center of the element, on each side (2 measurements). Horizontal

Fig. 7. a, b Test set-up with load on the outer and inner face of element 2.5-2, respectively.

deformations were measured at the center, +40 and −40 cm at the loaded sheet, and at the center of the other sheet (4 horizontal measurements). During testing, the vertical deformation was linear up to a load of 60 kN (or 6 times the design load), at which point the test was stopped as it was considered to have shown sufficient structural capacity. No noteworthy observations, visual or audial, were made during the test. No additional cracks occurred.

Subsequently, the specimen was moved in the set-up and the outer face was loaded, also to 60 kN. showing comparable behaviour. However, the vertical stiffness was significantly lower, which can be explained from the smaller concrete section (due to the sawing process, this face was almost 15% smaller), and the fact that this face had more cracks as a result of the previous bending test, which allow for some crushing resulting in the associated deformations.

It was concluded that compression would unlikely be the governing load case, given that an already damaged element could easily carry 6 times a conservative estimate of the design load (Fig. 8).

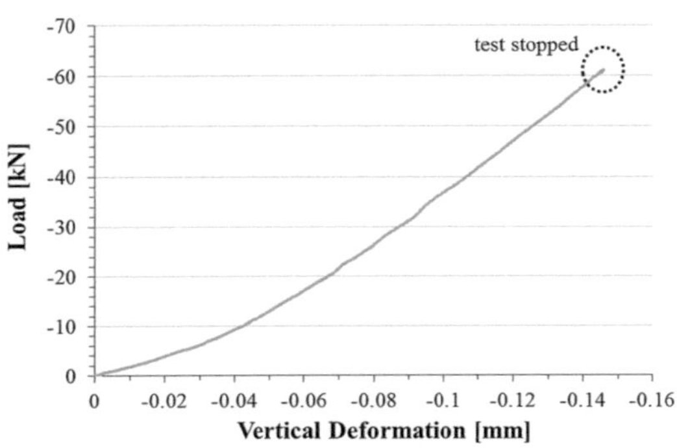

Fig. 8. Element 2.5-2 in compression test on inner face: average load displacement graph for left and right LVDT.

3.7 Impact Test

The 1.33 m element was subjected to pendulum impact loading in a set-up derived from EN 12600. Since the element shape, size and support do not correspond with the codified set-up, some adjustments were made (Fig. 9). The element stands on its base plate. Horizontally, it is supported back to the steel frame by two horizontal timber beams at the bottom and the top of the element (centre distance 1.24 m). The element and impact body were positioned so that the impact body would hit the element at mid height of the top of the middle wave on the front side. It is noted that a pendulum impact test is not usually required for one-storey building walls. The test was nevertheless performed to acquire an idea of its resistance to dynamic, localized loads and its sensitivity to damage.

a **b**

Fig. 9. a and b Pendulum impact test set-up, and element after testing with visible damage caused by multiple 1200 mm drop height impacts.

The EN 12600 predefines 3 consecutive drop heights: 190 mm, 450 mm and 1200 mm. The element was impacted from each drop height several times. No noticeable effect was observed after several impacts from 190 mm and 450 mm height. When the drop height was increased to 1200 mm, increasing damage occurred upon each impact, beginning with two horizontal cracks upon the first impact, spalling from the back side on the second impact and spalling at each side on the third impact. A drop height of 1200 mm corresponds to strict severe impact regulations for balustrades at height, for which damage is allowed but fall-through is not. The behaviour of the printed wall element was therefore considered safe for the intended use.

3.8 Overall Conclusion on Structural Performance

The test specimens met the requirements set for each loading condition with considerable margin, and the design was therefore considered safe and sufficiently strong. Nevertheless, some defects were encountered that called for an improved manufacturing design. However, before such studies could be performed, the project was discontinued due to a shift of priorities with the client.

4 Bicycle Bridge Gemert

Around the same period, another project was being developed at the TU/e: a 3D concrete printed bicycle bridge. This project was initiated by general contractor BAM as part of a larger infrastructure project. An extensive description of the project and related testing is given by Salet et al. [15, 16]. This section is largely taken from that publication.

4.1 Project and Design

To cross a small waterway in Gemert, the Netherlands, a 6.5 m span bicycle bridge with a width of 3.5 m was designed. It consists of elements printed vertically, that were subsequently rotated to form a horizontal element. The elements were prestressed together so that the section always remains in compression (Fig. 10a and b). This approach is similar to one adopted by Lim et al. [17]. The main advantage is that the tensile capacity of the printed concrete with hundreds of interfaces, becomes irrelevant.

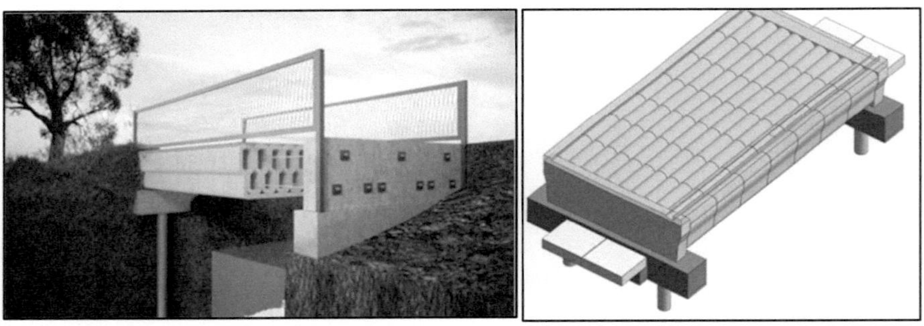

a b

Fig. 10. a, b 3DCP bicycle bridge, design impression (by BAM). Reproduced from [16].

The zig-zag cross-section of the bridge elements results in a significant material reduction while maintaining the required flexural and shear resistance. The prestress is introduced through two solid, cast RC bulkheads. The bulkheads have been placed on abutments on a pile foundation.

The final design also included zones with automatically embedded reinforcement cables [12, 13] to provide additional shear capacity and ductility in the lateral direction, even though the structural concept and dimensions are designed to work without them. Figure 11a and b show the reinforcement cables and post-tensioned tendons.

Thus, the structural principle of the bridge is quite different from that of the pavilion: the application of prestress and consequent compression loading in the entire section avoids a number of uncertainties associated with the tensile strength of the print mortar and the filament interfaces. It does, however, make the creep and shrinkage behaviour of the mortar highly relevant as they (partially) determine the prestress loss that may occur.

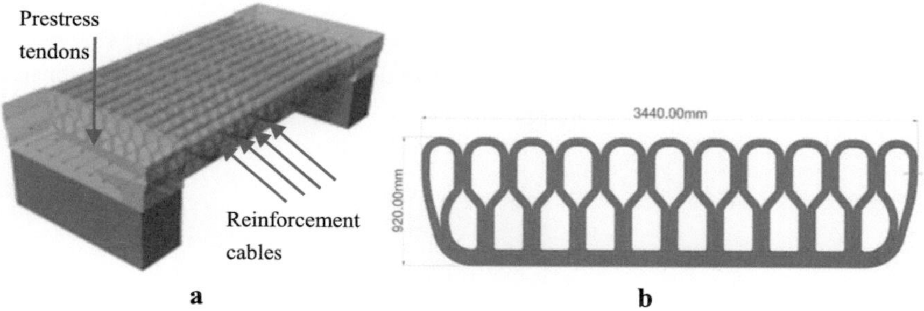

Fig. 11. a, b Schematic 3D image showing prestress tendons and reinforcement cable zones (left) and print element section geometry (right). Reproduced from [16].

4.2 Test Program

To validate the structural safety of the design, it was proposed first to test a 50% scaled mock-up of the bridge destructively in a 4-point bending test, and additionally to perform an a non-destructive in-situ test of the completed bridged to the design load. Furthermore, a long term observation would be started to monitor possible prestress losses. Considering the structural application in an uncontrolled outside environment, this regime was considered appropriate.

4.3 Printing

Like the pavilion project, a large mouth opening (10×60 mm^2) of the nozzle was required to obtain the appropriate layer width, and the same print speed of 80 mm/s was used. At 5:20 per layer, it took 48 h to print all bridge elements (Fig. 12). To allow continuous printing and avoid dry joints, two full wet systems (mixer-pump, hoses and print nozzle) were used.

Fig. 12. a, b Printing of a bicycle bridge element.

To avoid deterioration of the interlayer adhesion, while maintaining proper introduction of the reinforcement cable, a hybrid down/back flow nozzle was developed, shown in Fig. 13. The development of the nozzle design is discussed more extensively in [16].

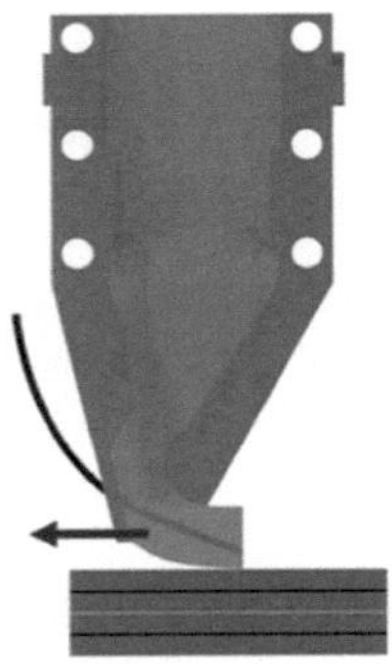

Fig. 13. Down/back-flow nozzle. Reproduced from [16].

4.4 4-Point Bending Test

A mock-up was printed at 50% of the bridge final size, resulting in global dimensions of l × b × h = 1720 × 500 × 460 mm. The mock-up elements were joined by three post-tensioning tendons at the top and six at the bottom (Dywidag, Ø15.7 mm, Y1860 steel grade). The prestress forces were measured with load cells on one top and one bottom tendon. The immediate prestress loss was around 20%, while the additional prestress loss after two weeks of prestressing was an additional 4%. After assembly, the mock-up was subjected to a load-controlled 4-point bending test until failure (Fig. 14). The load rate was 30 kN/min. After having reached a load of 120 kN, an alternating load-unload sequence was applied, at each increase of 30 kN load, until 300 kN. Finally, the bridge was loaded to 350 kN and the test was stopped. Figure 15 shows the load-displacement curve.

The load-unload sequence was applied to identify the point at which structural damage occurs, which is shown by a loss of stiffness at reloading. The test showed damage first occurred between 240 and 270 kN load. The first visible crack was noted at 300 kN load, in the mid-span, indicating a flexural failure. Thus, the mock-up easily passed the pass/fail criterium, which was set at a load carrying capacity of 176 kN (Fig. 16).

4.5 In-Situ Test

An on-site test was performed on the final bridge to ascertain the scaled-up production process had not introduced unexpected errors. Figure 17 shows the bridge loaded to the bending moment of the serviceability limit state. The deformations were minimal and

Fig. 14. Scale model in 4-point bending test set-up. Reproduced from [16].

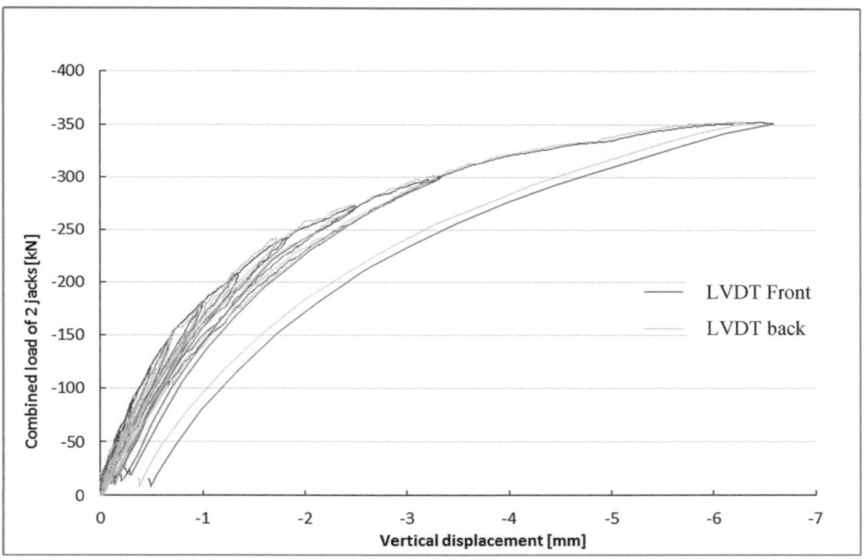

Fig. 15. Load-displacement curve of the 4-point bending test on 1:2 scale model. Reproduced from [16].

could not be measured – actually, no response of the bridge to the loading was observed at all. Consequently, the bridge was approved under the 'design by testing' option in the European and Dutch national regulations. The bridge was taken into use in October 2017 (Fig. 18).

a **b**

Fig. 16. a, b Bending crack, becoming visible at 300 kN total load. Reproduced from [16].

Fig. 17. In-situ testing of the bridge with water-filled containers.

Fig. 18. Completed bridge at the opening. (photo: Kuppens fotografie)

5 Discussion and Conclusion

The large scale experiments on both the pavilion and the bridge showed they were designed according to adequately robust structural principles as well as with sufficient margin. Furthermore, no unexpectedly premature failures occurred. Nonetheless, the two discussed projects yielded a variety lessons related to object scale and structural response.

Due to the non-static nature of the concrete, prolonged printing may cause failures due to hardening particles, as was experienced in one of the printed pavilion elements. Air entrapments or water flow inconsistencies may cause less catastrophic, but nonetheless undesirable, areas of reduced quality. Large scale printing should include quality control based on acceptance criteria to ensure no inferior quality printed elements are approved.

The projects also showed that even the effects of seemingly small alterations to the manufacturing process on the structural quality should be thoroughly considered. This was underlined by the experiences with the nozzle designs.

Even though the bridge section elements took over 5 min per layer to print, all elements stayed well within the limits with regard to deteriorating interface strength between layers. This is due to the long dormant state of the applied print mortar.

Another geometrical effect did occur, however. The plastic deformation behaviour of the material in the dormant stage resulted in dimensional deviations from the design that are not necessarily negligible. Due to the self-weight of the added layers in the pavilion elements, the face thickness increased to over 70 mm, while 60 mm was designed. This is an effect that is really particular to DFC technologies, as the final geometry is not well defined by the mold as with conventional cast concrete. In this case, the effect was beneficial in structural terms, but the correspondence of the actual geometry and the designed geometry properties should be considered as they in cases may also turn out to be detrimental or lead to inefficient material use.

A thoroughly robust design using prestress introduces relatively small risks to the structural performance. The structural response is relatively easy to predict as it is hardly dependent on more sensitive parameters like flexural tensile strength. However, when a composite element is designed, the compatibility with other materials is extremely important, as was experienced in the pavilion elements that cracked due to unexpected stresses and loads caused by the insulation foam and the wall ties (in combination with also unexpected reduced interface strength).

For DFC processes in general, some of the findings may be relevant and others may not. But the observations do indicate that large scale mock-ups and testing are highly beneficial to quality control and should form an integral part of approval processes when new DFC technologies are used for structural applications, as long as the relations between design, material, process, and product are not fully understood. These may serve not only to validate the feasibility of designs but also to avoid overly conservative approaches.

Acknowledgements. The following partners were involved in the Nyborg Pavilion project:
- Client: Nyborg municipality, Denmark,
- Architect: I. Moltke, Denmark,
- Structural engineering: Witteveen+Bos, the Netherlands,
- Research, print design and manufacturing of printed elements: Eindhoven University of Technology, the Netherlands,
- Material supplier print mortar: Saint-Gobain Weber Beamix, the Netherlands.
- In the 3D concrete printed bicycle bridge project, the partners were:
- Client: Province Noord-Brabant, the Netherlands,
- General contractor, initiator: BAM Infra, the Netherlands,
- Structural design and engineering: Witteveen+Bos consulting engineers, the Netherlands,
- Research, print design and manufacturing of printed elements: Eindhoven University of Technology, the Netherlands,
- Material supplier print mortar: Saint-Gobain Weber Beamix, the Netherlands,
- Prestress system and application: Dywidag, the Netherlands,
- Reinforcement cable supplier: Bekaert N, Belgium.

References

1. Le, T., et al.: Hardened properties of high performance printing concrete. Cem. Concr. Res. **42**, 558–566 (2011)
2. Panda, B., Paul, S.C., Tan, M.J.: Anisotropic mechanical performance of 3D printed fiber reinforced sustainable construction material. Mater. Lett. **209**, 146–149 (2017)
3. Sanjayan, J.G., Nematollahi, B., Xia, M., Marchment, T.: Effect of print-time interval on inter-layer strength of 3D printed concrete. Constr. Build. Mater. (accepted for publication)
4. Ahmed, Z.Y., Bos, F.P., Wolfs, R.J.M., Salet, T.A.M.: Design considerations due to scale effects in 3D concrete printing. In: Proceedings of 8th International Conference of the Arab Society for Computer Aided Architectural Design (ASCAAD 2016), London, UK
5. Nederlands Normalisatie Instituut: NEN-EN 1990:2002 Eurocode - Basis of structural design. Delft, The Netherlands (2002)
6. Bos, F., Wolfs, R., Ahmed, Z., Salet, T.: Additive manufacturing of concrete in construction: potentials and challenges of 3D concrete printing. Virtual Phys. Prototyp. **11**, 209–225 (2016). https://doi.org/10.1080/17452759.2016.1209867
7. Doomen, C.C.M.: The effect of layered manufacturing on the strength properties of printable concrete. MSc graduation thesis, TU/e, Eindhoven (2016)
8. Slager, G.J.: Influence of the interface between layers on the tensile properties of 3D-printed concrete. M.Sc. graduation thesis (A-2017.194). TU/e, Eindhoven (2017)
9. 3ders: http://www.3ders.org/articles/20170602-cybe-construction-completes-3d-printing-of-168-sq-m-rdrone-laboratory-in-dubai.html. Accessed 25 Mar 2018
10. 3ders: https://www.3ders.org/articles/20180312-a-3d-printed-house-for-4000-silicon-valleys-new-story-can-build-it.html. Accessed 25 Mar 2018
11. 3ders: http://www.3ders.org/articles/20150118-winsun-builds-world-first-3d-printed-villa-and-tallest-3d-printed-building-in-china.html. Accessed 25 Mar 2018
12. Bos, F.P., Ahmed, Z.Y., Wolfs, R.J.M., Salet, T.A.M.: 3D printing concrete with reinforcement. In: Luković, M., Hordijk, D.A. (eds.) High Tech Concrete: Where Technology and Engineering Meet, pp. 2484–2493. Springer, Cham (2017). https://doi.org/10.1007/978-3-319-59471-2_283

13. Bos, F.P., Ahmed, Z.Y., Jutinov, E.R., Salet, T.A.M.: Experimental exploration of metal cable as reinforcement in 3D printed concrete. Materials **10**, 1314 (2017). https://doi.org/10.3390/ma10111314
14. Vermue, F.: Numerical and experimental assesment of 3D Concrete Printed Sandwich Panels with different insulating core materials. MSc graduation thesis, TU/e, Eindhoven (2017)
15. Salet, T.A.M., Ahmed, Z.Y., Bos, F.P., Laagland, H.L.M.: 3D printed concrete bridge. In: Chua, C.K., Yeong, W.Y., Tan, M.J., Liu, E.J., Tor, S.B. (eds.) 3rd International Conference on Progress in Additive Manufacturing (Pro-AM 2018), Singapore, 14–17 May 2018
16. Salet, T.A.M., Ahmed, Z.Y., Bos, F.P., Laagland, H.L.M.: Design of a 3D printed concrete bridge by testing. Virtual Phys. Prototyp. **13**(3), 222–236 (2018)
17. Lim, S., Buswell, R., Le, T., Austin, S., Gibb, A., Thorpe, T.: Developments in construction-scale additive manufacturing processes. Autom. Constr. **21**, 262–268 (2012)

Method of Enhancing Interlayer Bond Strength in 3D Concrete Printing

Taylor Marchment(✉) and Jay Sanjayan

Swinburne University of Technology, Hawthorn, VIC 3122, Australia
tmarchment@swin.edu.au

Abstract. Additive manufacturing is predicted to revolutionize the way in which we construct our cities and structures. These technologies can create a big potential for freeform design whilst also providing reductions in cost, materials wastage and workplace injuries. 3D concrete printing (3DCP) is one technique that is being investigated. Although many benefits are evident, there are many technological issues that have yet to be explored, particularly that of the bonding strength in extrusion based 3DCP. Extrusion based 3DCP works on a layer by layer deposition of a stiff cementitious mix, forming a material interface. This interface essentially becomes a position of weakness, forming a weak bond.

Currently the interlayer bond is assumed to be related to either mechanical anchorage effects or chemical hydration effects. This paper hypothesis that the mechanical effects are predominant and to prove the hypothesis presents a series of experiments that were carried out to analyse and enhance the bond at this interface. The methodology employed in this study will focus on applying a cement paste to the top of an extruded substrate layer before the secondary layer is deposited.

We demonstrated that the application of a paste at the interlayer does show an increase in bond strength. The greatest bond strength was found in pastes mixed with additives to sustain flow characteristics, over a time gap. The increase in contact area on both layers is now verified to be a crucial factor in bond strength development.

Keywords: 3D printing · Interlayer bond · Additive manufacturing

1 Introduction

Digital fabrication is the method of using digital modelling technology combined with a robotic output to construct an object. Within the digital fabrication field, 3D printing has been gaining a lot of attention and is forecasted to become the 3rd industrial revolution [1]. 3D printing is a form of additive manufacturing, meaning that material is added and built up to create an object, rather than machined away. This type of manufacturing allows for greater cost savings, minimising material wastage and controlled manufacture of complex geometries. Typically 3D printing has been primarily utilised in industries such as military, aerospace, automotive and biomedical with a material application of various polymers and metals [2].

Concrete being one of the world's most utilised materials, with an estimated 2 billion tonnes used per year globally, is a primary construction material [1]. Better

T. Wangler and R. J. Flatt (Eds.): DC 2018, RILEM Bookseries 19, pp. 148–156, 2019.
https://doi.org/10.1007/978-3-319-99519-9_13

utilisation of this material could see dramatic advantages within the construction industry. Unlike other industries, construction is lacking technological innovation, there are high labour costs, high levels of material wastage and safety issues [3]. Adopting 3D concrete printing (3DCP) in construction has the ability to not only save time and money through efficient material usage, but also add the ability to create more complex, freeform architectural designs; without the exponential costs [4].

A small number of 3DCP techniques have been explored thus far with many issues and limitations arising. Extrusion based 3DCP is the most commonly investigated technique due to large scale construction applications [5]. Extrusion 3DCP involves a layer by layer deposition of a stiff cementitious mix. The layers must remain stiff enough to hold their shape and build more layers upon, but also must remain pumpable to be transported to the end effector. A major challenge arising from this is the application of conventional steel reinforcements which provide tensile strength in concrete structures. Another major issue that becomes more critical without the use of tensile reinforcement, is the bond strength between layers [6].

The interlayer bond is considered the weakest point in 3D printed elements and remains under investigation in 3DCP research. Currently there are two assumptions as to why the interface bond remains the weakest point of 3D printed elements. (1) Mechanical Bond: The interlayer contains voids due to the stiffness of the fresh concrete layers. This is mechanical bond problem of not being malleable enough to penetrate and anchor into the surface pores. (2) Chemical Bond: The chemical hydration process across layers is hindered and compromised. These theories would therefore be exacerbated, succumbed to the stiffening effects of cement subject to time gap intervals in layer depositions [7, 8].

This paper hypothesize that the mechanical bond is the primary issue and aims to increase bond strength by investigating the mechanical effects of increased contact area at the layer interface. Taking inspiration from the construction techniques of bricks and mortar; this experimental process will be performed through the application of malleable cementitious pastes atop 3D printed specimens prior to subsequent overlay layer deposition.

2 Materials and Methods

2.1 Mix Proportions

The mix used for 3D printing can be defined as a mortar mix consisting of general purpose (Type GP) ordinary Portland cement (OPC) conforming to AS 3972 and a mixture of two types of premium graded sand (supplied by Sibelco Pty Ltd., namely the 16/30 and the 30/60). Sand Type-1 (16/30) is a course sand with a particle size of 700 μm and sand Type-2 (30/60) is a finer sand with a particle size of 300 μm. The mix ratio in reference to weight was adopted as 1 (cement) : 1 (Type-1 sand) : 0.5 (Type-2 sand) : 0.36 (water).

Four various OPC paste mixtures with a cement to binder ratio of 0.36 were used as the bonding interface for the 3D printed layers. Of these four paste mixtures, three contained concrete additives (supplied by BASF Pty Ltd.). Paste-1 employed the use of a retarder (MasterSet RT 122), Paste-2 contained a viscosity modifying agent (MasterMatrix 362), and Paste-3 adopted a slump retainer (MasterSure 1008). Paste-1 and Paste-3 consisted of 0.3 ml of additive per 100 g of cement, while Mix-2 contained 0.9 ml of additive per 100 g cement.

2.2 3D Printing Process

3D Printing Apparatus

The 3D printing system used was a custom-made apparatus to simulate the 3D printing process of controlled extrusion in the x, y and z planes. The printing material is extruded by means of a piston through a Ø50 mm × 400 mm cylinder, out of a nozzle. The nozzle had an opening of 15 mm × 25 mm set at an angle of 45° to the build platform (see Fig. 1).

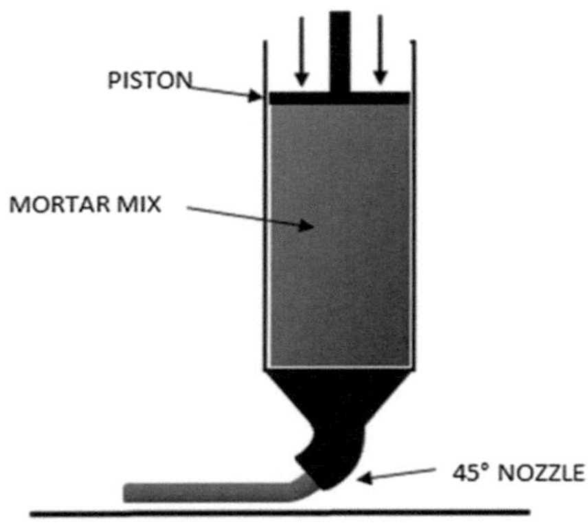

Fig. 1. Schematic illustration of the piston based extrusion 3D printing apparatus

Sample Preparation

Samples consisted of two printed layers, with an OPC paste applied in-between, with a deposition rate of 12 mm/s. The process begins by means of loading the 3D printing apparatus with the mortar mix, then depositing the first layer (substrate layer) for all samples. Immediately after deposition, the application of the relevant OPC paste mixture was applied by means of a 25 mm wide brush (ideally replicating the application of a second nozzle attached and coincident with the extruder) (see Fig. 2a). In creating the second layer (overlay layer) another batch of the mortar mix was loaded into the printing apparatus cylinder and deposited, maintaining a 15 min time gap interval from the substrate layer deposition. Samples are made in approximately 400 mm lengths and left to cure at a room temperature of 23 ± 3 °C, for 7 days in a conditioning room.

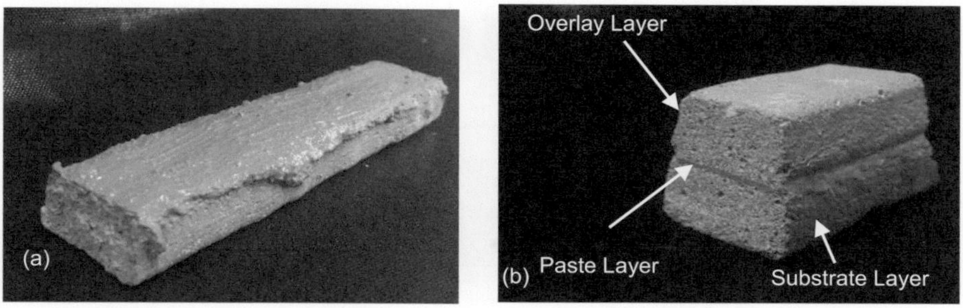

Fig. 2. (a) Application of the paste layer on top of substrate layer. (b) A cured and cut 50 mm long specimen.

2.3 Testing Methods

Cube Samples

The compressive strength of the paste mixtures and the mortar mix were measured by means of cast cubed samples. Samples were cast in 25 mm × 25 mm × 25 mm cubes and cured for 7 days, under the same conditions as the 3D printed samples. Testing was performed at 7 days curing subject to compressive load rate of 20 ± 2 MPa/min conforming to AS1012.9:2014. A minimum of 6 cube samples for each mix were tested.

Flow Table

The flow-ability of the mortar mix and the paste mixtures were measured on a flow table conforming to ASTM C230/C230 M. The test adopted conformed to ASTM C1437 standard test method for flow of hydraulic cement mortar. Using this standard procedure a measure of the flow at 3 min from the time of water addition combined with mixing was used. Material from the same batch was left to sit for 15 min before being subject to the same testing procedure. This was to measure the change in flow/stiffness after a 15 min delay time, replicating the 3D printing procedure outlined previously in the sample preparation outlined in Sect. 2.2.

Interlayer Bond Strength

The interlayer bond strength was measured by means of a uniaxial tensile test at 7 days curing. Samples for each of the various deposited types were cut into approximate 50 mm long specimens to fit in the test apparatus (Fig. 2b). The testing apparatus consisted of 2 pin connected clamps centrally loaded on the top and bottom faces of the specimen to eliminate eccentricity. The clamps grip the specimen with tapered claws at the interlayer, ensuring that an even distribution of stress is achieved, with no slippage (Fig. 3). The specimens were subject to a displacement rate of 1 mm/min, with a minimum of 6 specimens tested for each sample type.

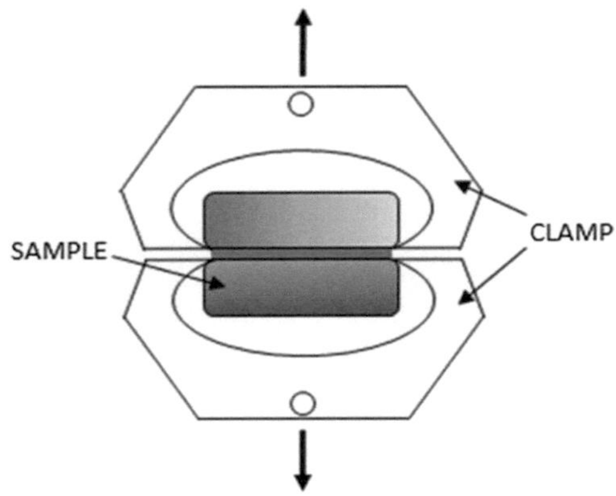

Fig. 3. Schematic illustration of the uniaxial test setup clamps and how they grip the specimen

3 Results and Discussion

3.1 Cube Tests

The w/c ratio for each mixture was maintained at 0.36 to allow for comparative strength consistency and negate adverse strength development among mixtures. To correlate any trends relating the strength of the pastes, compared to the interlayer bond strength, 7-days compressive strength of the pastes were measured in the form of cast cube samples. Although the bond strength is correlated to tensile strength and contact area, we can assume that the comparative strength of each mixture itself, independent of the layering method.

From the data presented in Fig. 4 the extruded mortar mix had an average compressive strength of 34 MPa which was the lowest among the samples tested. This can be drawn to being the only mix containing sand, thus assuming to have slightly different properties. The OPC paste and Pastes-1 & -2 had a similar compressive strength of 42 MPa to 45 MPa. Paste-3 was observed to have the highest compressive strength. From these results we can draw that there are some major strength differences between mixtures. First of all since the mortar mix is the weakest of the samples it can partly explain weakness at an interface based on its strength. In the case of the no paste mixture where the interface relies on the strength of the mortar mix we must also assume the strength as a factor of the lack of contact area due to the stiffness of the fresh mix. The second point is that the pastes exhibit similar compressive strengths, except Paste-3 which is comparatively higher. When assessing the range of standard deviation, variations between these pastes are minimal.

3.2 Flow Table Tests

Flow table tests were performed to measure the flow characteristics of mixtures, which is ideal for an approximate comparison. A comparison of the flow-ability of each mix at 3 min from the time of water addition, and 15 min was undertaken by measuring the

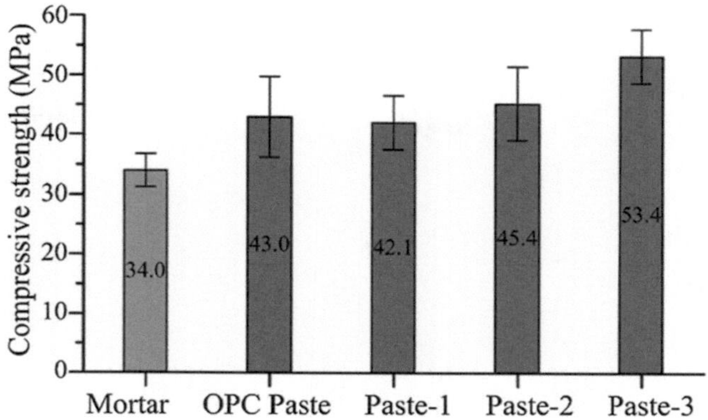

Fig. 4. Compressive strength results of the 25 × 25 × 25 mm cube samples of the paste mixtures

spread diameter obtained conforming to standard ASTM C1437. This allows a replication of the sample preparation process described in Sect. 2.2, comparing the relative stiffness of the paste, subject to layer deposition time gap interval.

Figure 5 presents the results from the flow table tests of the mortar and paste mixtures. The extruded mortar mix was the overall stiffest mix in the group, as was expected due to being the deposited 3D printed layer. It obtained a spread diameter of 166 mm when tested at 3 min, and a spread diameter of 151 mm when tested at 15 min. Amongst the paste mixtures and considering the varied accuracy of the testing regime, spread diameter results are relatively similar. Although what can observed is the OPC paste has a greater rate of stiffening after 15 min compared to the other paste mixtures (214 mm to 184 mm). Whereas the Pastes-1, -2 & -3 contain admixtures, these pastes express minimal change in flow after the 15 min time period, with Paste-3 exhibiting slightly higher flow. The main observation that can be drawn from these results confirm the aims of the experiment; that paste mixtures with additives slow the stiffening effects of cement, to maintain a malleable bond interface.

3.3 Interlayer Bond Strength

The uniaxial tensile tests undertaken to measure the interlayer bond strength are displayed in Fig. 6. The samples with no interlayer paste were observed to produce the lowest bond strength at an average of 0.44 MPa. The OPC paste samples attained a higher bond strength of 0.72 MPa, while the pastes containing admixtures saw an even greater increase in bond strength. Paste-1 generated the highest strength results of 1.26 MPa, while Paste-2 and Paste-3 mixtures presented similar results of 0.98 MPa and 1.00 MPa. All samples conformed to a uniform fracture through the interlayer, with paste samples fracturing from the overlay interface.

Out of the sample size it can be noted that a large scatter of results is present within the specimen range, which is common in tensile testing concrete [9]. Although there is a wide variation within the data range, a trend can be drawn between samples with a

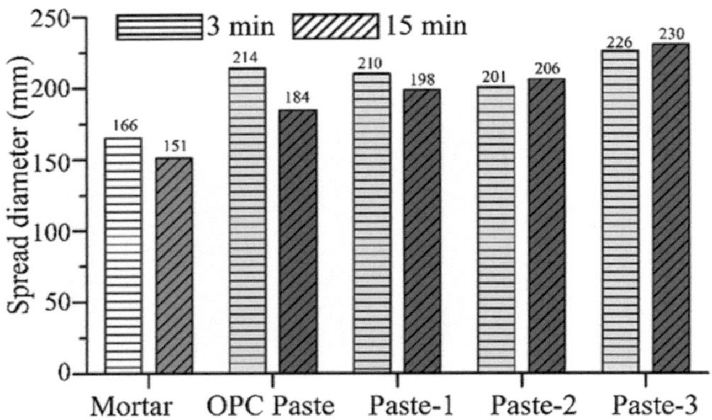

Fig. 5. Flow table test results for the paste mixtures

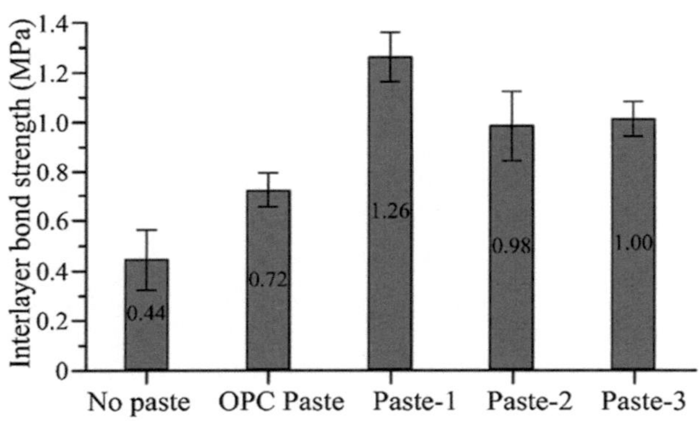

Fig. 6. Bond strength test results

paste layer at the interface and the control samples utilising typical 3D printed construction (no paste). Applying an OPC paste without admixtures provided a notable bond strength increase of approximately 60%. While pastes mixed with an additive were observed to produce a notably higher bond strength of approximately 120%–180% higher than the no paste samples. Although the addition of a paste layer has proved to provide an interlayer bond strength increase, the variations are investigated further through the testing of material properties such as paste strength and flow.

When making a comparison of the bond strength to the material strength we can see a trend within the data of Fig. 4 that the paste mixtures are stronger than the mortar mixture. This attains to one conclusion as to why the interlayer bond amongst paste samples is higher than samples without. Although Paste-3 had the strongest compressive strength, the bonding strength was not exhibited as the strongest. The strength of the material is only one component in the equation of bond strength. Beushausen [10] found through his study on concrete to concrete bonding, that the workability in the overlaying substance is a greater influencing factor in bonding than the material strength. Observing the trend in the mixture flow results shown in Fig. 5 on flow, the correlation of workability/flowability to bond strength has similarities. The paste samples exhibit higher bond strength to no paste samples due to overall workability. While comparing the OPC paste to the pastes containing admixtures, the paste with admixtures display a maintained flow-ability after a 15 min time gap. This confirms Beushausen's [10] findings as the paste mixtures containing additives attain a higher bond strength attributed to the higher degree of workability. The conclusions that cannot be drawn through these test results are to how Paste-1 samples achieved the highest interlayer bond strength, when Paste-3 displayed superior material properties regarding workability and strength.

4 Conclusion

Increasing interlayer bond strength by the application of cementitious paste mixtures, applied at the bonding interface between extrusion-based concrete 3D printed layers were investigated. Samples made manufactured into two layers with 15 min time intervals between layer depositions. Conventional 3DCP sample with no paste layer were compared with samples employing the use of cementitious pastes, mixed with and without concrete additives. These methods where undertaken to increase interlayer contact area and mechanical strength through surface pore anchorage. The effectiveness of these pastes were measured through uniaxial tensile tests, providing the ultimate interlayer bond strength. The material properties such as flow characteristics and compressive strengths were also measured to investigate any correlations into bond strength to further understand the bond mechanism. From this study the following conclusions can be drawn:

- Compressive strength of the mixtures was used to outline any strength differences in the layered mixtures. With a maintained w/c ratio of 0.36 used for all mixtures, it was observed that the paste mixtures were relatively similar in strength but were comparatively stronger than that of the mortar mixture.
- The flow characteristics of the mixtures show a similar trend to that of the strength characteristics. Although pastes containing additives maintained a higher degree of workability over a 15 min time interval compared to the OPC paste.

Considering the strength of the pastes and the correlating workability of each, a relationship can be drawn to the interlayer bonding strength. The use of a workable paste layer has shown to display a higher bond strength, compared to conventional layer on layer deposition. With the addition of additives to offset the stiffening effects of cement over time, the bond strength is further enhanced. Thus, we can conclude that our assumption of increasing surface area and mechanical anchorage is a prime factor in generating a strong interlayer bond. Although further research should be undertaken into the hydration and chemical effects across layer interfaces.

Acknowledgements. This paper gained support through an Australian Government Research Training Program Scholarship.

References

1. Wangler, T., et al.: Digital concrete: opportunities and challenges. RILEM Tech. Lett. **1**, 67 (2016)
2. Wohlers, T.: Wohlers report 2016. Wohlers Associates, Inc., Fort Collins (2016)
3. Harty, C.: Implementing innovation in construction: contexts, relative boundedness and actor-network theory. Constr. Manag. Econ. **26**(10), 1029–1041 (2008)
4. Agustí-Juan, I., et al.: Potential benefits of digital fabrication for complex structures: Environmental assessment of a robotically fabricated concrete wall. J. Clean. Prod. **154**, 330–340 (2017)
5. Sakin, M., Kiroglu, Y.C.: 3D printing of buildings: construction of the sustainable houses of the future by BIM. Energy Procedia **134**, 702–711 (2017)
6. Khoshnevis, B., Bekey, G.: Automated Construction Using Contour Crafting–Applications on Earth and Beyond. Nist Special Publication Sp, pp. 489–494 (2003)
7. Zareiyan, B., Khoshnevis, B.: Interlayer adhesion and strength of structures in Contour Crafting - effects of aggregate size, extrusion rate, and layer thickness. Autom. Constr. **81**, 112–121 (2017)
8. Zareiyan, B., Khoshnevis, B.: Effects of interlocking on interlayer adhesion and strength of structures in 3D printing of concrete. Autom. Constr. **83**, 212–221 (2017)
9. Beushausen, H., Alexander, M.G.: Bond strength development between concretes of different ages. Mag. Concr. Res. **60**(1), 65–74 (2008)
10. Beushausen, H.: The influence of concrete substrate preparation on overlay bond strength. Mag. Concr. Res. **62**(11), 845–852 (2010)

Exploiting the Potential of Digital Fabrication for Sustainable and Economic Concrete Structures

J. Mata-Falcón(✉), P. Bischof, and W. Kaufmann

Institute of Structural Engineering, Department of Civil, Environmental and Geomatic Engineering, ETH Zurich, Zurich, Switzerland
mata-falcon@ibk.baug.ethz.ch

Abstract. Digital technologies overcome typical constraints of traditional concrete construction processes caused by the high impact of labour costs and bring about many new possibilities to the conceptual design, dimensioning, detailing, and production of concrete structures. While the potential of geometric flexibility is being extensively explored, most digital technologies encounter difficulties in penetrating the market due to lacking compliance with structural integrity requirements. To maximise their impact, it is essential that digital concrete processes (i) integrate reinforcement resisting tensile forces and (ii) address conventional structures with geometric simplicity. This paper discusses the potential of digital concrete fabrication processes to reduce the quantity of reinforcement required in concrete structures. For example, "minimum reinforcement" can be tremendously reduced by (i) tailoring the concrete grade locally to the actual needs and (ii) ensuring small crack spacings and correspondingly reduced crack widths by means of crack initiators. An experimental study shows that the strength reduction in the interfaces between layers from extrusion processes can be quantified with reasonable accuracy, which allows using these weak interfaces as crack initiators. A mechanical model to quantify the corresponding potential for saving "minimum reinforcement" when using 3D printing is presented. It is found that weak interfaces in layer joints with 33% of the concrete tensile strength inside the layer allow reducing up to 80% the minimum reinforcement for a given maximum crack width requirement under imposed deformations.

Keywords: Digital fabrication · Concrete structures · Minimum reinforcement Sustainability · Durability · 3D printing

1 Introduction

Reinforced concrete is by far the most used building material today. Existing concrete and reinforcement technology and approaches for its dimensioning have been optimised for more than a century hand in hand with traditional construction methods. This has led to highly efficient conventional concrete construction processes, hard to beat by novel technologies.

T. Wangler and R. J. Flatt (Eds.): DC 2018, RILEM Bookseries 19, pp. 157–166, 2019.
https://doi.org/10.1007/978-3-319-99519-9_14

The implementation of digital technologies brings about many new possibilities for the conceptual design, dimensioning, detailing, and production of concrete structures, promising to revolutionise the concrete construction industry. Over the past years, the digital fabrication research community has developed diverse interesting digital fabrication technologies with strong focus on materials and geometrical aspects [1–5]. These technologies have the potential of overcoming typical constraints of traditional processes caused by the high impact of labour costs. Nevertheless, even the most promising processes are currently only competitive for special applications and encounter difficulties in penetrating the market due to high costs and lacking compliance with structural integrity requirements of clients and building codes. Therefore, digital fabrication for large-scale and mass-market concrete structures is a persisting challenge.

The difficulties of digital fabrication inventions to penetrate the market are analogous to the obstacles found by historical key developments in structural concrete. The graph in Fig. 1 shows that after invention of new concepts, several decades of research and small scale applications passed before the concepts reached maturity with ideas how to exploit their potential. Only then, the key developments of reinforced concrete, prestressed concrete and prefabrication were adopted by mass market.

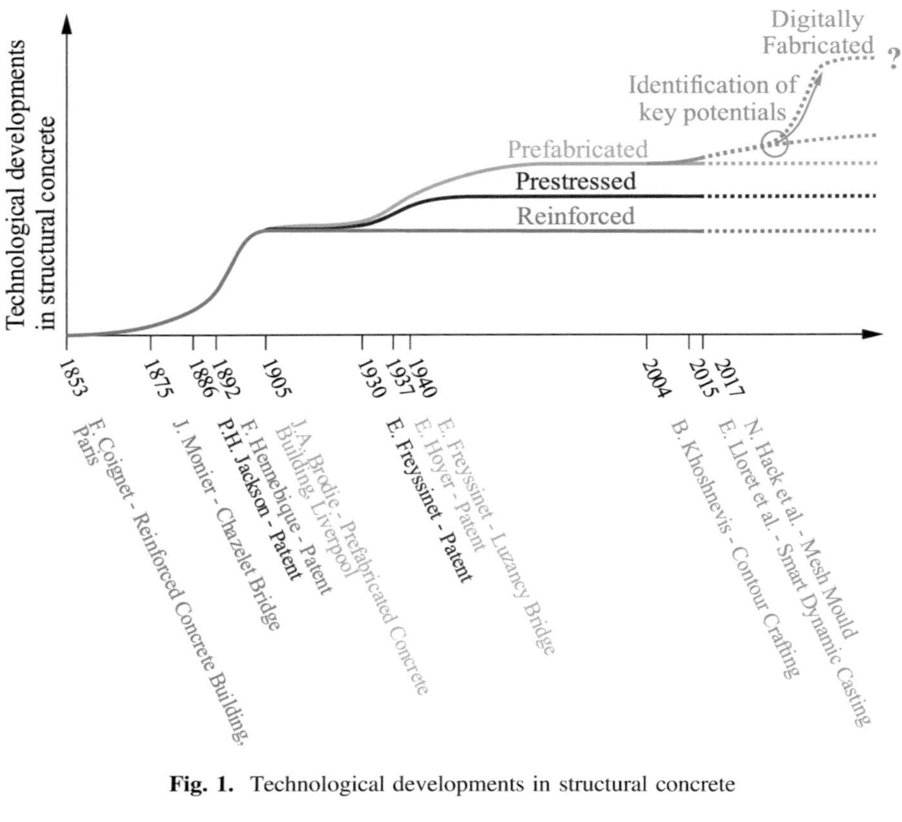

Fig. 1. Technological developments in structural concrete

In general, the conceptual design, dimensioning, detailing, and production of concrete structures should be systematically re-thought to assess the potential of digital fabrication. Benefits that digital fabrication brings to structural concrete need to be identified at early development stage in order to exploit the full potential offered by these new technologies to build economic, code-compliant and sustainable concrete structures. The present contribution discusses some intrinsic benefits of digital concrete fabrication related to reducing the required quantity of reinforcement.

2 Requirements of Mass-Market Concrete Structures

2.1 Necessity of Reinforcement

Current 3D printing by layered extrusion (e.g. "contour crafting") – the most popular digital fabrication technology with concrete today– typically does not allow adding reinforcement during the production. The use of reinforcement resisting tensile forces is, however, essential for the load bearing capacity of structural concrete, since concrete tensile strength is unreliable and therefore typically neglected in structural design. Moreover, reinforcement is not only required to provide strength: A "minimum reinforcement" is always required to (i) avoid brittle failures at cracking, (ii) ensure a sufficiently ductile behaviour to enable plastic redistribution, and (iii) limit deformations and crack widths in real-life structures. The first two functions ensure a reasonable deformation capacity, which is required to apply standard bearing capacity verifications according to design codes for structural concrete. The third function of minimum reinforcement addresses the behaviour under service conditions and durability, which governs the overall reinforcement quantity in many structures. In conclusion, 3D printed elements without reinforcement are essentially suitable only as a replacement of unreinforced masonry, which could be erected much more efficiently both conventionally as well as using robots [6].

While the combined use of external post-tensioned reinforcement and fibre-reinforced concrete can significantly improve the structural performance of 3D printed concrete elements, their strength and ductility will still be limited with respect to conventionally reinforced concrete structures. Hence, the integration of internal reinforcement in 3D printing is fundamental to produce load bearing concrete structures more efficiently than with traditional construction methods. In some digital fabrication techniques, such as Smart Dynamic Casting [3] and Mesh Mould [4] processes, this requirement has already been implemented.

2.2 Limited Use of Geometrical Complexity

The current building stock almost completely consists of rectangular buildings, as summarised by Steadman [7] from two exemplary studies. With digital fabrication technologies, complex geometries may be produced in some cases with a similar production effort than simpler geometries. Hence, digital concrete fabrication could overcome the restriction caused by rectangular building components in existing technologies. This flexibility brings form optimization design processes into a new dimension, allowing using structurally optimum shapes and minimizing the use of materials for a certain structural element at the same production cost. While this is a powerful benefit, its potential will most likely not be fully exploited in mass market concrete structures, because of other requirements of geometric simplicity independent of the building technology, as pointed out e.g. by Steadman [7] who elaborated reasons for vertical and horizontal rectangularity of buildings: Verticality is structurally efficient as typically dominant gravity loads are transferred directly to the ground, and horizontal flat slabs and straight walls serve to exploit the building space. Furthermore, flat floors permit easy manoeuvring and placing of, usually rectangular, furniture, and horizontal rectangularity provides unlimited possibilities to divide buildings and rooms and freedom to pack next to each other. While some of these requirements may not apply to all buildings (e.g. in single spaces such as theatres or museums non-rectangularity is frequent since close packing is not required), geometric simplicity will be still required in most cases. Hence, it is important to find efficient digital construction processes for these standard applications.

3 Benefits of Digital Fabrication for Structural Concrete

3.1 Key Potentials

Besides the geometric flexibility, whose use may be limited as outlined above, digital fabrication brings about many potentials that should be exploited while aiming at conforming to mass-market requirements.

The robotic placement of reinforcement permits providing reinforcement in any direction and length in the optimum directions and strictly in the statically required amount, significantly reducing the reinforcement quantities for ultimate limit state, especially in walls and slabs. Furthermore, it has the potential of reducing on-site adjustments to the planned reinforcement due to limitations on manual placement and hazards and health risks of construction workers (e.g. lifting heavy rebar and fall hazards of steel fixers).

Equally promising is a reduction of minimum reinforcement, which often makes up for more than 50% of the total reinforcement content. The required amount of minimum reinforcement to avoid brittle failures at cracking (see Subsect. 2.1) increases proportionally to the cracking load of the structure, which is directly dependent on the concrete tensile strength. Hence, by tailoring the concrete grade locally to the actual needs by digital fabrication processes, i.e. reducing the cement content where a low concrete strength is sufficient, the corresponding minimum reinforcement could be reduced very significantly (in addition to savings in cement content). Another process that could tremendously reduce the minimum reinforcement is the fabrication of crack initiators by digital technologies. These crack initiators would (i) reduce the cracking load and (ii) ensure small crack spacings and correspondingly reduced crack widths. Both aspects contribute to reducing the minimum reinforcement. Crack initiators could be easily produced by means of geometric or material discontinuities introduced in digital fabrication technologies. One example of intrinsic crack initiators caused by material discontinuities are the weak interfaces between concrete layers in 3D printing processes. These weak interfaces, intuitively perceived as a disadvantage of 3D concrete printing, can be used either to reduce the reinforcement content or to improve the durability of the structures as will be further discussed in the Subsect. 3.2.

3.2 Crack Initiators by 3D Printing Weak Interfaces

Relevance

Weak interfaces between layers cast at different time have been widely reported in the literature. For example, Roussel and Cussigh [8] reported on the formation of distinct layers (also designated as "cold joint"), which may occur in interfaces of Self Compacting Concrete (SCC) depending on the delay between the casting of layers. The occurrence of distinct layers "depends strongly on the thixotropic behaviour of the SCC, the thickness of the layers and on the roughness of the interface between the two layers" [8]. Zareiyan and Khoshnevis [9] experimentally analysed the bond strength between layers cast after each other by means of compression and splitting tests. This study showed that a casting interval of 12 min with layers of 50.8 mm thickness caused a reduction of the tensile strength (measured at an age 28 days) of about 10%.
Due to shrinkage and temperature strains, concrete is prone to cracking regardless of the amount of reinforcement in the structure and the presence of mechanical actions. In engineering practice, reduced crack widths are relevant for surfaces directly exposed to the environment for reasons of water tightness, durability and/or aesthetics (fair-faced concrete). Designers typically increase the reinforcement content to reduce the crack spacing and hence the crack widths. By using weak interfaces as crack initiators, the crack spacing and crack widths could be reduced without increased reinforcement content.

The following sections present (i) an experimental study to assess the reduction of the bending tensile strength that can be achieved in layered casting processes, as well as (ii) a mechanical model to quantify the potential saving of minimum reinforcement when using 3D printing weak interfaces.

Pilot Experiments

In a pilot experimental study by the authors, three-point bending flexural tests on 60 specimens of 80 mm × 80 mm × 270 mm with a span of 200 mm were conducted. The specimens were cast vertically in two steps (approx. 135 mm + 135 mm) at room temperature. The casting interval between layers was varied (min. 3 specimens per casting interval). The study aimed at examining whether it is possible to regulate the concrete tensile strength f_{ct} of a specific type of concrete in the interface between layers.

The concrete was similar to the one used for Smart Dynamic Casting projects [3], which is similar to the standard formulations used in 3D printing processes. The mix was slightly modified in order to increase the flowability and, hence, to easily level the layer surface. A level layer surface improves the repeatability of the interface tensile strength.

Figure 2 shows the bending tensile strength in average and in its scatter (minimum to maximum) of weak interfaces with varying casting intervals in relationship to the reference specimens (continuous casting) after 21 days. Furthermore, photos of the failed weak interfaces are shown, of the reference specimen and of the specimens with 12-min, 24-min and 1-h casting interval between layers. The results show that the loss of bending tensile strength in a weak interface greatly evolves between 10 min and 30 min for the used type of concrete. For intervals greater than 2 h, the two parts of the specimens had no adherence at all.

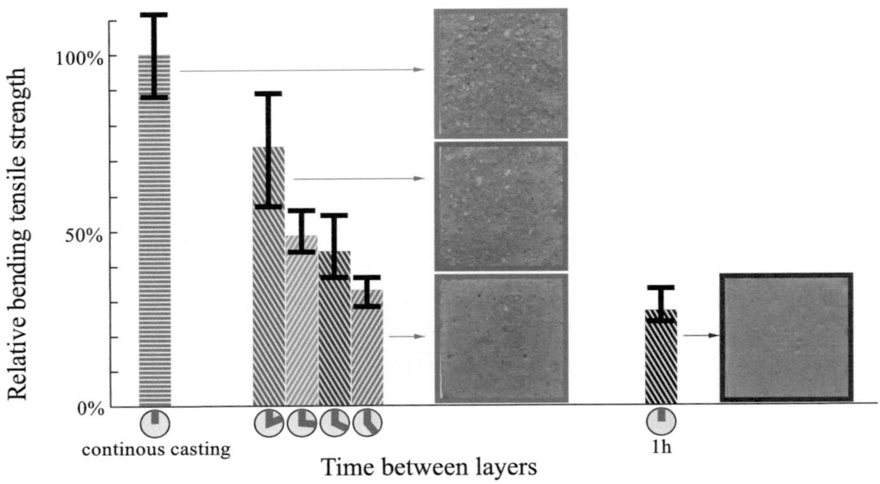

Fig. 2. Bending tensile strength of weak interfaces and its scatter for varying casting intervals normalised with respect to the reference specimens with continuous casting, and photos of failed interfaces.

Reduction of Minimum Reinforcement

Crack widths in concrete structures are typically calculated by multiplying the crack spacing with the average tensile strain of the reinforcement ($wr = s_r \varepsilon_{sm}$), neglecting the typically small reduction of the crack widths caused by the tensile strains of the concrete between the cracks. Crack spacings as well as tensile strains and stresses between the cracks can be determined from basic mechanical principles using e.g. the Tension Chord Model (TCM) [10]. In this section, the TCM is modified to consider the effect of weak interfaces between 3D printing layers in order to quantify the reduction of the crack width and/or the reduction of the reinforcement allowed by the presence of crack initiators (subscript *wi* used for "weak interfaces" in the following).

The main assumption of the TCM is the consideration of a stepped, rigid-perfectly plastic bond shear stress-slip relationship with bond strengths $\tau_b = \tau_{b0} = 2 f_{ct}$ prior to yielding, and $\tau_b = \tau_{b1} = f_{ct}$ after onset of yielding. Hence, at serviceability load levels where the reinforcement remains elastic, constant bond stresses are considered, similar to the simplification adopted by many design codes. While cracks are considered as stress free, tensile stresses are transferred from reinforcement to concrete between cracks by bond shear stresses. The maximum spacing between two cracks (s_{r0}) will be produced when the tensile strength of concrete (f_{ct}) is reached at the centre between two cracks (Fig. 3a). For the modified TCM presented here, considering crack initiators by 3D printing weak interfaces (Fig. 3b), the layering process is assumed to affect exclusively the concrete tensile strength in the weak interfaces ($f_{ct,wi}$), which is reduced as follows:

$$f_{ct,wi} = \xi \cdot f_{ct} \tag{1}$$

The authors are currently conducting an experimental study to evaluate whether the layering process could also influence the bond strength τ_{b0} and τ_{b1}.

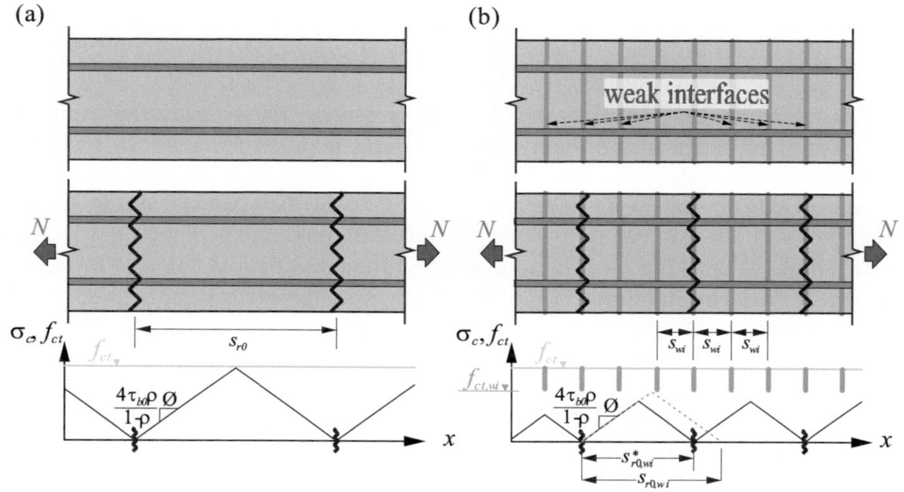

Fig. 3. Tension chords with maximum crack spacing and distribution of concrete tensile stresses: (a) TCM for a continuously cast element; (b) modified TCM considering weak interfaces.

Under these assumptions, and for layer thicknesses (s_{wi}) considerably smaller than the crack spacing without weak layers (as usually happens for standard 3D printing processes, e.g. s_{wi} = 20 mm, $s_{wi}/s_{r0} \approx 0.1$) the maximum crack spacing $s_{r0,wi}$ of a tension chord incorporating weak interfaces is determined by:

$$s_{r0,wi} = \frac{f_{ct,wi} \emptyset (1 - \rho)}{2\tau_{b0}\rho} = \xi \cdot s_{r0} \tag{2}$$

where Ø is the diameter of the reinforcement, ρ is the geometric reinforcement ratio of the tension chord and s_{r0} is the maximum crack spacing in an element without crack initiators.

The presence of weak interfaces reduces the maximum crack spacing proportionally to the reduction of the concrete tensile strength in the interfaces between layers (ξ).

The cracking load and the steel stresses at the crack at the onset of cracking ($\sigma_{sr,cr}$) are also reduced proportionally to the relative strength in the interfaces:

$$\sigma_{sr,cr,wi} = \xi \cdot f_{ct}\left(\frac{1}{\rho} - 1 + \frac{E_s}{E_c}\right) = \xi \cdot \sigma_{sr,cr} \tag{3}$$

where E_s and E_c are the Young's moduli of steel and concrete respectively.

The maximum crack width at the onset of cracking ($w_{max,cr}$), relevant for durability verifications in elements subjected to imposed deformations, can be estimated as follows in the presence of weak interface crack initiators:

$$w_{max,cr,wi} = s_{r0,wi}\left(\frac{\sigma_{sr,cr,wi}}{E_s} - \frac{\tau_{b0} s_{r0,wi}}{\emptyset E_s}\right) = \xi s_{r0}\left(\frac{\xi \sigma_{sr,cr}}{E_s} - \frac{\tau_{b0} \xi s_{r0}}{\emptyset E_s}\right) = \xi^2 \cdot w_{max,cr} \tag{4}$$

The effect of weak interfaces in layering processes allows reducing either the maximum crack width or the reinforcement content required for a certain maximum crack width requirement. As Fig. 4 shows for an example tension chord with reinforcement of Ø18 mm, weak interfaces with a 33% of the concrete tensile strength inside the layer allow reducing by 67% the minimum reinforcement required to avoid brittle failures at cracking (reduction quasi-proportional to the reduction of the tensile strength in the weak interfaces, ξ). At the same time, these weak layer joints would allow reducing by around 90% the maximum crack with a constant reinforcement content (reduction proportional to the square of the reduction of the tensile strength in the weak interfaces, ξ^2). Hence, the possibility to introduce controlled crack initiators has a huge potential to reduce very significantly the amount of minimum reinforcement and/or improve the durability of many reinforced concrete structures.

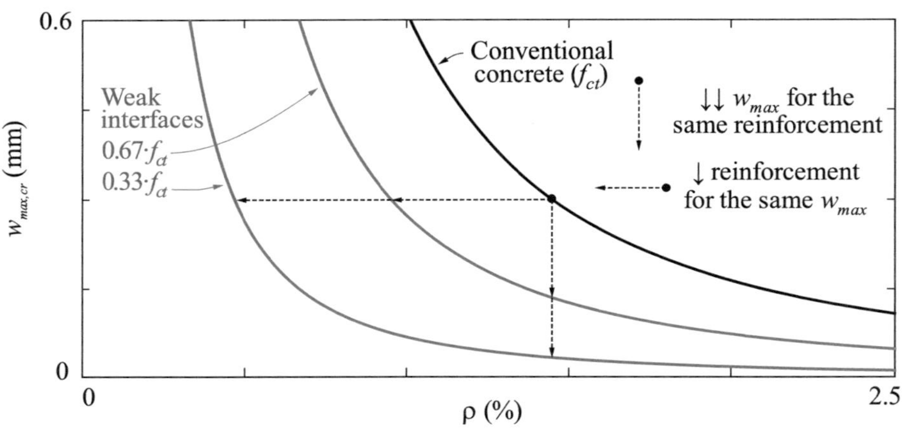

Fig. 4. Impact of weak joints between layers of extrusion processes on the amount of reinforcement required to control the maximum crack width caused by imposed deformations, assuming thin layers (Ø = 18 mm; f_{ct} = 4.8 MPa).

In real-life structures, a reduction of the amount of minimum reinforcement has additional benefits for the crack control since it allows reducing the diameter of the reinforcement when the reinforcement spacing is fixed, e.g. because of constructive reasons. In such cases, using the assumptions of the TCM, the crack width is directly proportional to the diameter of the reinforcement. Hence, the amount of minimum reinforcement can be further reduced for a certain crack width requirement. This effect is shown in Fig. 5 with a fixed longitudinal reinforcement spacing of 100 mm. In this case, weak joints with 33% of the concrete tensile strength inside the layer allow reducing by 80% the minimum reinforcement. Since this example considers a layer thickness (s_{wi}) in the same order of magnitude as the crack spacing, the simplification considered in Eq. (2) cannot be directly adopted and the maximum crack spacing has to be rounded to an odd number of layers ($s^{*}_{ro,wi}$), as shown in Fig. 3b.

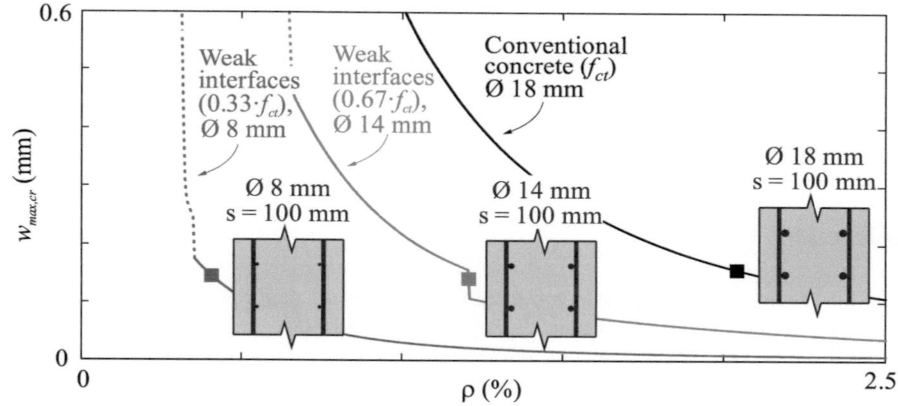

Fig. 5. Example of the required longitudinal reinforcement content for a maximum crack width requirement of 0.15 mm considering a conventional concrete wall and concrete walls with weak interfaces between layers (f_{ct} = 4.8 MPa; s_{wi} = 100 mm). The dashed line indicates yielding of reinforcement at cracking.

4 Conclusion and Outlook

Digital technologies are extensively exploring the potential of geometric flexibility. Nevertheless, most of them encounter difficulties in penetrating the market due to lacking compliance with structural integrity requirements. To maximise their impact, is essential that digital concrete processes (i) integrate reinforcement resisting tensile forces and (ii) address conventional structures without geometrical complexity.

Digital concrete fabrication processes have a great potential to reduce the quantity of reinforcement required in concrete structures even for standard geometries. The robotic placement of reinforcement permits providing reinforcement in the statically required amount. Moreover, minimum reinforcement can be tremendously reduced by

(i) tailoring the concrete grade locally to the actual needs and (ii) ensuring small crack spacings and correspondingly reduced crack widths by means of crack initiators.

The results of an experimental pilot study show that the strength reduction in the interfaces between layers from extrusion processes can be quantified depending on the casting interval of the layers. This allows using these weak interfaces as crack initiators. A mechanical model to quantify the potential for saving minimum reinforcement when using these crack initiators generated in 3D printing is presented. It is found that weak interfaces in layer joints with 33% of the concrete tensile strength inside the layer allow reducing up to 80% the minimum reinforcement for a given maximum crack width requirement against imposed deformations.

Current research of the authors is addressing beneficial effects of weak interfaces and other intrinsic characteristics of digitally fabricated concrete structures aiming at exploiting them for developing sustainable and economic products and production processes able to penetrate the mass market.

Acknowledgements. This research is supported by the National Centre for Competence in Research in Digital Fabrication in Architecture, funded by the Swiss National Science Foundation (project number 51NF40_141853).

The authors would like to thank Lex Reiter (IFB, ETH Zürich) for his indispensable support for the planning and the conduction of the presented experiments.

References

1. Khoshnevis, B.: Automated construction by contour crafting—related robotics and information technologies. Autom. Constr. **13**(1), 5–19 (2004)
2. Buswell, R.A., Soar, R.C., Gibb, A.G.F., Thorpe, T.: Freeform construction: mega-scale rapid manufacturing for construction. Autom. Constr. **16**(2), 224–231 (2007)
3. Lloret, E., et al.: Complex concrete structures: merging existing casting techniques with digital fabrication. Comput.-Aided Des. **60**, 40–49 (2015)
4. Hack, N., et al.: Mesh mould: an on site, robotically fabricated, functional formwork. Presented at the Second Concrete Innovation Conference (2nd CIC), Tromsø, Norway (2017)
5. Wangler, T., et al.: Digital concrete: opportunities and challenges. RILEM Tech. Lett. **1**, 67–75 (2016)
6. Helm, V., Ercan, S., Gramazio, F., Kohler, M.: In-situ robotic construction: Extending the digital fabrication chain in architecture. In: Proceedings of the 32nd Annual Conference of the Association for Computer Aided Design in Architecture (ACADIA), San Francisco, pp. 169–176 (2012)
7. Steadman, P.: Why are most buildings rectangular? Archit. Res. Q. **10**(2), 119–130 (2006)
8. Roussel, N., Cussigh, F.: Distinct-layer casting of SCC: the mechanical consequences of thixotropy. Cem. Concr. Res. **38**(5), 624–632 (2008)
9. Zareiyan, B., Khoshnevis, B.: Interlayer adhesion and strength of structures in Contour Crafting - Effects of aggregate size, extrusion rate, and layer thickness. Autom. Constr. **81**, 112–121 (2017)
10. Marti, P., Alvarez, M., Kaufmann, W., Sigrist, V.: Tension chord model for structural concrete. Struct. Eng. Int. **8**, 287–298 (1998)

Alternative Reinforcements for Digital Concrete Construction

Viktor Mechtcherine[1(✉)], Venkatesh Naidu Nerella[1], Hiroki Ogura[2], Jasmin Grafe[1], Erik Spaniol[1], Martin Hertel[1], and Uwe Füssel[1]

[1] Technische Universität Dresden, 01052 Dresden, Germany
mechtcherine@tu-dresden.de
[2] Institute of Technology, Shimizu Corporation, Tokyo 135-8530, Japan

Abstract. Applications of structural concrete require use of reinforcement in one form or another. The known reinforcement concepts in additive concrete construction typically rely on conventional reinforcement approaches, which provide a solid basis for structural design, since existing guidelines and codes can be used. However, the use of conventional steel reinforcement poses serious limitations to the digitalization and automation of fabrication techniques. The article at hand presents two alternative approaches of reinforcing 3D-printed concrete structures: (1) additive manufacturing of steel reinforcement elements, (2) use of high-performance microfiber to achieve strain-hardening behavior of printed concrete. For both approaches materials and manufacturing techniques are briefly described followed by the results of mechanical testing and complimentary microscopic investigations. The printed steel bars showed similar mechanical performance in comparison to ordinary steel bars of the same cross-section area and comparable bond behavior to concrete too, as observed in pull-out experiments. The addition of 1% to 1.5% high-density polyethylene microfiber to fine-grained matrix enabled for printable strain-hardening cement-based composites (SHCC) with a tensile strain capacity of up to 3.2%.

Keywords: Additive concrete construction · 3D-printing · Steel bars Polymer fiber · ECC · SHCC · Strain-hardening cement-based composites

1 3D-Printed Steel Reinforcement

1.1 Introduction

Additive concrete technologies and corresponding research activities have been developing very rapidly in many countries. Most of such technologies deal with the placement of concrete, whereas the proposed concepts for introducing reinforcement are less advanced. Several methods for reinforcing concrete in the context of additive construction were suggested, e.g. (1) integrating conventional steel reinforcement between printed layers and in cavities [1–3], (2) enveloping conventional steel reinforcement by printable concrete [4], and (3) placing concrete between two layers of reinforcement with narrow mesh (Mesh Mold) [5]. These approaches must be followed and promoted further, but also different new solution strategies are needed in order to bring the additive manufacturing in construction on a new level [6, 7]. The approach

T. Wangler and R. J. Flatt (Eds.): DC 2018, RILEM Bookseries 19, pp. 167–175, 2019.
https://doi.org/10.1007/978-3-319-99519-9_15

suggested here is to manufacture steel reinforcement, specifically such elements of complex geometry as installations or fastenings, additively. Since such technology would enable high freedom with respect to shaping reinforcement elements, the *shape following force* concept can be easily implemented opening new opportunities for structural design.

In this part of the article, the manufacture technique for steel bars is briefly presented followed by the experimental characterization of the new reinforcement and its comparison to common steel bars. Finally, the bond behavior of novel reinforcement to printable concrete is described. The more detailed presentation of the concept and the results can be found in [8].

1.2 Materials, Manufacture and Testing Techniques

Under consideration of the productivity, robustness and costs, the application of arc- and wire-based technologies of additive manufacture appear to be most adequate with respect to 3D-printing of steel reinforcement. In the work at hand, gas-metal arc welding approach with a continuously delivered wire electrode was applied [8].

Steel bars manufactured using this techniques are shown in Fig. 1a. In order to control the position of the bar failure in tension tests, the bar diameter was slightly reduced in the middle region of the specimens (Fig. 1b).

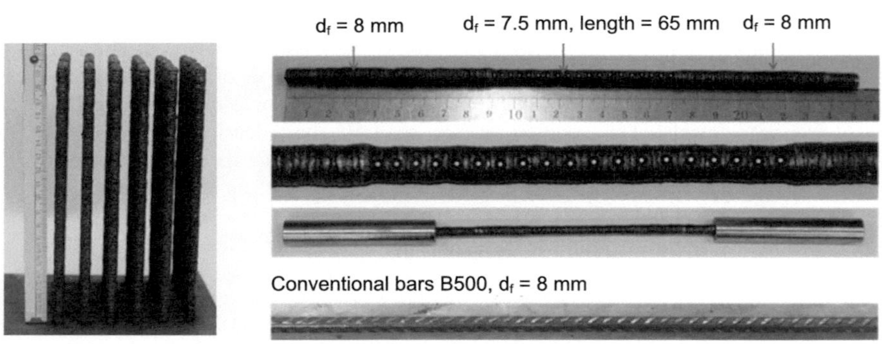

Fig. 1. (a) Steel bars produced by 3D printing, (b) Specimens for uniaxial tension tests [8].

The printed bars were subjected to tensile loading under displacement controlled regime. The deformations were measured by means of a high resolution camera and evaluated using digital image analysis algorithms [8]. To investigate the bond of reinforcement bars to concrete, pullout tests according to [9] were carried out.

1.3 Experimental Results

Figure 2 shows experimental results yielded by tension tests on the printed steel bars in comparison to common steel bars of the same diameter. The additively manufactured reinforcement had lower values of both yield stress and tensile strength than those measured on common bars (28% and 16%, respectively). Also Young's modulus was 28% lower in case of 3D-printed specimens. However, the printed reinforcement exhibited a more pronounced deformability, reaching approximately 250% higher values of the ultimate strain.

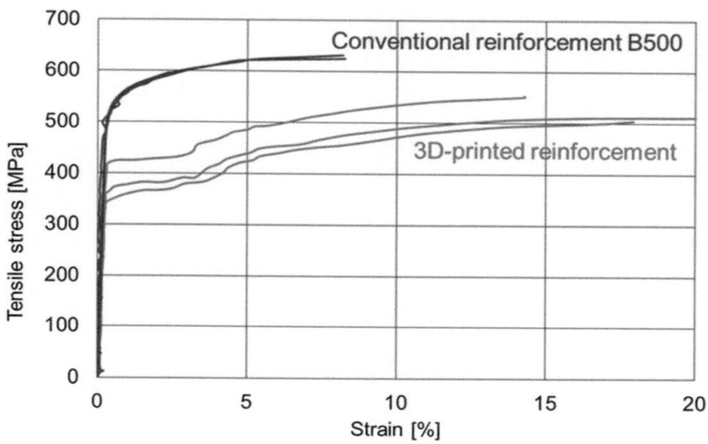

Fig. 2. Results of tension tests on bars made by 3D-printing and on commonly used rebar [8].

The increase in ductility for the additively manufactured steel reinforcement was well decipherable by visual inspection of failed specimens as well. While commonly used steel rebar showed shear failure with moderate plastic deformations, 3D-printed bars exhibited an obvious necking.

Additionally, the condition of fractured bars was studied by means of an ESEM to characterize the effect of the additive manufacture by gas-metal arc welding on micro-structure of the novel reinforcement. Figures 3a and c show that a necking failure results in a stairway-like crack, which can be probably traced back to heterogeneity introduced by additive crafting process. With a higher resolution trans-crystalline failure becomes recognisable by comb-like micro-structure, characteristic for ductile fracture, cf. Fig. 3b. In contrast, the failure surface of the common steel bars appeared more regular and evener, Fig. 3e.

The pull-out tests (see [8] for details) demonstrated that the bonding behaviour of printed steel bars to concrete was comparable to that of common steel bars. The bond strength was approximately 20–25% lower in case of additively manufactured bars. It should be stressed here that no specific measures were undertaken to improve the bond in the cited experiments, i.e. the printed bars exhibited merely "natural" roughness. However, the bond can be surely considerably increased, if necessary, by printing the reinforcement with particular surface morphology like ribs. From the point of view of manufacturing such profiling is well possible, as demonstrated in [8].

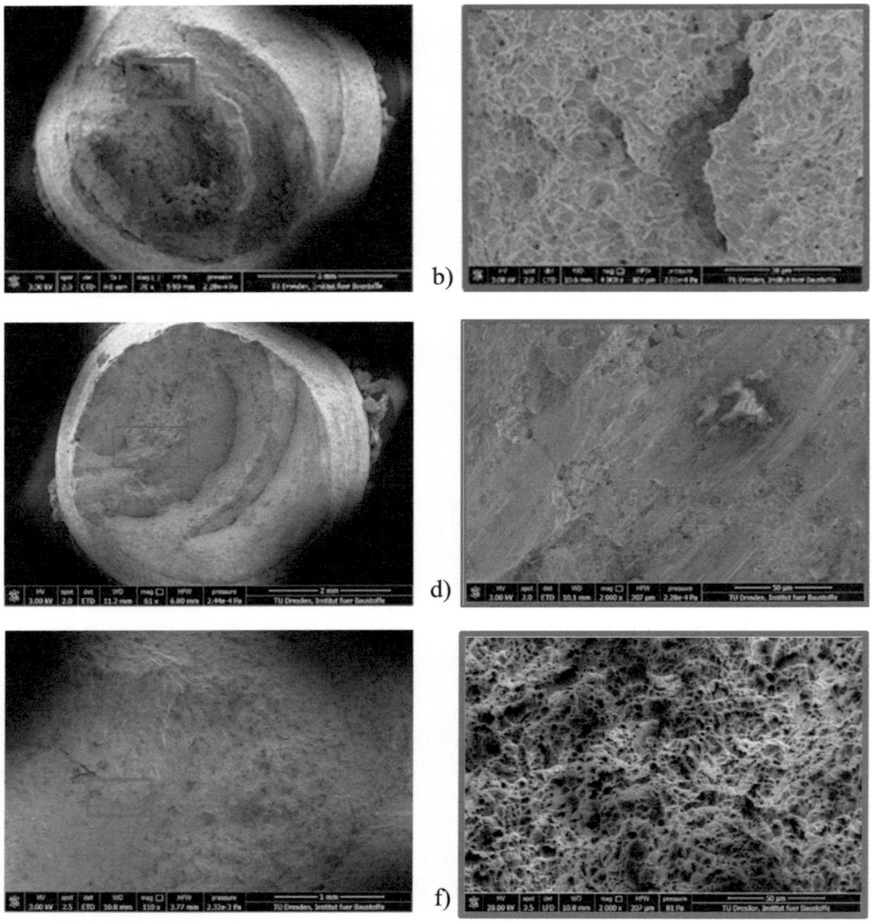

Fig. 3. (a–d) surfaces of fractured printed steel bars; heterogeneity can be observed at higher resolution: (c) comb-like microstructure typical for ductile fracture and (d) inter-crystalline failure characteristic for more brittle, shear failure mechanism, (e) common steel bars after failure showing a more uniform microstructure with dominating trans-crystalline shear failure.

2 3D-Printed Strain-Hardening Cement-Based Composites

2.1 Introduction

The possibility of using dispersed-fibre reinforcement is seemingly capable of being most easily integrated into the 3D-printing process by using fibre-reinforced concrete instead of plain concrete. This straightforward approach has been pursued by only few researchers as yet. Hambach and Volkmer [10] used basalt fiber of 3 to 6 mm length as well as alternatively dispersed glass or carbon fibers. They reported a pronounced orientation of fiber due to extrusion process and a significant increase in flexural strength. Panda et al. [11] worked with various percentages of glass fibers of 3 mm to 8 mm length. Anisotropy and increase in flexural strength were observed also in that study. Taking into account tremendous advances in the field of high-performance fiber-reinforced cement-based composites, see e.g. [12, 13], there is a very high potential for further improvement in mechanical properties of printable concrete especially in terms of ductility.

With this respect, strain-hardening cement-based composites (SHCC) are a particularly promising group of fiber-reinforced concretes. They develop multiple, fine cracks when subjected to increasing tensile loading, exhibiting strain capacities of up to several percent. Additionally to extremely high mechanical performance under quasi-static and dynamic loading [13], SHCC have a number of further advantages such as: (1) narrow cracks typical of SHCC are very favorable with respect to the durability of structural elements [14], (2) fine polymeric fibers used in SHCC help to mitigate negative consequences of plastic, autogenous and drying shrinkage, thus, preventing formation of cracks [15]. Since freshly printed elements are exposed to drying right from the beginning, if no specific curing measures are taken, the mitigation of shrinkage seems to be an issue very important for printed concrete materials.

This part of the article presents a printable SHCC developed at the Institute of Construction Materials of the TU Dresden and describes its production technique and mechanical behavior. Further details may be found in [16].

2.2 Materials, Manufacture and Testing Techniques

Two printable SHCC mixtures will be presented here, designated as Mixture B and Mixture C. Both mixtures had a water-to-binder ratio (W/B) of 0.22. The binder comprised 75% cement CEM II/A-M (S-LL) 52.5R, 15% silica fume and 10% fly ash. Silica fume and fly ash were used to increase viscosity and cohesion of mixtures. Silica fume was added as slurry containing 50 wt% solid matter. As aggregate, a fine sand with maximum size of 1.0 mm was used for both mixtures, however, in differing quantity. While Mixture B had a sand-to-binder (S/B) proportion of 0.5, the proportion needed to be decreased to 0.2 in the case of Mixture C, since this mixture had a higher fiber content. High-density polyethylene (HDPE) microfibers were chosen for this study based on positive experience of pervious work on conventionally casted SHCC [12]. The length and diameter of the fiber were 6 mm and 0.012 mm, respectively. Density of HDPE fibers was 0.97 g/cm^3 and tensile strength of the fibers was approximately 3000 MPa. Mixtures B and C contained 1.0% and 1.5% fiber, respectively, by volume of the composite. For both mixtures 2.0% polycarboxylate-based superplasticizer by mass of binder was used. The mixing was performed with pan-type mixer having a capacity of 20 L.

The cone spread diameter measured according to EN 1015-3 was for Mixture B 120 mm before strokes and 135 mm after strokes. For Mixture C the corresponding values were 119 mm and 133 mm, respectively. Further details on rheological propertiesand extrudability measurements can be found in [16].

3D-printing device used in this study consisted of an eccentric screw pump, a conveying pipe and a nozzle for material deposition and forming. The nozzle had a cross-section of 18,7 mm by 30 mm, see Fig. 4. A linear printing speed of 50 mm/s was selected. Wall-shaped samples with length of 1000 mm, width of 30 mm and height of 120 mm were fabricated starting at 20 min after mixing. The time gap between two subsequent layers was one minute.

Fig. 4. Printing with SHCC Mixture B.

For uniaxial tensile tests, prism specimens with dimensions of 250 mm × 24 mm 40 mm were saw-cut from the printed wall. The longitudinal axis was in horizontal direction, thus, parallel to interfaces between the layers. The ends of the specimens were strengthened by gluing pieces of SHCC on both sides, see Fig. 6a, to ensure that fracture localization occurred in the middle part of the specimen. For comparison, some SHCC specimens having the same compositions (Mixtures B and C) were casted in a conventional manner and tested as well. All the specimens were cured under water for 7 days and then stored until test day in controlled climate with 20 °C temperature and 65% relative humidity.

2.3 Experimental Results

The stress-strain diagrams obtained from uniaxial tension tests on printed SHCC are shown in Fig. 5b. The courses of the curves are characteristic for SHCC as observed in numerous earlier investigation on cast material. They show a pronounced increase in load-bearing capacity after the formation of the first crack, marked by transition from linear elastic regime to strain-hardening regime. The strain-hardening is accompanied by multiple cracking, while formation of cracks causes the certain unsteadiness in the curve (sudden drops followed by regain and further increase in stress level). As expected Mixture C with the higher fiber content of 1.5% exhibited considerably higher strain capacity with average values of approximately 3.2%, while Mixture B with 1% HDPE fiber provided an average strain capacity of about 0.9%. The first crack stress was nearly the same for both mixtures and amounted approximately 4.2 MPa. The average tensile strength of Mixture C was with 5.7 MPa slightly higher than that of mixture B with 5.3 MPa.

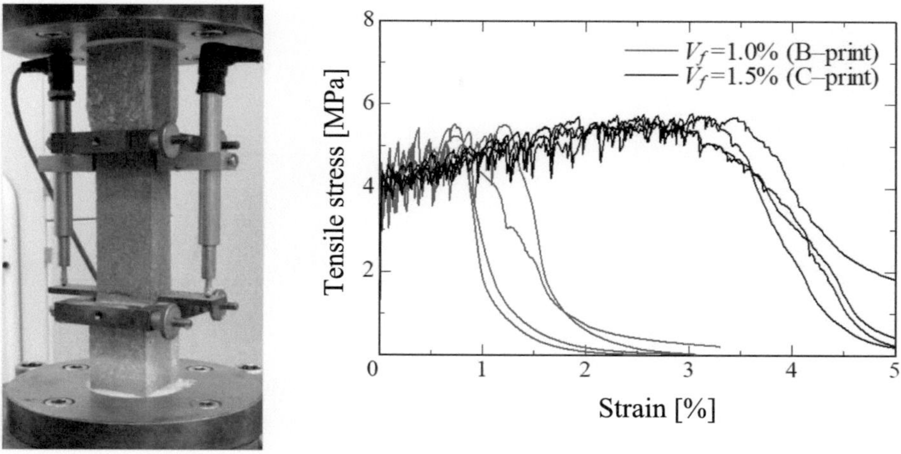

Fig. 5. (a) Setup of uniaxial tension tests, (b) stress-strain diagrams obtained from uniaxial tension tests on printed SHCC.

Tensile tests performed on cast SHCC specimens yielded similar results as for printed SHCC with respect to the first crack stress and tensile strength. However, the values of strain capacity measured for cast specimens were significantly lower than those for printed specimens: the average strain capacity amounted around 0.6% for Mixture B and 1.7% for Mixture C. A probable reason for a better performance of printed SHCC in terms of ductility can be likely traced back to a more favourable orientation of fiber due to extrusion process. It has been often observed that extrusion causes a pronounced, nearly unidirectional fiber orientation parallel to the longitudinal axis. As a result most fibers cross cracks under nearly straight angles which is favourable for force transfer across cracks. In case of cast specimens the fiber orientation is less pronounced. The compressive strength was determined on cast specimens only; its average values were 105 MPa and 104 MPa for Mixtures B and C, respectively.

Figure 6 shows a typical crack pattern of SHCC specimens as observed on their surfaces after uniaxial tension tests. Fine parallel cracks are nearly uniformly distributed over the length of the specimen, which is characteristic for high-performance SHCC. Pronounced fiber pullout can be observed on the fracture surface, see Fig. 6. The fiber pullout is a common failure mode for HDPE fiber, it can be traced to its smooth, hydrophobic surface and subsequently modest and only frictional bond to cementitious matrix of SHCC.

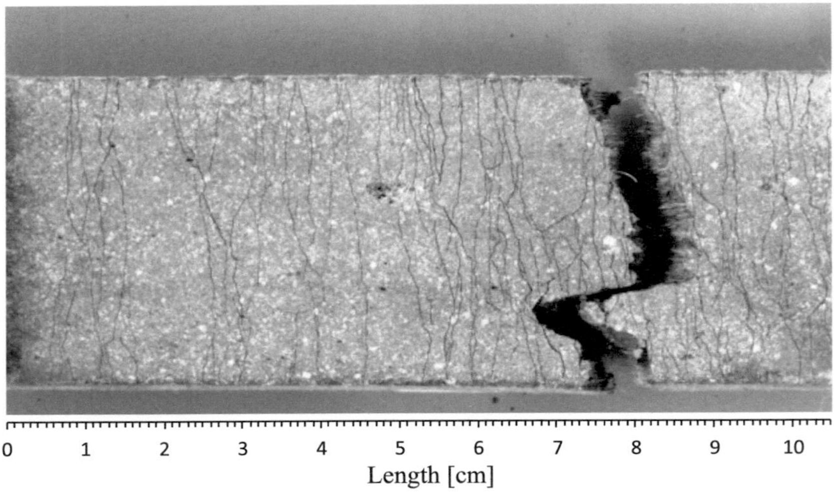

Fig. 6. Typical crack pattern as observed on the surfaces of SHCC specimens after uniaxial tension tests, here Mixture C.

3 Summary

This article was dedicated to two new types of reinforcement for 3D-concrete-printing.

The first approach was to print the steel reinforcement. The development of a 3D-printing process based on gas-metal arc welding with fully automatic, adaptive process control enables the production of steel reinforcement bars with adequate geometrical precision and geometric freedom at reasonable production speeds. 3D-printed steel bars exhibited approximately 20% lower values of the yield stress and tensile strength in comparison to conventional reinforcement bars. However, in contrast to the conventional bars they showed pronounced yielding and higher strain capacity. Microscopic investigation of fracture surfaces confirmed the ductile mode of failure in case of printed steel bars, showing additionally some particularities of the micro-structure due to additive production process. Printed steel bars showed a satisfactory bond to printable fine-grained concrete, thus being comparable with the bonding performance of conventional steel.

The second approach was to use dispersed fiber reinforcement for achieving a strain-hardening behavior of printable concrete. Two fine-grain compositions of printable strain-hardening cement-based composites were presented, one containing 1% and another 1.5% of HDPE fiber. The mechanical properties of 3D-printed SHCC were characterized using tension tests performed on prism specimens saw-cut out of printed scaled-down wall. Printed SHCC exhibited higher strain capacity than cast SHCC made of selfsame compositions, which can be explained by favorable fiber orientation due to extrusion process. Printed SHCC showed a well distributed fine multiple cracking typical for SHCC materials.

Further research on both types of reinforcement is in progress.

References

1. Valencia, N.: https://www.archdaily.com/804596/worlds-first-3d-printed-bridge-opens-in-spain. Accessed 14 Aug 2018
2. Apis-cor Homepage: http://apis-cor.com/en/faq/texnologiya-stroitelstva. Accessed 29 Dec 2017
3. TU Eindhoven: https://www.tue.nl/en/university/departments/built-environment/news/17-10-2017-worlds-first-3d-printed-reinforced-concrete-bridge-opened/. Accessed 14 Aug 2018
4. Scott, C.: Chinese Construction Company 3D Prints an Entire Two-Story House On-Site in 45 Days, 20 March 2017. https://3dprint.com/138664/huashang-tengda-3d-print-house/
5. Gramazio Kohler Research: Mesh Mould and In situ Fabricator, 2016–2017 in the DFAB House, 10 December 2017. http://dfab.arch.ethz.ch/web/e/forschung/324.html
6. Mechtcherine, V., Nerella, V.N.: Incorporating reinforcement in 3D-printing with concrete. Beton- und Stahlbeton **113** (2018). https://doi.org/10.1002/best.201800003
7. Mechtcherine, V., Nerella, V.N.: 3D printing with concrete: state-of-the art, trends, challenges. Bautechnik **95**, 275–287 (2018)
8. Mechtcherine, V., Grafe, J., Nerella, V.N., Spaniol, E., Hertel, M., Füssel, U.: 3D-printed steel reinforcement for digital concrete construction – manufacture, mechanical properties and bond behavior. Constr. Build. Mater. **179**, 125–137 (2018)
9. Rehm, G.: Über die Grundlagen des Verbundes zwischen Stahl und Beton, vol. 138. DAfStb, Wilhelm Ernst & Sohn, Berlin (1961)
10. Hambach, M., Volkmer, D.: Properties of 3D-printed fiber-reinforced Portland cement paste. Cem. Concr. Compos. **79**, 62–70 (2017)
11. Panda, B., Paul, S.C., Tan, M.J.: Anisotropic mechanical performance of 3D printed fiber reinforced sustainable construction material. Mater. Lett. **209**, 146–149 (2017)
12. Curosu, I., Liebscher, M., Mechtcherine, V., Bellmann, C., Michel, S.: Tensile behavior of high-strength strain-hardening cement-based composites (HS-SHCC) made with high-performance polyethylene, aramid and PBO fibers. Cem. Concr. Res. **98**, 71–81 (2017)
13. Müller, S., Mechtcherine, V.: Fatigue behaviour of strain-hardening cement-based composites (SHCC). Cem. Concr. Res. **92**, 75–83 (2017)
14. Altmann, M., Mechtcherine, V.: Durability design strategies for new cementitious materials. Cem. Concr. Res. **54**, 114–125 (2013)
15. Serpukhov, I., Mechtcherine, V.: Early-age shrinkage of ordinary concrete and a strain-hardening cement-based composite (SHCC) in the conditions of hot weather casting. In: CONCREEP 2015 Proceedings, pp. 1504–1513 (2015)
16. Ogura, H., Nerella, V.N., Mechtcherine, V.: Developing and testing of strain-hardeningcement-based composites (SHCC) in the context of 3D-printing. Materials **11**, 1375 (2018)

Additive Manufacturing and Characterization of Architectured Cement-Based Materials via X-ray Micro-computed Tomography

Mohamadreza Moini[1(✉)], Jan Olek[1], Bryan Magee[2], Pablo Zavattieri[1], and Jeffrey Youngblood[1]

[1] Purdue University, West Lafayette, IN 47907, USA
mmoini@purdue.edu
[2] Ulster University, Newtownabbey BT370QB, UK

Abstract. There is an increasing interest in the fabrication of cement-based materials via additive manufacturing (AM) techniques. However, the processing-induced heterogeneities and interfaces represent a major challenge. The role of processing in creating interfaces and their characteristics requires understanding of the microstructure of 3D-printed hardened cement paste (hcp). This work investigates the microstructural features of architectured cement-based materials, including processing-induced heterogeneous patterns, interfacial regions (IRs), and pore network distributions with respect to the architectural patterns. A 3D printer was modified and merged with an extrusion system and specimens were 3D-printed using a layer-wise direct ink writing (DIW) process capable of fabrication of 'lamellar' architectures of materials. A lab-based X-ray microscope (XRM) was used to perform X-ray micro-computed tomography (micro-CT) evaluations to explore the microstructural characteristics of 3-day old intact (i.e. not tested) 3D printed and cast specimens at two levels of magnification: 0.4X and 4X. CT scans of printed specimen revealed a patterned pore network and several microstructural features, including: (a) macropores (visible during printing), (b) micropores at interfacial regions (IRs), (c) accumulation of anhydrous cement particles near macropores, and (d) rearrangement of filaments away from their designed toolpath. In comparison, microstructural investigation of cast specimen at 4X scan revealed randomly distributed pores with no connectivity throughout the specimen. The aptitude of micro-CT as a non-destructive technique for microstructural characterization of architectured cement-based materials is discussed. The role of processing to induce and to pattern heterogeneities such as IRs in materials is demonstrated and the role of architecture in controlling such heterogeneities and their directionality through the interface is discussed.

Keywords: 3D-printing · Cement paste · Micro-CT · Interfacial Region (IR)

T. Wangler and R. J. Flatt (Eds.): DC 2018, RILEM Bookseries 19, pp. 176–189, 2019.
https://doi.org/10.1007/978-3-319-99519-9_16

1 Introduction

Properties of hardened cement paste (hcp) are influenced by its microstructure and the way in which the material is cast and placed [1]. Specifically, characteristics of the pore network (i.e., the size and distribution of internal flaws), the morphology of microstructural components, and heterogeneities in the microstructure, all affect mechanical properties of hardened cement-based materials [1, 2]. Over the last decades, the use of advanced characterization techniques, such as scanning electron microscopy (SEM), has significantly advanced the understanding of microstructure of cement-based materials. However, the effectiveness of SEM is limited in terms of obtaining three-dimensional information about connectivity and size distribution of pore network [1, 3].

X-ray micro-computed tomography (micro-CT) has been previously applied to cement-based mortar and concrete materials to characterize pore network [4] and microstructure [5] and their relationships with a variety of key properties including: fracture properties [5–7], damage mechanisms [2, 8, 9], mass transport [10], and evolution of cement hydration [11]. Micro-CT is a non-destructive technique that captures three-dimensional images of materials without the need for destructive preparation processes such as drying, surface treatments, and vacuuming; all of which are commonly required when preparing specimens for typical microstructural characterization techniques such as MIP, gas sorption, and SEM [1]. As such, the use of micro-CT for microstructural characterization of 3D-printed hcp is advantageous as it does not alter the microstructure while allowing for imaging of elements as large as 10ths of millimeters.

During this research, a laboratory-based (i.e. not requiring synchrotron facilities) X-ray microscope was used to explore the microstructural characteristics of intact (i.e. not tested) printed and cast 3-day old cement paste specimens at two magnifications: 0.4X and 4X corresponding, respectively, to the resolution of 32.24 μm and 4.04 μm.

3D-printing via direct-ink-writing (DIW) of colloids, slurries, and pastes allows for control of the architecture of the element and can give rise to a variety of microstructural features [12–14]. In this work, four microstructural features observed in 3D-printed lamellar architecture are discussed. All four features were qualitatively detected at 0.4X scans, and further verified at 4X scan. Finally, the microstructure of the printed specimen was compared to a microstructure of conventionally cast specimen.

2 Methods

2.1 3D-Printing Setup

To establish a 3D printer capable of printing cement paste via DIW technique, two separate units were combined. A gantry-based 3D printer (Ultimaker 2 Extended+ used for printing thermoplastic materials) was merged with a stepper motor-driven extrusion system (Structur3d Discov3ry Paste Extruder) to serve as a paste extruder (see Fig. 1(a)). These two units were connected with plastic tubing and luer locks. The resulting system is capable of printing pastes with high yield stress and viscosity and can fabricate elements at prototyping scale (mm). The extruder is capable of applying desirable displacements (extrusion rates) via mounted 75 mL ink-charged syringes as depicted in Fig. 1(b). The 3D printer was modified to allow for mounting of a nozzle holder assembly on the printer gantry rods (Fig. 1(c)). The nozzle holder consists of two lightweight aluminum parts designed and custom-fabricated specifically for this printer. To continuously feed the paste from the syringe to the printer nozzle, a polyethylene tube for paste delivery (with an internal diameter of 4.3 mm and 450 mm long) was passed through the nozzle holder assembly. A female luer lock was inserted into the nozzle holder to connect the tube to the nozzle (Fig. 1(d)). To connect the other side of the tube onto the syringe a male luer lock was used (Fig. 1(e)). The luer locks and the standard Nordsen nozzle (gauge 15 with 1.36 I.D.) are shown Fig. 1(f).

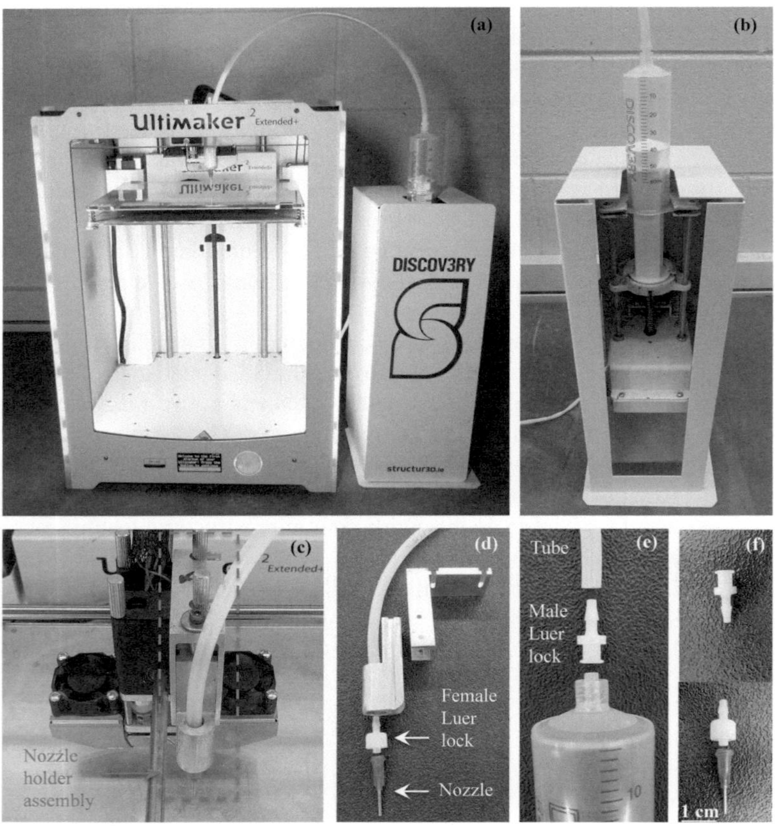

Fig. 1. 3D printer setup: (**a**) Gantry 3D printer (Ultimaker 2 Extended+) and stepper motor-based extrusion (Discov3ry) system; (**b**) Syringe and plunger mounted onto the stepper motor system; (**c**) Nozzle holder assembly mounted onto the 3D printer gantry guide rods; (**d**) Nozzle holder assembly and the female luer lock used to connect the tube to the nozzle; (**e**) male luer lock used to connect syringe to the tube; (**f**) nozzle and luer locks shown separately

2.2 Ink Design, Mixing Procedure, and Curing

An iterative mix design process was employed to identify cement paste inks with flow properties (i.e., yield stress, and viscosity) suitable for the DIW process. Combinations of high-range-water-reducing-admixtures (HRWRA) and viscosity modifying admixtures (VMA) are widely used in self-consolidating concrete to mitigate opposing effects of segregation at rest and high fluidity during pumping [15]. In DIW process of cement paste, it is also critical to achieve yields stresses that are high enough to allow the materials to be self-supporting (shape-holding). In the process of DIW of ceramic slurries (on the mm scale) the increase in the content of solids results in both, the increase in yield stress and in viscosity [16]. Moreover, the yield stress of cement paste is known to decrease with the addition of superplasticizers and the apparent viscosity is typically known to increase with the addition of VMAs [15]. In this work, the successful ink was designed with yield high enough for shape-holding and viscosity suitable for extrusion. A low water to cement ratio of 0.27, corresponding to a solid content of 53% (by mass), was used in this work. This percentage of solid content is found to provide suitable in ceramic paste as well [16]. In the case of cement paste, such low water to cement ratio can produce very stiff mixtures that can experience bleeding while undergoing extrusion. As such, a HRWRA was used to lower yield stress (and to ensure extrudability) and VMA was used to reduce bleeding. The use of VMA is highly desirable for DIW process. Specifically, it enhances the stability of the ink during extrusion and that of the specimen upon deposition [15].

The final ink used in this work consisted of the sub 150 μm fraction of commercially available Type I cement (ASTM C150 [17]) obtained from Buzzi Unicem, USA; the deionized water; HRWRA (MasterGlenium 7700), and VMA (MasterMatrix 362). Both chemical admixtures met the requirements of the ASTM C494 [18]. Optimal dosages were established based on findings of a related previous study [19]. For each 250.0 g of cement, the mix contained 65.2, 1.1 and 3.0 grams of deionized water, HRWRA and VMA, respectively. HRWRA and VMA were added to the water consecutively and stirred until they could not be visually observed. The liquid phase was then added to cement. A Twister Evolution Venturi vacuum mixer was used for mixing the paste to eliminate entrapped air as the presence of the air bubbles will degrade the quality of the ink. The pre-mixing mode of the mixer was used during the first 25 s of the mixing to process the paste at slow speed while subjected to a 70% of vacuum level provided by the mixer. This was followed by mixing at 400 rpm for 90 s at 70% vacuum level. The paste was mixed for a second time at 400 rpm for 90 s at 100% vacuum level of the mixer. The paste was then loaded into the syringe. The syringe was then outfitted with the plunger and mounted on the extruder as depicted in Fig. 1(b). The mixing process was performed within 5 min after combining cement and liquid. The specimens were cast and printed in lab environment at 18 ± 3 °C and 45 ± 5% relative humidity. Immediately after printing (or casting), the specimens were placed in a sealed curing box which maintained constant relative humidity of 93 ± 2% (by using saturated solution of potassium nitrate). The box was kept at constant temperature of 18 ± 3 °C.

2.3 Slicing and Design

In order to generate a toolpath required for lamellar architecture, a commercially available slicer (Simplify3D) was used to generate the G-code commands. A cubical 3D object was introduced to the slicer and geometrical parameters were assigned to it to achieve desired printing path and architecture. The G-code command included 5 axis of control: Point cloud coordinates (X, Y, Z axis) to control the movements of the nozzle and the bed; extrusion (E axis) to control the amount of extrusion relative to the nozzle movement; and printing speed (F axis) to control the speed of the nozzle movement. E and F axis were controlled via an extrusion rate multiplier and printing speed in the slicer. Several other printing parameters, including the location, the amount, and the speed of retraction, were also scripted to the G-code in the slicer. To generate the tool path, a 25 × 25 × 25 mm cube was introduced to the slicer and a continuous printing path in each layer was designed to create lamellar architectures (Fig. 2(a)). To achieve a solid specimen, 100% infill was used for lamellar architectures (Fig. 2(b)). A layer height (filament height) of 1.00 mm and the internal diameter of the nozzle of 1.36 mm were specified in the slicer. The specified printing speed was 250 mm/min, resulting in a speed of 87 s/layer. These printing and geometrical parameters were established by trial and error to obtain a suitable print quality (i.e., filament width and height close to that specified in the slicer). A schematic cross-section of the lamellar architecture considered, and a resulting specimen, are shown in Fig. 2(c) and (d).

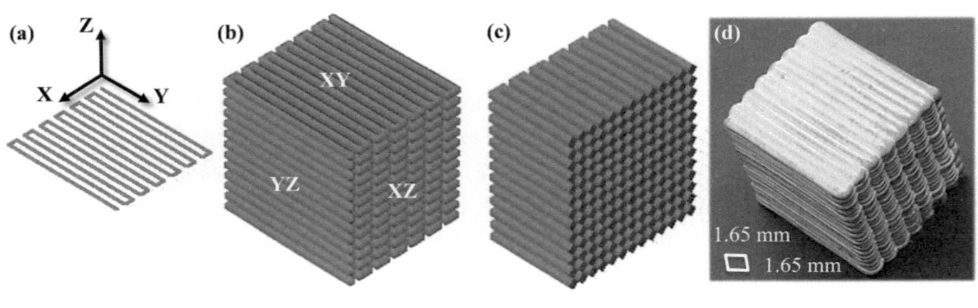

Fig. 2. Schematics of lamellar architectures: (**a**) Printing path of individual layers; (**b**) Printing path of a cube specimen and; (**c**) Cross-section of the specimen with lamellar architecture; (**d**) 3D lamellar architecture cube printed via DIW

2.4 Micro-CT and Scanning Specimens

X-ray micro-computed tomography (micro-CT) is an imaging technique that involves the recording of series of 2D radiographs (images), taken at various angles around a rotating object, to mathematically reconstruct a spatial map and digitally render the entire volume (i.e., three-dimensional appearance) of an object [2]. Resulting 3D renditions are typically presented as a series of 2D (i.e., sliced images) with intensities corresponding to X-ray absorption and material density at each voxel [20]. The resulting variations in intensity allow for identification of various phases and features of the microstructure and their 3D distribution.

Conventionally, X-ray microscopes employed for micro-CT characterization technique use a flat panel detector and thus rely on single-step (i.e., geometric) magnification. As a result, resolution degrades with increasing sample size and working distance. In this study, an X-ray microscope (XRM), Zeiss Xradia 510 Versa was utilized, which allows of an increase in the resolution of scans through dual-stage magnification process. In the first stage, the field of view (FOV) desirable to scan the entire volume of the specimen was established via geometric magnification process, which involved setting distances between the source, detector, and specimen (as in conventional micro-CTs). In a second stage, additional optical magnification was enabled at the detector system through objective lenses. The detector is equipped with scintillator and objective lens which converts X-rays to light rays and thus allows for optical magnification and higher resolution. The initial (i.e. the 0.4X) scan allowed a large FOV and thus facilitated the scan of the entire specimen (32.24 μm pixel size). This was followed by a 4X scan, allowing higher resolution (4.04 μm pixel size) at regions of interest (ROI). A beam energy of 150 keV, a power of 10 W, exposure times of 0.94 s and 4 s, and full 360° rotation were used for 0.4X and 4X scans of printed specimen and a beam energy of 140 keV, a power of 9 W, exposure times of 1 s, and full 360° rotation were used for 0.4X scans of cast specimen respectively. Dragonfly software was used for post-processing of the data. One cast and one 3D-printed hcp cubes ($25 \times 25 \times 25$ mm $\pm$ 1 mm) were used in this experiment.

3 Results and Discussion

3.1 3D-Printed Lamellar Architecture Micro-CT (0.4X and 4X Scans)

0.4X Scan. The 0.4X magnification CT scan of the intact specimen revealed the presence of four microstructural features: macropores, micropores, rearrangement of filaments and accumulation of anhydrous cement grains. These features are illustrated in Figs. 3(a), (b) and (c), for XZ, YZ, XY planes respectively, and in Fig. 3(d) for the 3D rendition of the interior of the lamellar architecture specimen. In micro-CT images of hcp, darker intensities represent pores filled with air or water, with greyscale intensities corresponding to hydrated cement paste products and brighter regions corresponding to anhydrous cement grains [1]. The previously mentioned four microstructural features are analyzed in more details below:

Macropores. As seen in Figs. 3(a) and (d), there are several regions that contain large pores (macropores) located between adjacent filaments of the same layer. These pores are aligned along the filament in the direction of the Y axis as can be seen in the regions outlined by white rectangles in Figs. 3(a) and (d). They are typically wider than 100 μm and range in length from few to 10ths of millimeters. These macropores are the result of variability in the width of the filament as it is being extruded from the tip of the nozzle. That variability is, in turn, likely the result of inconsistencies in the local properties of ink under extrusion.

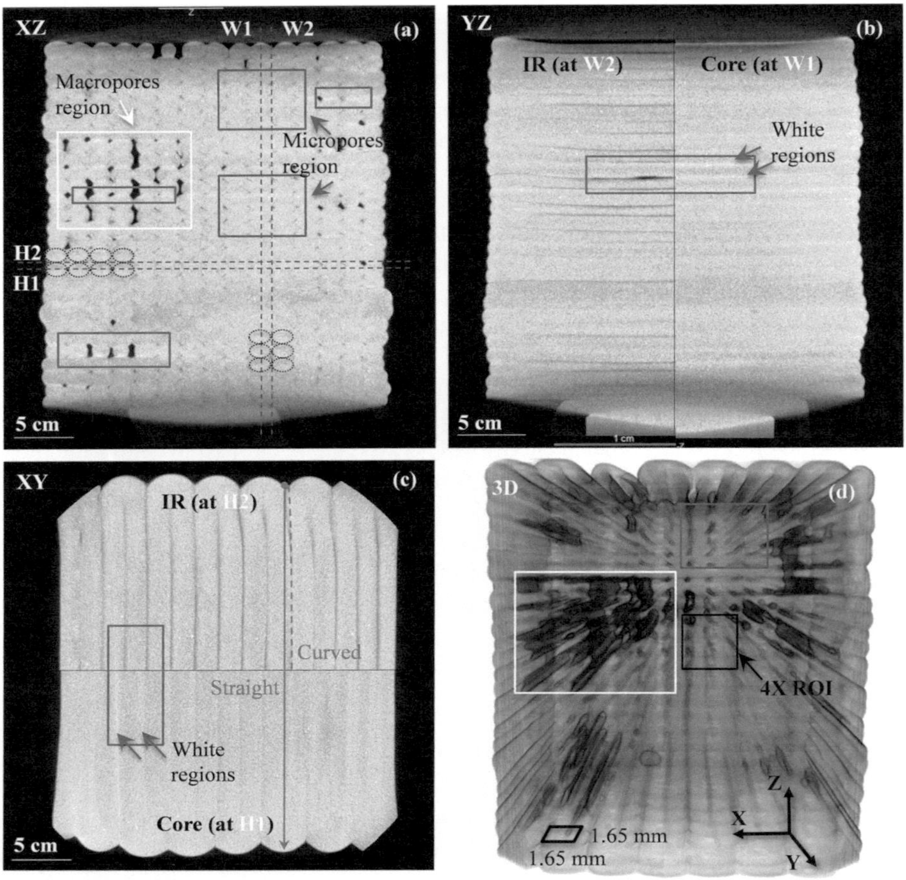

Fig. 3. X-ray micro-CT images of the microstructure of 3-days old paste specimen with lamellar architecture collected during 0.4X scan: (**a**) 2D projection on the XZ plane; (**b**) 2D projection on the YZ plane; (**c**) 2D projection on the XY plane, and; (**d**) Three-dimensional rendition of the entire volume of the specimen

Micropores (Micro-channels). The second type of pores was observed to exist between adjacent filaments of the adjacent layers (see the regions outlined by blue rectangles shown in Figs. 3(a) and (d). These micropores (or micro-channels) are smaller than 100 μm in diameter but can be 10ths of millimeters long. They are located between filaments along the Y axis (i.e., they are present in the interfacial regions (IRs) of the filaments). The horizontal (i.e., XY) slices H1 and H2 shown in Fig. 3(c), correspond to, respectively, the core (i.e., through the center) and interfacial regions of the filaments as indicated in cross-sectional view (XZ) in Fig. 3(a). Similarly, vertical (YZ) slices W1 and W2 shown in Fig. 3(b) also correspond to the core and interface regions of the filaments indicated in Fig. 3(a). Analysis of Fig. 3(b) and (c) indicate that micro-channels (appearing in these Figures as darker regions within the matrix) are only present in the images representing slices through interfaces (i.e. they are absent from images representing slices through the cores). It should be noted, however, that these micro-channels are more pronounced in horizontal than in the vertical planes. This can be clearly seen by comparing images of the IR H2 (shown in Fig. 3(b) and IR W2 (shown in Fig. 3(c)).

Finally, one would expect that the gap formed between four semi-circular filaments will have a diamond-shaped cross-section (when viewed along the axes of the filaments). However, as illustrated in Figs. 4 and 5, this was not the case for the lamellar architecture specimen prepared for this study as the gaps have been found to be triangular in shape. The next sections describe two other characteristics of the printed microstructure (rearrangement of the filaments and accumulation of anhydrous cement grains near macropores), which are considered to be responsible for the formation of these triangular gaps.

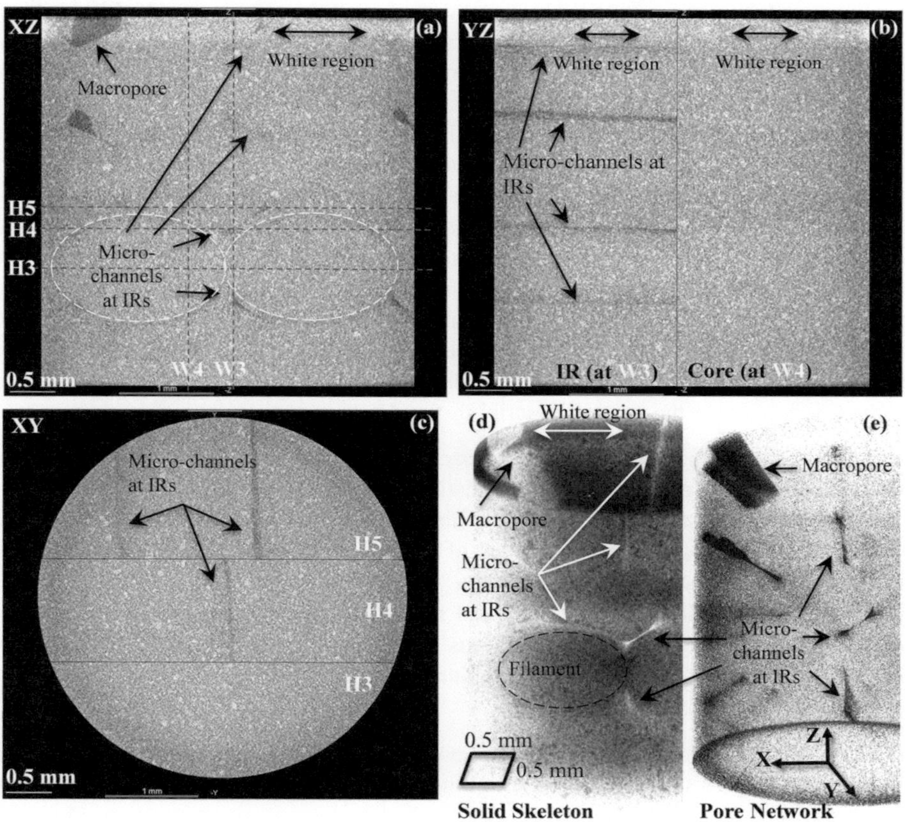

Fig. 4. X-ray micro-CT images of the microstructure of 3-days old intact specimen with lamellar architecture collected during 4X scan: (**a**) XZ; (**b**) YZ; (**c**) XY planes; (**d**) 3D image of the solid skeleton; (**e**) 3D image of pore network

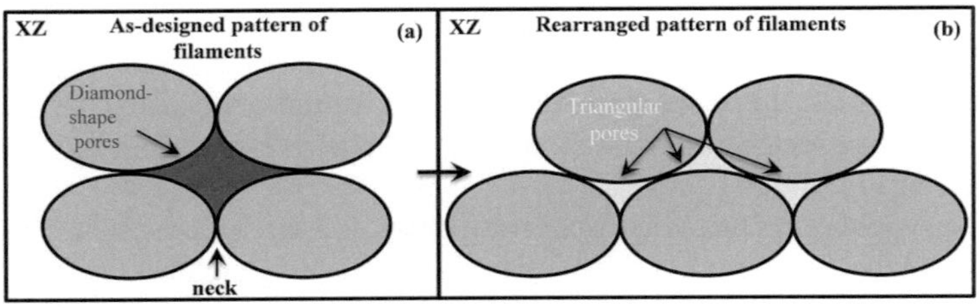

Fig. 5. Schematic illustration of arrangements of filaments and shapes of the pores along the Y direction: (**a**) As-designed pattern in lamellar architecture containing diamond-shaped pore and; (**b**) Rearranged filament pattern showing triangular-shaped pores (similar to those depicted in 4X CT image (Fig. 4(e))

Self-Drifting of Filaments from the Programmed Toolpath. The examination of CT images revealed that the internal filaments of the lamellar architecture self-drifted (to an extent of about half width of the filament) from their targeted (programmed) toolpath. From the top view (i.e., XY plane), the filaments in all interior layers were observed to follow slightly curved, rather than straight, paths. This is in spite of the fact that the nozzle was programmed to move along a linear tool path parallel to the Y axis. This observation was made possible due to the differences in the gray level intensities between the signals from the core sections and interfaces of the filaments. A typical curvature is highlighted in Fig. 3(c) by the set of two orange lines: the solid line on the left showing the location where a straight interface should have been observed and the dashed line on the right that shows the actual (slightly curved) path of the actual interface. In the consecutive layers of the microstructure, the direction of this curvature successively changes to left and right. This has to do with the opposite direction of printing path in successive layers. The causes of this self-drifting of the filaments are further elaborated in the section describing the 4X scan.

Accumulation of Anhydrous Cement Grains near Macropores ('White Regions'). The 0.4X CT scans also revealed the accumulation of the unhydrated (anhydrous) cement grains in the IRs near the macropores. This phenomenon was observed in both, the horizontal and vertical interfacial planes of lamellar architecture. In addition, the 4X scan demonstrated the presence of a subtle, 'brighter' zone at the horizontal interfaces, typically also near the macropores. These brighter regions of the microstructure are referred to in this paper as 'white regions' and are outlined by green rectangles in Figs. 3(a), (b) and (c). The more in-depth analysis of these white regions is presented in the section describing the 4X scan.

4X Scan. The 3D scan of the specimens resulting from the 0.4X scan shown in Fig. 3 (d) has been used to select an internal region of interest (ROI) to be evaluated at higher (i.e., 4X) magnification. That ROI is outlined by a black square rectangle in Fig. 3(d) and the details are presented in Fig. 4. The examination at the magnification 4X was performed to further explore the four types of microstructural features discussed during the analysis of the results from the 0.4X scan. Specifically, the 4X CT scan provided additional information on the shape of the micro-channels and their connectivity in the

IRs, the rearrangement of the paths of the filaments, and the nature of white regions. The 2D projections of individual XZ, YZ, XY planes are shown in, respectively, Figs. 4(a), (b) and (c); Fig. 4(d) shows the 3D rendition of the interior of the specimen. The higher resolution of these images (compared to the resolution obtained during the 0.4X scans), allowed for the identification of larger cement grains (they appear as bright spots against the gray matrix of the hydrated products) and pores (they appear as dark spots).

Macropores. The previously mentioned ROI was selected to capture the bottom edge of a typical macropore between two filaments of the same layer. An example of such pore is shown in the top left corner of Fig. 4(a). This macropore is also shown in the image of the solid skeleton (Fig. 4(d)) and that of the pore network (Fig. 4(e)).

Micropores (Micro-channels). This section provides further analysis of the shape of the micro-channels as observed in 4X images. Specifically, it can be observed that the triangular micro-channels ((Fig. 4(a)) are commonly connected to one another and, when viewed in 3D, they form an inclined dog-bone shape as illustrated in Fig. 4(d) and (e). As filaments rearrange upon deposition and move closer together, the connectivity between the micro-channels can be facilitated. Thorough investigation of the 4X scanned volume qualitatively indicated that this connectivity between micro-channels occurs through IRs and form pattern of pore network aligned with the filament architecture (Fig. 4(d) and (e)). The typical three horizontal (XY) slices of H3, H4, H5 (indicated in Fig. 4(a)) and shown in Fig. 4(c)), demonstrate the homogeneous characteristics of the microstructure along the 'cores' (H3), compared to heterogeneous characteristic along IRs where micro-channels are present (H4, H5). Similarly, the vertical (YZ) slices of W3 and W4 (indicated in Fig. 4(a)) and shown in Fig. 4(b)), demonstrate the homogeneous characteristic of the microstructure along the 'cores' (W4), compared to the heterogeneous characteristic of the microstructure along 'IRs' (W3).

Self-Drifting of Filaments from the Programmed Toolpath. As discussed in the section presenting the 4X scan, the triangular cross-sectional shape of the micro-channels is correlated with the rearrangement of the filaments. This indicates that the filaments are shifted upon deposition towards an adjacent filament deposited before them. This was confirmed via visual observations during printing. Rearrangement of the filament toward a higher packing arrangement shown in Fig. 4(e) further confirms this observation. As the filaments rearrange from their designed path, the shape of the pores changes from diamond to triangle as schematically shown in Fig. 5(a) and (b). The similarity between the triangular cross-sectional shapes of micro-channels shown in Fig. 4(e) and the illustration presented in Fig. 5(b) confirms that the formation of the triangular-shape pores is the result of self-drifting of the filaments.

A variety of mechanisms for interaction and driving force for filament rearrangements can be hypothesized. As an example, during deposition, each filament interacts with the adjacent filament and therefore is shifted towards it. As each filament is deposited, it is designed to overlap and make contact with its adjacent filaments. In addition, the phenomenon such as die swell and relaxation of the filament due to gravity upon deposition may facilitate this contact. Once the filaments make contact, the viscous flow can be driven by the differences in surface curvature between the body of the filament and its neck regions (see Fig. 5(a)), and can bring the adjacent filaments closer. In addition, the rearrangement may cause lamellar architecture to achieve a higher packing density and possibly establish a densifying mechanism of viscoelastic materials upon deposition.

Accumulation of Anhydrous Cement Grains near Macropores ('White Regions'). The subtle trace of 'white region' at the horizontal interfaces discussed in connection with the 0.4X scan is captured in greater details in 4X (see the top parts of Fig. 4(a) and (b)). A thorough investigation of this region demonstrates that the white region contained a higher fraction of anhydrous cement grains, which explains why they appear brighter as cement grains have higher density relative to the density of the hydration products and pores. Similarly, the images of the solid skeleton (shown in Fig. 4(d)) and that of the pore network (shown in Fig. 4(e)), both illustrate the accumulation of anhydrous cement grains in the white region. The fact that white regions are only seen in the horizontal planes suggests that these locations offer preferential drying environment during the printing of each layer. The presence of the macropores may have additionally enhanced drying. The use of ink with a low water to cement ratio of 0.275 can also be contributing to the increased rate of drying at the horizontal IRs due to longer exposure time to air drying compared to vertical IRs.

3.2 Cast Specimen in 0.4X Micro-CT (0.4X Scan)

0.4X Scan. The 0.4X scan of cast specimen showed the random distribution of the pore network. The XY plane of 0.4X scan of intact cast specimen is illustrated in Fig. 6 (a) whereas Fig. 6(b) shows the 3D rendition of the interior of the specimen. The darker regions in both images represent pore regions. The random distribution of pores in cast elements and the lack of patterned heterogeneous network compared to those observed in lamellar architecture (Fig. 3), highlight the differences between the microstructure of traditionally cast and 3D-printed hcp specimen.

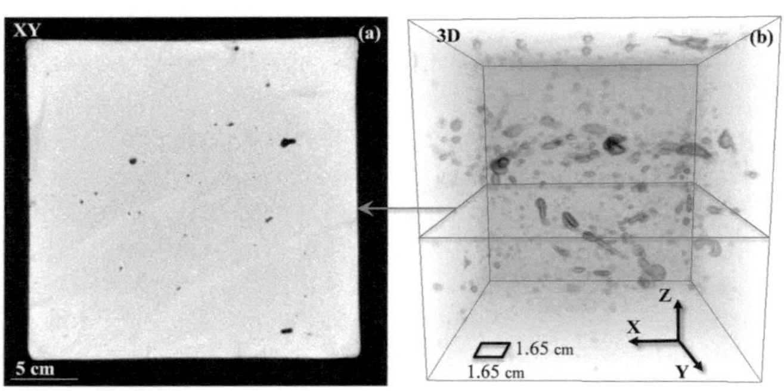

Fig. 6. X-ray micro-CT images of the 3-days old intact cast specimen collected during the 0.4X scan: (**a**) 2D projection on the XY plane and; (**b**) 3D rendition of the entire specimen

Further Discussion. The alignment of macropores and micro-channels with the pattern of the filaments in the lamellar architecture (Figs. 3(d) and 4(e)) suggests that the pore network is inherently associated with the specific architecture of the specimen. However, it is also possible that the location and the amount of the macropores are linked to the specific printing parameters used in this study.

From 0.4X scan, it is evident that the characteristics of the microstructure are different between the solid cores and IRs, induced by the presence of interfacial porosity and the differences in its distribution. It could be hypothesized that the heterogeneous micropore network is present due to the initial presence of extrusion-induced lubricating layer (i.e. layer containing water that surrounds the filament upon deposition) which makes the outer region of the filaments more prone to evaporation, thus resulting in the creation of the micropores at IRs. The evaporation and the subsequent wetting at the interface due to deposition of successive layers could also give rise to anisotropic properties of the microstructure and 3D-printed elements. It is also considered that the presence of triangular micro-channels, as shown in Fig. 5(b), could be induced during deposition. Given the round shape of the nozzle, air and lubricating water surrounding the filaments can be trapped in between filaments upon deposition of each layer. The presence of lubricating layer followed by drying mechanisms during the print and continuing hydration of cement paste can leave micropores at the IRs across the circumferential zone of the filament as indicated by the dashed oval in Fig. 4 (e). Horizontal interfaces were exposed to air for a longer period of time than vertical interfaces during the printing process.

The higher amount of horizontal (dark) IRs compared to (also dark) vertical IRs (Fig. 3(c) and (b)) indicates the higher amount of drying and resulting micropores in horizontal planes (XY). This observation indicates the existence of the correlation between exposure time and formation of heterogeneous IRs in different planes, which can result in anisotropy of mechanical properties of the 3D-printed specimens. The accumulation of anhydrous cement grains near macropores in horizontal IRs could be the result of the higher amount of drying at the horizontal IRs. These IRs are shown to be connected to one another through the micro-channels in the microstructure of 3D-printed lamellar architecture as shown in Fig. 4(e). This heterogeneously patterned pore network, together with the presence of macropores, causes anisotropy in the 3D-printed elements. This anisotropy can lead to differences in mechanical properties between the 3D-printed specimen and the cast specimens, depending on the architecture of materials and the interfacial (bond) strength of 3D-printed element in different directions. The presence of the interface is very important as IRs follows the pattern of the filament and overall layered architecture of the element. The architecture can then be designed to allow damage and micro-cracking to be promoted at the weak IR to achieve higher fracture resistance without sacrificing the strength [24].

It was previously discussed that in 3D-printed hcp elements, a wide variety of heterogeneous features could exist over a broad range of scales [21–23]. Application of micro-CT characterization technique to 3D printed lamellar architecture demonstrated the presence of the weak IRs induced by the processing and patterned via architecture. In addition, rearrangement of the filaments and the resulting change in morphology of the pores was revealed. Connectivity of micropores was discovered in the 4X scan in 3D-printed architecture. As revealed by comparing the 0.4X scans of the printed and cast specimens, their microstructure is very different. These observations reveal the role of processing in determining the microstructure of materials.

4 Summary

- Micro-CT images of 3D-printed lamellar architecture collected during this study revealed four characteristic features of the microstructure not observed in the cast specimens, thus indicating these features resulted from the processing of material.
- The previously mentioned microstructural features included the following: (a) macropores (i.e., visible gaps formed during printing), (b) micropores at interfacial regions (IRs) of filaments in the form of micro-channels smaller than 100 μm, (c) self-rearrangement of filaments from their designed toolpath, and (d) high accumulation of anhydrous cement particles near the large pores. All of these four features were qualitatively detected at the 0.4X scans, and further verified at the 4X scan. The presence of these features could play a role in determining the overall mechanical response of architectured hcp specimen.
- Pore network (at both macro and micro scale) appeared to be aligned with respect to the direction of the filaments in the lamellar architecture of the printed specimen.
- Micro-CT demonstrated great aptitude (as non-destructive technique) for capturing spatial heterogeneities of the microstructure of 3D-printed elements as large as 10ths of mm. The dual-stage magnifications system used in this study facilitated higher resolutions at large working distance, thus allowing CT of larger specimens.

Acknowledgements. The authors gratefully acknowledge generous support of this research by the National Science Foundation (CMMI Grant 1562927).

References

1. Gallucci, E., Scrivener, K., Groso, A., Stampanoni, M., Margaritondo, G.: 3D experimental investigation of the microstructure of cement pastes using synchrotron X-ray microtomography (μCT). Cem. Concr. Res. **37**(3), 360–368 (2007)
2. de Wolski, S.C., Bolander, J.E., Landis, E.N.: An in-situ X-ray microtomography study of split cylinder fracture in cement-based materials. Exp. Mech. **54**(7), 1227–1235 (2014)
3. Scrivener, K.L.: Backscattered electron imaging of cementitious microstructures: understanding and quantification. Cem. Concr. Compos. **26**(8), 935–945 (2004)
4. Lu, S., Landis, E.N., Keane, D.T.: X-ray microtomographic studies of pore structure and permeability in Portland cement concrete. Mater. Struct. **39**(6), 611–620 (2006)
5. Diamond, S., Landis, E.: Microstructural features of a mortar as seen by computed microtomography. Mater. Struct. **40**(9), 989–993 (2007)
6. Landis, E.N., Nagy, E.N., Keane, D.T.: Microstructure and fracture in three-dimensions. Eng. Fract. Mech. **70**(7–8), 911–925 (2003)
7. Landis, E.N., Nagy, E.N., Keane, D.T., Nagy, G.: Technique to measure 3D work-of-fracture of concrete in compression. J. Eng. Mech. **125**(6), 599–605 (1999)

8. Landis, E.N., Keane, D.T.: X-ray microtomography for fracture studies in cement-based materials. In: Developments in X-Ray Tomography II, pp. 105–114. International Society for Optics and Photonics (1999)
9. Landis, E.N., Zhang, T., Nagy, E.N., Nagy, G., Franklin, W.R.: Cracking, damage and fracture in four dimensions. Mater. Struct. **40**(4), 357–364 (2007)
10. Sugiyama, T., Promentilla, M.A.B., Hitomi, T., Takeda, N.: Application of synchrotron microtomography for pore structure characterization of deteriorated cementitious materials due to leaching. Cem. Concr. Res. **40**(8), 1265–1270 (2010)
11. Gastaldi, D., Canonico, F., Capelli, L., Boccaleri, E., Milanesio, M., Palin, L., Croce, G., Marone, F., Mader, K.: In situ tomographic investigation on the early hydration behaviors of cementing systems. Constr. Build. Mater. **29**, 284–290 (2012)
12. Lewis, J.A.: Direct ink writing of 3D functional materials. Adv. Funct. Mater. **16**(17), 2193–2204 (2006)
13. Duoss, E.B., Weisgraber, T.H., Hearon, K., Zhu, C., Small, W., Metz, T.R., Vericella, J.J., Barth, H.D., Kuntz, J.D., Maxwell, R.S., Spadaccini, C.M.: Three-dimensional printing of elastomeric, cellular architectures with negative stiffness. Adv. Funct. Mater. **24**(31), 4905–4913 (2014)
14. Raney, J.R., Compton, B.G., Mueller, J., Ober, T.J., Shea, K., Lewis, J.A.: Rotational 3D printing of damage-tolerant composites with programmable mechanics. Proc. Natl. Acad. Sci. **115**(6), 1198–1203 (2018)
15. Aïtcin, P.C., Flatt, R.J. (eds.): Science and Technology of Concrete Admixtures. Woodhead Publishing, Sawston (2015)
16. Rueschhoff, L., Costakis, W., Michie, M., Youngblood, J., Trice, R.: Additive manufacturing of dense ceramic parts via direct ink writing of aqueous alumina suspensions. Int. J. Appl. Ceram. Technol. **13**(5), 821–830 (2016)
17. ASTM C150 – Standard Specification for Portland Cement (2017)
18. ASTM C494 – Standard Specification for Chemical Admixtures for Concrete (2016)
19. Moini M.: Master's thesis, University of Wisconsin-Milwaukee (2015)
20. Landis, E.N., Keane, D.T.: X-ray microtomography. Mater. Charact. **61**(12), 1305–1316 (2010)
21. Le, T.T., Austin, S.A., Lim, S., Buswell, R.A., Law, R., Gibb, A.G., Thorpe, T.: Hardened properties of high-performance printing concrete. Cem. Concr. Res. **42**(3), 558–566 (2012)
22. Spangenberg, J., Roussel, N., Hattel, J.H., Stang, H., Skocek, J., Geiker, M.R.: Flow induced particle migration in fresh concrete: theoretical frame, numerical simulations and experimental results on model fluids. Cem. Concr. Res. **42**(4), 633–641 (2012)
23. Choi, M., Roussel, N., Kim, Y., Kim, J.: Lubrication layer properties during concrete pumping. Cem. Concr. Res. **45**, 69–78 (2013)
24. Moini, M., Olek, J., Youngblood, J.P., Magee, B., Zavattieri, P.D.: Additive manufacturing and performance of architectured cement-based materials. Adv. Mater. https://doi.org/10.1002/adma.201802123

Hardened Properties of 3D Printable 'One-Part' Geopolymer for Construction Applications

Behzad Nematollahi(✉), Ming Xia, Shin Hau Bong, and Jay Sanjayan

Centre for Sustainable Infrastructure, Faculty of Science, Engineering and Technology, Swinburne University of Technology, Hawthorn, VIC 3122, Australia
bnematollahi@swin.edu.au

Abstract. This paper reports the hardened properties of an extrusion-based 3D printable 'one-part' geopolymer for construction applications. To date, all of the 3D printable geopolymers reported in the literature had 'two-part' mix formulations, made by using liquid activators. In contrast, the 3D printable geopolymer developed in this study has a 'one-part' (just-add-water) mix formulation, made by using a small amount of solid activator instead of the commonly used liquid activators. Handling a small amount of solid activators instead of large quantities of user-hostile liquid activators significantly enhances commercial viability and large-scale application of the 3D printable geopolymers in construction industry. Effects of print-time interval on the inter-layer strength, along with compressive and flexural strengths of the developed 3D printed 'one-part' geopolymer in different directions were investigated. Specimens were printed with 2 and 15 min delay times (print-time intervals). Compressive, flexural and inter-layer strengths of the 3D printed 'one-part' geopolymer were measured. The results showed that the print-time interval had a significant effect on the inter-layer strength of the 3D printed 'one-part' geopolymer. However, the effect of the print-time interval on the compressive and flexural strengths of the 3D printed 'one-part' geopolymer was negligible. The results also showed that the compressive and flexural strengths of the 3D printed 'one-part' geopolymer depended on the loading direction.

Keywords: 3D concrete printing · One-part geopolymer · Solid activator Inter-layer strength · Extrusion · Print-time interval

1 Introduction

Implementation of additive manufacturing techniques in construction industry such as 3D concrete printing (3DCP) opens many thresholds into the future of this industry. The 3DCP process is an additive, layer-by-layer based, manufacturing technique to build concrete components with complex geometries without the use of expensive formwork, which hugely enhances geometrical freedom, while drastically reducing construction time [1].

T. Wangler and R. J. Flatt (Eds.): DC 2018, RILEM Bookseries 19, pp. 190–199, 2019.
https://doi.org/10.1007/978-3-319-99519-9_17

In recent years, different techniques have been developed to adopt 3DCP in the construction industry, one of which is extrusion-based 3DCP. This technique is analogous to fused deposition modelling (FDM) method, which extrudes cementitious material from a nozzle mounted on a gantry, crane or a robotic arm to print a concrete component layer-by-layer. Example of such technique is Concrete Printing developed by Le et al. [2, 3].

One of the main limitations of the extrusion-based 3DCP is the limited scope of printable concretes. Conventional concrete in its current form is not suitable for extrusion-based 3DCP. Setting characteristics of ordinary Portland cement (OPC) limit its use for extrusion-based 3DCP. Therefore, researchers are trialling other types of cementitious materials such as sulphur aluminate cement [4]. 3D printable high-performance fibre-reinforced cement composite (HPFRCC) [2, 3] and ultra-high performance concrete (UHPC) [5] have also been explored. Nevertheless, the developed 3D printable HPFRCC and UHPC may not be considered as environmentally friendly, as they both contain a considerable amount of high early strength cement [2, 3, 5]. It is well established that OPC manufacturing is highly energy and emissions intensive. As such, it is urgently needed to expand the current severely limited range of printable concretes.

To tackle this limitation, the researchers at Swinburne University of Technology, Australia have been working to develop 3D printable geopolymers for construction applications. Geopolymer is a sustainable alternative to OPC. It is synthesized by alkali activation of aluminosilicate source materials such as fly ash and slag, which are industrial by-products of coal power stations and iron manufacture, respectively [6]. Previous studies reported that production of fly ash-based geopolymer has at least 80% less CO_2 emissions, and requires about 60% less energy, as compared to manufacture of OPC [7, 8]. A 3D printable fly ash-based geopolymer has recently been developed by the authors of this study [9, 10]. The effect of several mixture parameters such as type of activator, type of sodium silicate solution, mass ratio of sodium silicate to sodium hydroxide solutions, and activator to fly ash ratio on the fresh and hardened properties of the developed 3D printable fly ash-based geopolymer were investigated [9]. The authors of this study also investigated the effects of type of fiber on the inter-layer and flexural strengths of the developed 3D printable fly ash-based geopolymer [10].

However, one of the main obstacles for commercialization and widespread application of the developed 3D printable fly ash-based geopolymer is the use of corrosive and often viscouse liquid activators for its manufacture. Conventionally, geopolymer is synthesized from a 'two-part' mix formulation, including solid aluminosilicate source materials and liquid activators. The most important drawback with regards to the 'two-part' mix formulation is that in commercial and mass production of the 'two-part' geopolymer, handling large quantities of corrosive liquid activators is difficult [6].

This study aims to overcome the aforementioned obstacle by developing a 3D printable 'one-part' geopolymer, which as a 'just-add-water' mix formulation uses a small amount of solid activator instead of the commonly used liquid activators. The hardened properties of the developed 3D printable 'one-part' geopolymer were reported. Influences of print-time interval on the compressive, flexural and inter-layer strengths of the developed 3D printable 'one-part' geopolymer were investigated.

Effect of testing direction on the compressive and flexural strengths of the developed 3D printable 'one-part' geopolymer was also evaluated.

2 Experimental Procedures

2.1 Materials and Mixture Proportions

A low calcium (Class F) fly ash supplied from Gladstone power station in Queensland, Australia and a ground granulated blast furnace slag (henceforth referred to as slag) supplied from Independent Cement and Lime Pty Ltd., Australia were used in this study. Two types of silica sands with different particle sizes were used. The relatively coarser sand denoted as "CS" with an average particle size of 898 μm was supplied by Sibelco Australia Ltd. The relatively finer sand denoted as "FS" with an average particle size of 172 μm was supplied by TGS Industrial Sand Ltd. Anhydrous sodium metasilicate powder was used in this study as the solid activator. It was supplied by Redox, Australia with a chemical composition of 51 wt% Na_2O and 46 wt% SiO_2 (balance H_2O).

The mix proportions of the 3D printable 'one-part' geopolymer are given in Table 1. Several trials were conducted before identifying the final mix proportions given in Table 1, which satisfied the extrudability and buildability criteria of extrusion-based 3DCP.

Table 1. Mix proportions of 3D printable one-part geopolymer.

One-part geopolymer binder			Sand		Water
Fly ash	Slag	Solid activator[a]	"FS"	"CS"	
0.50	0.50	0.08	0.50	1.0	0.34

Note: All numbers are mass ratios of the precursor weight (fly ash+slag).
[a]Composed of anhydrous sodium metasilicate powder.

2.2 Mixing, Printing, Curing and Testing of Specimens

To prepare the 'one-part' geopolymer mixture, fly ash, slag and solid activator were added to a Hobart mixer and dry mixed for approximately two minutes. Tap water was then gradually added to the mix and the mixing was continued for another ten minutes to obtain a consistent fresh state.

Figure 1a presents a schematic illustration of the extrusion-based 3DCP process. The extrusion-based 3DCP process was simulated by using a small scale custom-made 3D printer.. A 3D printed 45° nozzle with a rectangle opening of 30 mm × 15 mm was connected to the end of a piston-type extruder, in which the fresh material was extruded through a metallic cylinder with the dimensions of 50 mm × 600 mm (diameter × length). The first layer measuring 200 mm (L) × 30 mm (W) × 15 mm (H) was printed, on top of which the second layer was printed after the selected delay

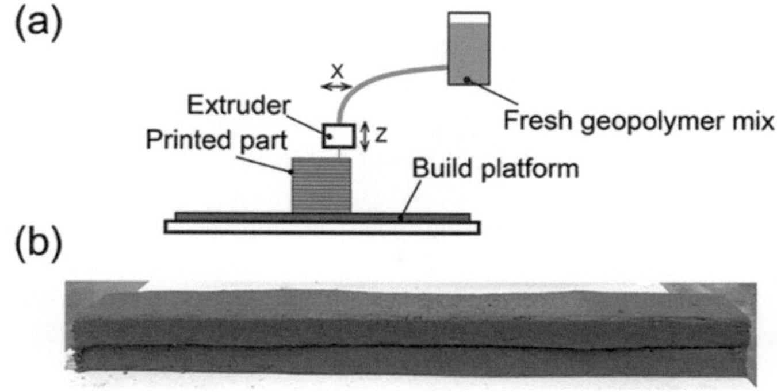

Fig. 1. (a) Schematic illustration of the extrusion-based 3DCP process, (b) a two-layer printed filament with the dimensions of 200 mm (L) × 30 mm (W) × 30 mm (H).

times (i.e. the print-time intervals between the layers), namely 2 min and 15 min. Figure 1b shows a two-layer printed filament of the developed 3D printable 'one-part' geopolymer.

Heat curing was adopted in this study to accelerate the curing process. It should be noted that the developed mixture shown in Table 1 can also be cured at ambient temperature. For the heat curing, at the end of printing process all printed filaments were placed in a container and sealed to minimize excessive moisture loss and placed in an oven at 60 °C for 24 h. At the end of heat curing period, the filaments were removed from the oven and stored in the laboratory at ambient temperature (23 °C ± 3 °C) until being cool. Previous studies reported that strength of geopolymer after completion of the heat curing does not change significantly over time [11–14]. Thus, in this study all specimens were tested after completion of the heat curing (i.e. one-day after printing).

The compressive strength of 3D printed 'one-part' geopolymer was measured by extracting 30 mm cube specimens from the 200 × 30 × 30 mm printed filaments and testing them in perpendicular, lateral and longitudinal directions (Fig. 2a). For each direction, at least six specimens were tested. The load rate was 20 MPa/min.

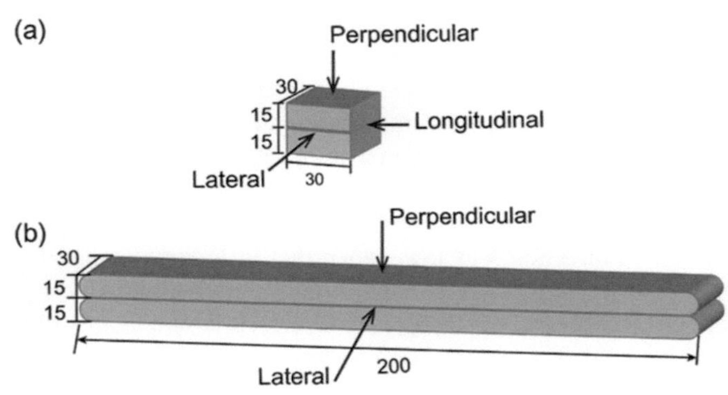

Fig. 2. Schematic illustration of specimens and testing directions for (a) compression and (b) flexural tests of 3D printed 'one-part' geopolymer.

The flexural strength of 3D printed 'one-part' geopolymer was measured by testing at least six specimens measuring 200 × 30 × 30 mm in perpendicular and lateral directions (Fig. 2b). A three-point bending test setup with a span of 150 mm was used to test the specimens under displacement control at the rate of 1.0 mm/min. Surfaces of all specimens tested prepared for compression and flexural tests were ground to have a smooth and flat surface.

Figure 3 shows the test setup and the specimen prepared for the inter-layer strength measurement. Specimens measuring 50 × 30 × 30 mm were extracted from the 200 × 30 × 30 mm printed filaments. For each delay time, at least six specimens were prepared. Two metallic T-sections were glued on top and bottom of the specimen using epoxy resin. A small notch with an approximate depth of 5 mm was made at the interface of the layers on both cross sections of the specimen to assure failure of the specimen at the interface. The test rate was 1 mm/min. Care was taken to ensure the specimen is aligned in the testing machine to prevent any eccentricity.

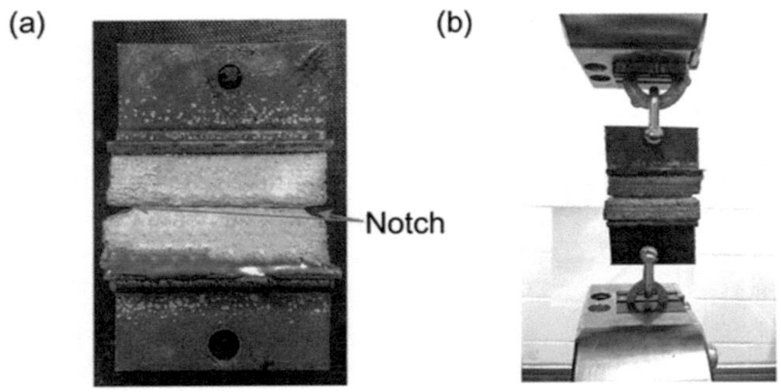

Fig. 3. (a) Test specimen for measuring the inter-layer strength, and (b) Test setup.

3 Results and Discussion

3.1 Compressive Strength

The compressive strength results of the 3D printed 'one-part' geopolymer are presented in Fig. 4. The compressive strength of the printed 'one-part' geopolymer exhibited an anisotropic phenomenon depending on the loading direction. The previous studies (Le et al. [3], Marchment et al. [15] and Sanjayan et al. [16]) also reported similar anisotropic behavior for the compressive strength of the 3D printed OPC concrete. Irrespective of the delay time, the mean compressive strength of the 3D printed 'one-part' geopolymer in lateral direction was lower than that in longitudinal and perpendicular directions. The highest compressive strength was in the longitudinal direction because the fresh geopolymer material was subjected to high pressure in this direction during the printing process [16]. The compressive strength in the perpendicular direction was in between the longitudinal and lateral directions. This is because the fresh geopolymer material was subjected to some level of pressure during the setting process due to the self-weight of the material. The lowest compressive strength was obtained in the lateral direction. Unlike the longitudinal and perpendicular directions, no pressure was exerted on the fresh geopolymer in this direction. Thus, the fresh geopolymer material was free to expand and settle in this direction, which resulted in lower compressive strength in this direction [16]. The trends observed in this study regarding to the compressive strength of the printed 'one-part' geopolymer in different testing directions was consistent with those reported by Sanjayan et al. [16] for 3D printed OPC concrete.

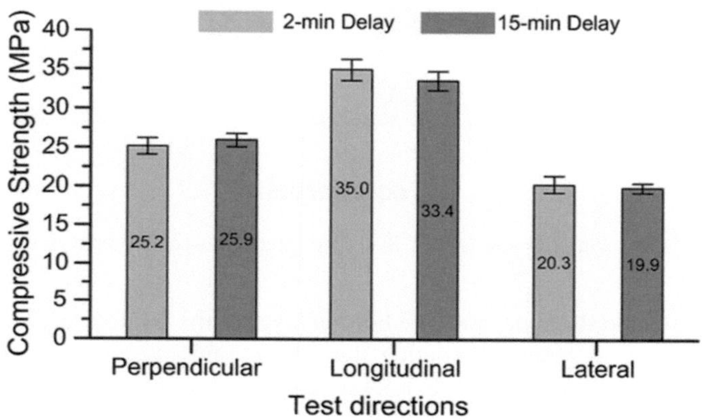

Fig. 4. Compressive strength of the 3D printed 'one-part' geopolymer.

According to Fig. 4, the mean compressive strength of the samples printed with 15 min and 2 min delay times were comparable. Therefore, it can be said that the delay time did not have a significant effect on the compressive strength of the printed 'one-part' geopolymer, regardless of the testing direction.

3.2 Flexural Strength

The flexural strength results of the 3D printed 'one-part' geopolymer are presented in Fig. 5. Similar to the compressive strength results (Fig. 4), the flexural strength of the printed 'one-part' geopolymer also showed the anisotropic behavior depending on the testing direction. The previous studies (Le et al. [3] and Sanjayan et al. [16]) also reported similar anisotropic behavior for the flexural strength of the 3D printed OPC concrete. The anisotropic behaviour of the compressive and flexural strengths of the printed 'one-part' geopolymer are considered to be an inherent characteristic of the layer-by-layer printing process [17]. It should be noted that the extent of this effect quantitatively is quite dependent on the specifics of the print process, material and facility.

According to Fig. 5, regardless of the delay time, the mean flexural strength of the 3D printed 'one-part' geopolymer in perpendicular direction was significantly higher than that in lateral direction. This result is in good agreement with the results obtained by Sanjayan et al. [16] for 3D printed OPC concrete. When testing in perpendicular direction, the flexural strength is governed by the mid-span of the bottom layer where

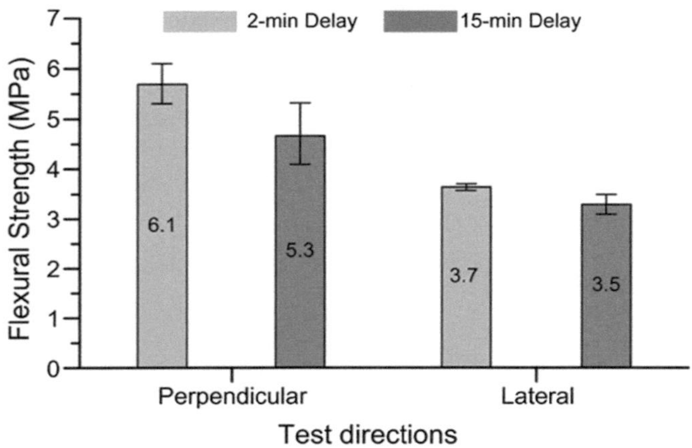

Fig. 5. Flexural strength of the 3D printed 'one-part' geopolymer.

the maximum tensile stress occurs. The bottom layer of the printed specimen was most probably subjected to good compaction due to the self-weight of the second layer, leading to higher load capacity of the bottom layer. This resulted in the higher flexural strength in perpendicular direction than lateral direction.

As can be seen in Fig. 5, the mean flexural strength of the printed 'one-part' geopolymer with 2 min delay time was slightly higher (6–15% depending on the testing direction) than that with 15 min delay time. Thus, it can be said that the delay time did not have a significant effect on the flexural strength of the printed 'one-part' geopolymer. This is consistent with the compressive strength results.

According to Fig. 5, the mean flexural strength of the printed 'one-part' geopolymer in perpendicular direction was 51–65% higher than that in lateral direction depending on the delay time. However, as shown in Fig. 4, the mean compressive strength of the printed 'one-part' geopolymer in perpendicular direction was only 24–30% higher than that in lateral direction depending on the delay time. Thus, it can be concluded that the extrusion-based 3DCP process adopted in this study introduced relatively little anisotropic behavior on the compressive strength of the printed 'one-part' geopolymer as compared to the flexural strength.

3.3 Inter-layer Strength

The inter-layer strength results of the 3D printed 'one-part' geopolymer are presented in Fig. 6. The inter-layer strength of the samples printed with 2 min delay time was 63% higher than that of the samples printed with 15 min delay time. This was expected because increasing the delay time reduces the adhesion between the top and bottom layers. This is consistent with the results reported by Le et al. [3], where the inter-layer strength of printed concrete reduced when delay time increased. It should be noted that all specimens failed at the interface between the layers, regardless of the delay time. In addition, it should also be pointed out that the inter-layer strength of 0.8–1.3 MPa obtained in this study was sufficiently high to prevent interfacial shear failure. This is supported by the mode of failure during flexural tests, where the flexural failures of the specimens tested in perpendicular direction were governed by the tensile strength of the bottom layer rather than the inter-layer shear strength.

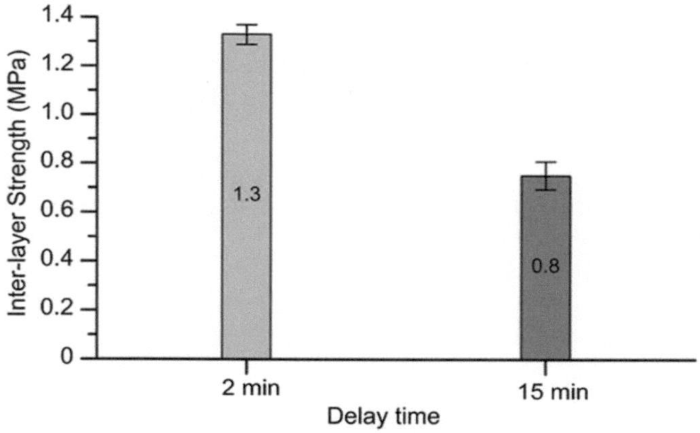

Fig. 6. Inter-layer strength of the 3D printed 'one-part' geopolymer.

4 Conclusions

A novel 3D printbale 'one-part' geopolymer suitable for extrusion-based 3D concrete printing for construction applications was developed in this study. As a 'just-add-water' mix formulation, a small amount of solid activator was used to make the developed 3D printable 'on-part' geopolymer to overcome the difficulties associated with handling large quantities of user-hostile liquid activators. This significantly increases the commercial feasibility of using 3D printable geopolymers in the construction industry. The effects of delay time and testing direction on the hardened properties of the developed 3D printable 'one-part' geopolymer including compressive, flexural and interlayer strengths were investigated. The following conclusions are drawn:

(1) The compressive strength of the 3D printable 'one-part' geopolymer exhibited an anisotropic strength behavior, depending on the loading direction. The maximum mean compressive strength was obtained in the longitudinal direction, followed by the perpendicular and lateral directions. This trend was true regardless of the delay time.
(2) Similar to the compressive strength, the flexural strength of the 3D printable 'one-part' geopolymer measured in the perpendicular and lateral directions also exhibited an anisotropic strength behavior. The flexural strength in the perpendicular direction was higher than in the lateral direction.
(3) The anisotropic strength behavior was more pronounced in the flexural than the compressive strength. The extrusion-based 3DCP process adopted in this study

introduced relatively more anisotropic behavior on the flexural strength than the compressive strength of the 3D printed ‘one-part’ geopolymer.

(4) The inter-layer strength of the 3D printable ‘one-part’ geopolymer was considerably decreased by increasing the delay time from 2 min to 15 min. However, the measured inter-layer strength of 0.8–1.3 MPa was sufficiently high to prevent interfacial shear failure. This is supported by the mode of failure during flexural tests, where the flexural failures of the specimens were governed by the tensile strength of the bottom layer rather than the inter-layer shear strength.

Acknowledgements. Authors acknowledge the support by the Australian Research Council Discovery Grant DP170103521, Linkage Infrastructure Grant LE170100168, and Discovery Early Career Researcher Award DE180101587.

References

1. Nematollahi, B., Xia, M., Sanjayan, J.: Current progress of 3D concrete printing technologies. In: 34th International Symposium on Automation and Robotics in Construction (ISARC 2017), Taiwan, pp. 260–267 (2017). https://doi.org/10.22260/ISARC2017/0035. Proceedings published by Tribun EU, s.r.o., Brno, 2017, ISBN: 978-80-263-1371-7
2. Le, T.T., Austin, S.A., Lim, S., Buswell, R.A., Gibb, A.G., Thorpe, T.: Mix design and fresh properties for high-performance printing concrete. Mater. Struct. **45**, 1221–1232 (2012)
3. Le, T.T., Austin, S.A., Lim, S., Buswell, R.A., Law, R., Gibb, A.G.F., Thorpe, T.: Hardened properties of high-performance printing concrete. Cem. Concr. Res. **42**, 558–566 (2012)
4. Lin, X., Zhang, T., Huo, L., Li, G., Zhang, N., Juan, L., Ji, W., Wang, B.: Preparation, properties and application of cement-based building 3D printing materials. Concr. Aust. **42**(3), 59–67 (2016)
5. Gosselin, C., Duballet, R., Roux, P., Gaudillière, N., Dirrenberger, J., Morel, P.: Large-scale 3d printing of ultra-high performance concrete—a new processing route for architects and builders. Mater. Des. **100**, 102–109 (2016)
6. Nematollahi, B., Sanjayan, J., Shaikh, F.U.A.: Synthesis of heat and ambient cured one-part geopolymer mixes with different grades of sodium silicate. Ceram. Int. **41**, 5696–5704 (2015)
7. Duxson, P., Provis, J.L., Lukey, G.C., Van Deventer, J.S.: The role of inorganic polymer technology in the development of ‘green concrete’. Cem. Concr. Res. **37**, 1590–1597 (2007)
8. Li, Z., Ding, Z., Zhang, Y.: Development of sustainable cementitious materials. In: International Workshop on Sustainable Development and Concrete Technology, Beijing, pp. 55–76 (2004)
9. Nematollahi, B., Vijay, P., Sanjayan, J., Xia, M., Nerella, V.N., Mechtcherine, V.: Systematic approach to develop geopolymers for 3D concrete printing applications. Arch. Civ. Mech. Eng. (2018, submitted)
10. Nematollahi, B., Xia, M., Sanjayan, J., Vijay, P.: Effect of type of fiber on inter-layer bond and flexural strengths of extrusion-based 3D printed geopolymer. In: Proceedings of the 2nd International Conference on Advanced Manufacturing and Materials (ICAMM), Tokyo (2018)
11. Hardjito, D., Wallah, S.E., Sumajouw, D.M.J., Rangan, B.V.: On the development of fly ash-based geopolymer concrete. ACI Mater. J. **101**, 467–472 (2004)

12. Nematollahi, B., Qiu, J., Yang, E.-H., Sanjayan, J.: Microscale investigation of fiber-matrix interface properties of strain-hardening geopolymer composite. Ceram. Int. **43**, 15616–15625 (2017)
13. Nematollahi, B., Sanjayan, J., Qiu, J., Yang, E.-H.: Micromechanics-based investigation of a sustainable ambient temperature cured one-part strain hardening geopolymer composite. Constr. Build. Mater. **131**, 552–563 (2017)
14. Nematollahi, B., Sanjayan, J., Qiu, J., Yang, E.-H.: High ductile behavior of a polyethylene fiber-reinforced one-part geopolymer composite: a micromechanics-based investigation. Arch. Civ. Mech. Eng. **17**(3), 555–563 (2017)
15. Marchment, T., Xia, M., Dodd, E., Sanjayan, J., Nematollahi, B.: Effect of delay time on the mechanical properties of extrusion-based 3D printed concrete. In: 34th International Symposium on Automation and Robotics in Construction (ISARC 2017), pp. 240–245. Taiwan (2017). https://doi.org/10.22260/ISARC2017/0032. Proceedings published by Tribun EU, s.r.o., Brno, 2017, ISBN: 978-80-263-1371-7
16. Sanjayan, J., Nematollahi, B., Xia, M., Marchment, T.: Effect of surface moisture on interlayer strength of 3D printed concrete. Constr. Build. Mater. **172**, 468–475 (2018)
17. Oxman, N., Tsai, E., Firstenberg, M.: Digital anisotropy: a variable elasticity rapid prototyping platform. Virtual Phys. Prototyp. **7**, 261–274 (2012)

Bond Strength in 3D Printed Geopolymer Mortar

Biranchi Panda([✉]), Nisar Ahamed Noor Mohamed, Yi Wei Daniel Tay, and Ming Jen Tan

Singapore Centre for 3D Printing, School of Mechanical and Aerospace Engineering, Nanyang Technological University, Singapore, Singapore
biranchi001@e.ntu.edu.sg

Abstract. 3D printing is getting significant attention in the construction industry. This technology, which has been talked for years, is now delivering tangible results, however it is not yet ready for mass production in mainstream construction. Thixotropy *material*, *3D printer* together with *3-dimensional computer model* are the key elements required for robust concrete printing. For a freshly printed material, the increase in structural buildup at rest, prior to the placement of a successive layer can result poor interlayer bond strength and therefore, in this paper, we aimed to investigate the *structuration effect* of a nano-clay modified geopolymer, used for 3D concrete printing. Different structuration rate was achieved by changing molar ratio of activator and experimental results conclude that, there exists an optimum printing zone for different molar ratios, beyond which interlayer bond strength will be very weak.

Keywords: Digital construction · Geopolymer · Structural buildup Bond strength

1 Introduction

With the rise of digital tools and technologies in the wider world around us, innovation in terms of digitalizing our construction process has come true with the advent of "3D concrete printing". This technology has proven its potential in redefining the traditional moulding process into layer by layer printing of concrete, without the need of human intervention [1]. Printable material development is one of the most challenging task for 3D printing since it involves the material characteristics to be highly stiff (yield stress) at rest and low viscous during flowing [2]. In cement science, thixotropy material is well known to possess such unique properties that can be directly used in concrete printing. However, the stiffness i.e. associated with material yield stress is not a constant parameter, rather it increases with time due to hydration and ongoing colloidal reaction. For regular Portland cement, it has been found that, the increase in yield stress is linear in dormant period, followed by an exponentially rise till setting of the material [3, 4]. This change of yield stress can cause serious problem in multi-layer casting, if the rate of increment i.e. structuration rate is too fast compare to delay time between subsequent layers [5]. In hot atmospheric condition, this effect is more significant due to rapid increase in structuration rate by the outside temperature. Roussel et al. [6] also

T. Wangler and R. J. Flatt (Eds.): DC 2018, RILEM Bookseries 19, pp. 200–206, 2019.
https://doi.org/10.1007/978-3-319-99519-9_18

found similar consequences for distinct-layer casting of SCC. They reported that there exists a critical delay between layers of casting, above which, it generates loss of mechanical strength.

In 3D concrete printing, similar phenomena like distinct-layer casting, was noticed where extruded material is deposited in layer by layer manner and therefore, it is interesting to investigate the effects of structuration rate on bond strength of the printed samples, processed by any extrusion-based technique. It is important to note that bond strength between two layers also depends on some other factors like interface adhesion, friction, aggregate interlock, and surface condition of the samples which is not studied in this paper [7]. Alternatively, we used geopolymer binder to explore the bond strength phenomena with three different molar ratios (MRs) of activator. Unlike Portland cement, geopolymer does not have any kind of colloidal interaction [8]. Therefore, there is a lack of yield stress which sometimes cause the printed bead to deform, upon addition of more layers on top of it. To resolve this issue, we introduced highly purified nano-clay in to the geopolymer mix and its rheology was improved for 3D concrete printing application. We preferred to use less fine aggregates in our mix and achieve the required yield stress with the help of this nano-clay. The nano-clay carries a negative charge on the faces and a positive charge on the ends and so they tend to associate with each other by electrical attraction between positively charged edges and negatively charged surfaces. Hence, the resultant structure leads to a higher initial viscosity, while under strong pre-shearing the viscosity decreases as the structure breaks down [9].

Following section will describe all the raw materials, used in this research to produce geopolymer and the experimental methods to characterize its thixotropic property, structuration rate of three differently activated (MRs: 2.0,1.85 and 1.60) mortars. Section 3 will discuss the results of tensile bond test carried out after 7 days of ambient curing and the effects of MR on the respective strength values. Finally, in the last part, we recommend to use continuous mixer for achieving high buildability and stronger printed sample in concrete printing process.

2 Materials and Methods

Class F fly ash (FA), provided by Sembcorp EOSM (India), Pvt. Ltd., was used in this paper for producing geopolymer binder. Ground granulated blast-furnace slag (GGBS) from Engro Pvt. Ltd was mixed with FA for ambient curing of the geopolymer. As suggested in [10], we added some micro silica (SF) to FA + GGBS mix for increasing cohesiveness of the paste, which will help in smooth extrusion of the mortar. In terms of aggregates, fine river sand (max particle size 2 mm) was used as per the calculation shown in Table 1. We added 2.5% of nano-clay as rheology modifier, especially, for 3D printing purpose to the designed geopolymer mix. Alkaline activating solutions were formulated by blending a commercial sodium silicate solution (MR = 2.0) with 45 wt% NaOH solution, to reach the desired MR of 1.85 and 1.60.

The developed geopolymer mortar was first characterized for its thixotropy property prior to use in 3D printing process. We followed similar methodology described in [11], for measuring *structural break down* and *buildup*, immediately after the mixing

Table 1. Mix design of geopolymer mortar.

FA to binder ratio	GGBS to binder ratio	SF to binder ratio	Sand to binder ratio	Activator to binder ratio
0.75	0.15	0.10	1.5	0.46

process. The input protocols are shown in Fig. 1. In the both cases, pre-shearing was done for 60 s to maintain uniformity and then for structural breakdown, we measured the difference between the initial and final shear stress as reported in [11]. The buildup property of the material was evaluated by mimicking the 3D printing process [12], i.e. (after Pre-shearing) (1) shear at low rate (that represents material at rest (ii) high shear (extrusion process) and (iii) low shear rate (again at resting stage after the extrusion).

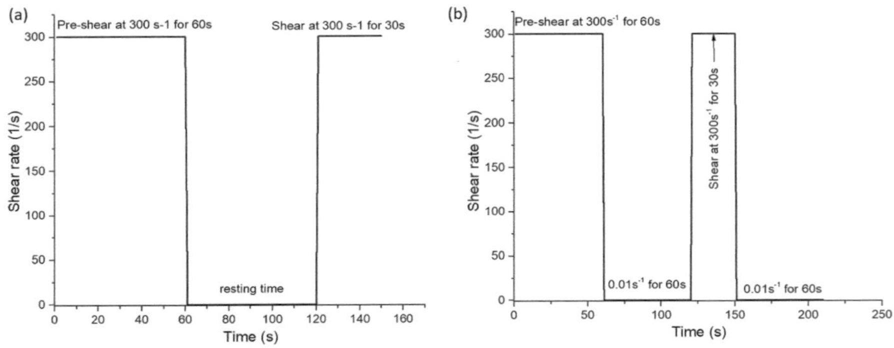

Fig. 1. Input protocols for measuring (a) structural break down and (b) buildup properties

The structuration rates of three different geopolymer, was evaluated by Anton Par MCR 102 rheometer in oscillation mode. Time sweep test, with in linear visco-elastic range at 10 rad/s frequency, was conducted to capture the increasing trend of storage modulus (G').

Two layers of geopolymer was printed with ten minutes time gap interval for the tensile strength measurement using Instron tensile test machine at 0.5 mm/min loading rate. Three samples of each MRs were tested, and their average value was reported in the following section. During printing, the extruder height was kept at the same height as of nozzle opening size, i.e. 30 × 15 mm and printing speed was same as the extruding velocity of the mortar for a constant flow rate of screw pump. 4 cm length samples were later extracted from a 400 mm long printed geopolymer and glued with two metal plates before placing under the load cell of machine. A schematic picture of geopolymer printing and test set up for bond strength is shown in Fig. 2.

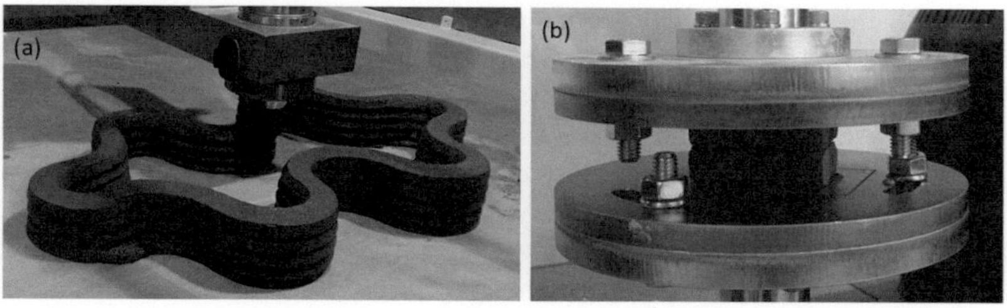

Fig. 2. (a) 3D printing of geopolymer mortar and (b) test setup for tensile bond strength

3 Results and Discussion

The obtained results of the geopolymer rheology in terms of *structural break down* and *build up* is shown in Fig. 3.

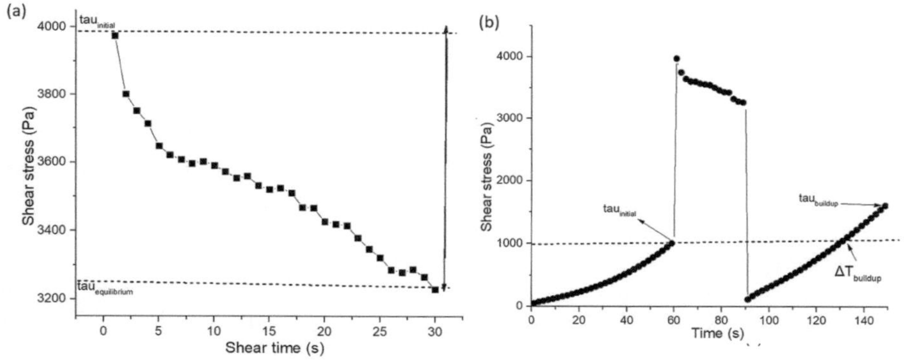

Fig. 3. (a) Structural break down and (b) buildup properties of a sample geopolymer mortar

It can be seen from the above figure that the geopolymer is behaving like shear thinning material with the shear stress going to down at a constant shear rate. The difference in $tau_{initial}$ and $tau_{equilibrium}$ (Fig. 3(a)) is considered as thixotropy index which is 0.23 here and with progress of time, it keeps on increasing, due to ongoing poly-condensation reaction in the material. Also, the area captured under this curve represents, easiness of breaking the material under applied shear force and in future, it can be considered as a test for easy of extrusion, if few mix designs are compared for the best results. From the Fig. 3(b), the time required to reach the initial $tau_{initial}$ was found to be around 40 s after the material came to rest, which indicates that its worth depositing a second layer after 40 s, to avoid deformation in the bottom layer. This quick recovery may be attributed to the flocculation nature of the nano-clay, since in geopolymer, there is no evidence of colloidal reaction like Portland cement [13, 14].

Results of 7 days bond strengths are shown in Table 2 along with their standard deviation values. As expected, the interface bond strength is decreasing with increase in

MR of the alkaline activator and depending on the printing quality (good or bad), the variations in the result can be more or less subjected to printer parameter settings. In geopolymer, MR of the activator plays a critical role on accelerating the hardening process [15], by rapidly enabling condensation reaction between Si and Al oligomers. Lower MR usually reacts faster with the binder and lowers the setting time of the material, thus making the open time very short for 3D concrete printing. If for such rapid setting material, the delay time between layers is not optimized then the contact (bond) will never strong due to lack of enough *moisture content* in the interface zone. Previous study by Roussel et al. [6], also confirmed similar behaviour by linking the change of static yield stress with mechanical strength of bonded SCC. Based on A_{thix} index, an expression of the critical delay was also formulated in their research, which declares, the critical delay can be of the order of 20 to 30 min for 0.3–0.5 Pa/s structuration rate.

Table 2. Tensile bond strengths of 3D printed geopolymer samples.

Samples Nos./Strength	MR 2.0	MR 1.85	MR 1.60
Tensile bond strength (MPa)	0.734	0.441	0.338
Standard deviation	0.04	0.02	0.02

According to plots in Fig. 4, it can be assumed that, due to faster poly-condensation reaction (sharp increase of G' curve) of MR 1.60, the surface moisture level has become less before deposition of another layer, which caused the poor interface bonding. However, in case of MR 2.0, due to more open time of the material, the bond became stronger compare to 1.85 and 1.60 MR. To support it, we have captured fracture surface images to relate the extent of bonding and it is obvious from Fig. 5 that the bonding in case of MR 2.0 was stronger (see rough surface), whereas for other MRs, it broke down at lower tensile load.

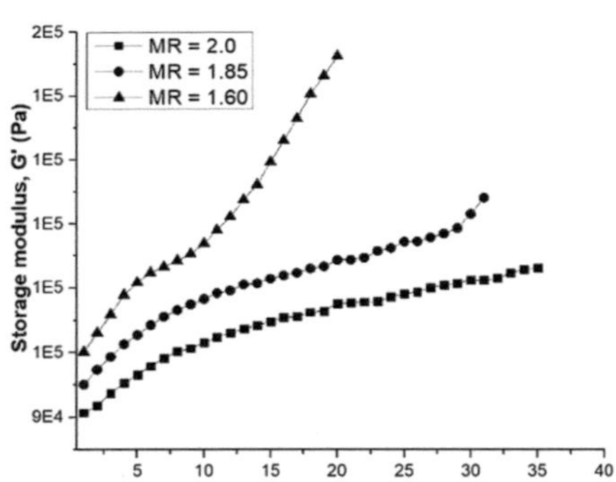

Fig. 4. Evolution of storage modulus (G') for the three molar ratios based geopolymer

Fig. 5. Surface roughness of interface (a) MR = 2.0 (b) MR = 1.85 (c) MR = 1.60 geopolymer

It is interesting to note that, due to viscous nature of the activator, geopolymer mortar after printing induced rough surface, which did not cause any kind of distinct layer problem. Also, it resulted strong bond strength in 7 days due to effective mixing (at interface) while being sheared by the top layer. Apart from these physical effects, the main controlling factor, i.e. *structuration rate* of the material always have a predominate effect which need to be control according to the delay time and printing speed of 3D concrete printer.

In the opinion of the present authors, if there is a need of printing large scale object, where second layer is getting deposited after 3 to 5 min of delay time, the material structuration rate need to be slow down to avoid distinct layer problem and inversely, for a quick printing, structuration rate can be faster, since the subsequent layers will be deposited at faster rate and the bottom layer need to have sufficient yield stress to hold these deposited layers. In case geopolymer mortar, the structuration rate is mostly governed by chemical reaction, which is not reversible, Therefore, decreasing the structuration rate (by using low reactive activator), may affect the overall strength of the printed sample. So, it is a wise alternative, to use continuous mixer, instead of batch mixer, that can exact mix the amount of material to be deposited in one or two layers. The advantage of using continuous mixer is that rapid hardening material can be printed quickly and at the same time, it can hold multiple layers for a large-scale printing. Batch mixing with slow structuration rate is not preferable as it may adversely affect the mechanical performance of the geopolymer samples.

4 Conclusions

We have defined from experimental results that a higher structuration rate can cause distinct-layer if the delay time is longer. Therefore, material must be designed considering the total printing time and the delay time to avoid the consequences of thixotropy. Since geopolymer has an inherent property of rapid setting and early mechanical strength, it's advisable to use continues mixer with controlled structuration rate, so that the final structure can be made with higher buildability and high mechanical strength. Though, in this research, we did not quantify the structuration index, but we managed to investigate the performance of different *structuration rates* on bond strength of printed geopolymers. In our future work, we would like to study the coupled effect of different *structuration rates* and *delay time*, and based on the obtained information, a model or data base can be created for the future 3D printing user, to digitally control their mix compositions, while ensuring stronger interface joint.

Acknowledgement. The authors would like to acknowledge Sembcorp Architects & Engineers Pte Ltd and National Research Foundation (NRF), Singapore for their funding and support in this research project.

References

1. Tay, Y.W.D., Panda, B., Paul, S.C., Noor Mohamed, N.A., Tan, M.J., Leong, K.F.: 3D printing trends in building and construction industry: a review. Virtual Phys. Prototyp. **12**(3), 261–276 (2017)
2. Bos, F., Wolfs, R., Ahmed, Z., Salet, T.: Additive manufacturing of concrete in construction: potentials and challenges of 3D concrete printing. Virtual Phys. Prototyp. **11**(3), 209–225 (2016)
3. Perrot, A., Rangeard, D., Pierre, A.: Structural built-up of cement-based materials used for 3D-printing extrusion techniques. Mater. Struct. **49**(4), 1213–1220 (2016)
4. Mettler, L.K., Wittel, F.K., Flatt, R.J., Herrmann, H.J.: Evolution of strength and failure of SCC during early hydration. Cem. Concr. Res. **89**, 288–296 (2016)
5. Megid, W.A., Khayat, K.H.: Bond strength in multilayer casting of self-consolidating concrete. ACI Mater. J. **114**(3), 467–476 (2017)
6. Roussel, N., Cussigh, F.: Distinct-layer casting of SCC: the mechanical consequences of thixotropy. Cem. Concr. Res. **38**(5), 624–632 (2008)
7. Panda, B., Paul, S.C., Mohamed, N.A.N., Tay, Y.W.D., Tan, M.J.: Measurement of tensile bond strength of 3D printed geopolymer mortar. Measurement **113**, 108–116 (2018)
8. Favier, A., Hot, J., Habert, G., Roussel, N., de Lacaillerie, J.B.D.E.: Flow properties of MK-based geopolymer pastes. A comparative study with standard Portland cement pastes. Soft Matter **10**(8), 1134–1141 (2014)
9. Ma, S., Qian, Y., Kawashima, S.: Experimental and modeling study on the non-linear structural build-up of fresh cement pastes incorporating viscosity modifying admixtures. Cem. Concr. Res. **108**, 1–9 (2018)
10. Panda, B., Tan, M.J.: Experimental study on mix proportion and fresh properties of fly ash based geopolymer for 3D concrete printing. Ceram. Int. (2018). https://doi.org/10.1016/j.ceramint.2018.03.031
11. Ouyang, J., Tan, Y., Corr, D.J., Shah, S.P.: The thixotropic behavior of fresh cement asphalt emulsion paste. Constr. Build. Mater. **114**, 906–912 (2016)
12. Li, H., Tan, Y.J., Leong, K.F., Li, L.: 3D bioprinting of highly thixotropic alginate/methylcellulose hydrogel with strong interface bonding. ACS Appl. Mater. Interfaces. **9**(23), 20086–20097 (2017)
13. Panda, B., Paul, S.C., Hui, L.J., Tay, Y.W.D., Tan, M.J.: Additive manufacturing of geopolymer for sustainable built environment. J. Clean. Prod. **167**, 281–288 (2017)
14. Favier, A., Hot, J., Habert, G., de Lacaillerie, J.D.E., Roussel, N.: Rheology of geopolymer: comparative study between Portland cement and metakaolin based geopolymer. In: 1st RILEM International Conference on Rheology and Processing of Construction Materials, Paris, vol. 1, p. 4794 (2014)
15. Duxson, P., Provis, J.L.: Designing precursors for geopolymer cements. J. Am. Ceram. Soc. **91**(12), 3864–3869 (2008)

Potentials of Steel Fibres for Mesh Mould Elements

P. Pfändler(✉), T. Wangler, J. Mata-Falcón, R. J. Flatt, and W. Kaufmann

ETH Zurich, Zurich, Switzerland
patrick.pfaendler@ifb.baug.ethz.ch

Abstract. Mesh Mould is a digital fabrication technique developed at ETH Zurich in which the reinforcement and formwork production are unified in a robotically controlled system. An industrial robot fabricates a dense, three-dimensional, double-sided, welded reinforcement mesh that is infilled with a special concrete mix that achieves sufficient compaction without flowing out the mesh, which acts as porous formwork. Since the project started in 2012, the actual generation of robot end-effector is capable of bending and welding conventional steel reinforcement of 6 and 4.5 mm in diameter. Due to the process, the load-bearing capacity of these Mesh Mould elements is not equal in both directions due to geometrical restrictions in the end-effector. This study aims to increase the load-bearing capacity in the weaker direction by using steel fibre reinforced concrete (SFRC), which orients the fibres during flowing in this direction and in addition prevents the leakage of the concrete by enhancing jamming. A total of 10 specimens with 540 × 210 × 80 mm dimensions were tested in a displacement controlled symmetric four-point bending test. By combining SFRC with a mesh, the bending strength increased significantly with respect to the samples without fibres. The capacity is higher than the capacity of the individual parts, which are evaluated in separate material tests. Nonetheless, the bending strength in this study was limited by the weld strength, which was considerably lower than the one achieved by the robot. Higher weld strength would lead to better performance than in this first study, which is a part of an ongoing research effort.

Keywords: Mesh Mould · Steel fibre reinforced concrete SFRC
Digital reinforcement assemblies · Digital construction

1 Introduction

1.1 Background About the Mesh Mould Project

The Mesh Mould project started in 2012 at ETH Zurich to explore the possibilities of digitally fabricating concrete structures with high geometric complexity without conventional moulds. The project started with a customized tool-head to build a polymer network through a spatial extrusion process with a rather low tensile strength of the polymer. At that time, the focus was on the shape rather than on the structural strength and led to a patent [1]. After this initial conceptual development, the project addressed

T. Wangler and R. J. Flatt (Eds.): DC 2018, RILEM Bookseries 19, pp. 207–216, 2019.
https://doi.org/10.1007/978-3-319-99519-9_19

the structural behaviour of the built elements and the polymer mesh was replaced by a steel reinforcement mesh in order to reach load-bearing capacity. The end-effector was adjusted in two steps to be able to build meshes with conventional steel reinforcement of Ø4.5 and Ø6.0 mm [2]. The reinforcement mesh is composed of continuous Ø6 mm rebars connected by short welded Ø4.5 mm segments at well-defined distances controlled by the robot. The mesh acts as a porous formwork, and with a proper concrete mix design with aggregate in the order of the size of the mesh spacing, retention of the concrete can be achieved through jamming [3]. The outer surface is sprayed with an additional layer of concrete for the purpose of appearance and durability [4] (Fig. 1).

(a) (b)

Fig. 1. (a) Concreting process of a Mesh Mould wall using a pump and (b) Finishing the wall [4]

The ongoing full digitalization of civil engineering design and construction will make it possible to design concrete structures with high geometrical complexity in the individual parts of the structure. This has further possible advantages by designing concrete structures in a structurally optimal shape in favour of saving material. The invention of the Mesh Mould system might lead in the future to significantly lower costs and less waste for non-standard shapes in the building industry [5]. The first prototype of a Mesh Mould wall as the main load-bearing element was built as a part of the DFAB house in Switzerland in 2017 [6].

1.2 Mechanical Behaviour of Mesh Mould Elements

First exploratory experiments on the mechanical behaviour of the Mesh Mould mesh under bending loads were made by Mata-Falcón [7] on hand-made reinforcement meshes. The main observation was a very close and thin crack distribution caused by shear deformation in the continuous direction of the reinforcement mesh by the load transfer in the perpendicular direction. Figure 2 shows clearly the observation during and after the bending test and displays the mesh size of 30 × 30 mm. The final design of the robot end-effector cannot produce this mesh size due to geometrical constraints and as a consequence, the reinforcement content decreases substantially in the discontinuous direction [2].

(a) (b)

Fig. 2. (a) Notched specimen under a four-point bending test with the crack pattern and (b) View from the bottom after the bending test with shear deformation in the continuous reinforcement bars [7]

The limitation of the robotic tool-head is a minimum layer distance of 30 mm in the continuous vertical direction and a minimum distance of 60 mm between two welded segments in the horizontal discontinuous direction, which is equal to a mesh size of 30 × 60 mm. The key problem is the reinforcement content in the discontinuous direction which is much weaker than in the perpendicular direction, which due to the mentioned constraints could not simply be improved by building a denser welded reinforcement mesh by the robot. In addition, this would have affected the production time for the mesh significantly.

The achievable reinforcement content could be insufficient to fulfil the requirements for a load bearing wall or even for small Mesh Mould elements subjected to bending in the weak direction. With a vertical reinforcement of Ø6 @ 30 mm, minimum reinforcement requirements in bending are satisfied for a thickness up to 250 mm. In the horizontal direction, however, the thickness is limited to 150 mm thick elements to ensure a ductile failure in bending, assuming failure of the rebars in tension.

The lack of adequate continuous reinforcement in the horizontal direction limits the structural application of Mesh Mould elements to less demanding structural applications. This study aims at overcoming these limitations by the use of steel fibres oriented in the horizontal direction. During the filling process of a wall, the fibres tend to align in the horizontal direction by flowing with the concrete, which strengthens the weak direction without requiring any changes in the robot end-effector. Furthermore, the fibres reduce the leakage of concrete during the casting process thanks to jamming of the mesh.

The present study is part of an ongoing research to expand the mechanical capabilities of the Mesh Mould system in the weak direction with a view to more structurally demanding applications in the future. This paper presents a first experimental study analysing the mechanical performance of thin Mesh Mould elements with steel fibre reinforced concrete (SFRC) under flexural actions.

2 Experimental Program

2.1 Test Specimens

The objective of the experimental program was to explore the influence of steel fibres on the mechanical behaviour of Mesh Mould elements. A total of 10 specimens (see Table 1) with 540 × 210 × 80 mm dimensions were tested in a four-point bending configuration. The behaviour of concrete with and without fibres was tested for specimens reinforced with a welded mesh, which was strengthened in some cases with additional reinforcing bars. Besides these tests, additional experiments without reinforcing bars were performed in order to characterize the concrete and the SFRC material properties [8].

Table 1. Main parameters of test specimens

Codification	Number of specimens	Fibres content [%]	Reinforcement
C+M	2	0.0	Mesh
C+sM	2	0.0	Mesh+2Ø6
SFRC+M	4	1.2	Mesh
SFRC	2	1.2	–

The welded reinforcement mesh is composed of (i) Ø6 mm rebars with 30 mm spacing in the direction transversal to bending and (ii) Ø5 mm diameter welded segments with a 60 mm spacing for the direction of tension. Hence, the element contains 3 to 4 reinforcing bars of Ø5 mm depending on the section (see Fig. 3). The mesh was welded at both ends to a steel profile to guarantee the anchorage of the reinforcement. The strengthening rebars were welded positioned between two discontinuous rebars and welded only to the steel profiles at the ends.

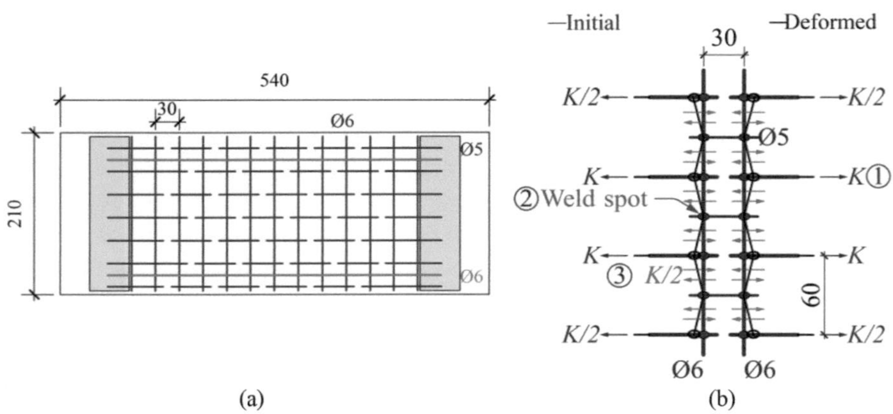

Fig. 3. (a) Tailor-made reinforcement mesh with additional strengthening and (b) Sketch of a section of the Mesh with the three possible failure modes and the shear deformation (all dimension in [mm])

2.2 Experimental Setup

All specimens were tested until rupture in a displacement controlled four-point bending test with 400 mm span and 200 mm distance between the loads (Fig. 4). The applied force (*F*), the deflection (*d*), and the crack mouth opening displacement (*CMOD*) at the notched midspan section were measured during the test. Moreover, pictures were taken at 0.1 Hz to capture the crack pattern of the specimens at different load steps. The loading speed was 0.005 mm/s before the peak load and 0.08 mm/s in the post-peak phase.

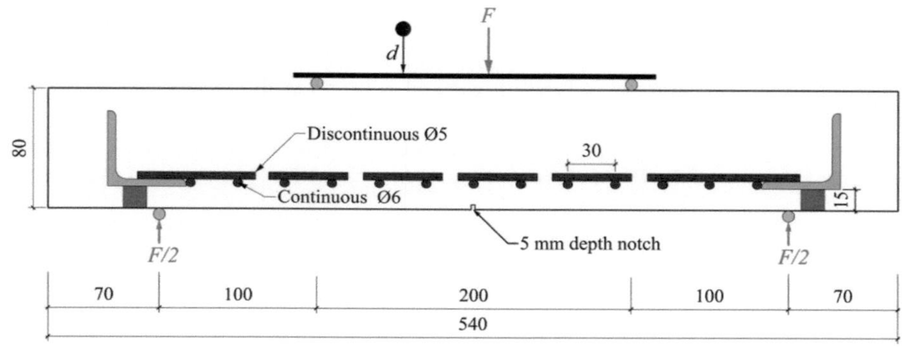

Fig. 4. Geometry and test setup (dimensions in [mm])

The fluid self-compacting mixture influenced the fibre dispersion by the direction of the casting flow after the SFRC was poured in the middle of the mould. Moreover, the production method oriented the fibres in the direction, in which the fibres are structurally more useful. The fibre distribution along the depth of the specimens was not uniform since the concrete mix was not perfectly non-segregating, but affects the structural performance in the weaker direction again in a positive way.

2.3 Material Properties

Concrete. The mixture of all samples was the same despite the additional steel fibres in the SFRC mix. The concrete mixture, initially optimized for casting into 3D-sand printed elements [10], was slightly adapted for this study. The concrete had a w/c of 0.33 and a water-to-binder ratio of 0.29, considering the amount of liquid in the superplasticizer and the reacting silica fume. The steel fibres used were straight OL 13/.16 fibres produced by Bekaert. The fibres were 0.16 mm in diameter and 13 mm long, with an ultimate tensile strength of 2'600 N/mm^2 according to the producer.

For the two mixes, the experimental setup described above was used to determine the bending strength. The mean bending capacity without any reinforcement was 1.00 kNm ($f_{ct,fl}$ 4.0 N/mm^2) and 1.95 kNm for SFRC with a volumetric content of 1.2% of fibres and a strain-hardening behaviour. Guided by DIN EN 14651 (Table 2), however using a different specimen geometry as well as the four-point test in this study,

Table 2. Overview of the results of the four-point bending test of the SFRC, where $f_{R,j}$ is the residual flexural strength at *CMOD* of 0.5, 1.5, 2.5 and 3.5 mm and height h_{sp} which is the distance between the notch and the top of the specimen [9]

Sample Name	$f_{r,1}$ [N/mm²]	$f_{r,2}$ [N/mm²]	$f_{r,3}$ [N/mm²]	$f_{r,4}$ [N/mm²]
SFRC#1	9.8	8.8	7.6	5.2[a]
SFRC#2	11.5	8.1[a]	5.5[a]	3.3[a]

[a]Value calculated through the linear relationship between the *CMOD* and the deflection *d*

the flexural tensile strength was calculated at different CMODs to characterize the behaviour of the SFRC in bending.

In addition, compression tests on a total of 8 cubes with dimensions 150 × 150 150 mm were conducted to obtain the compressive strength between 23 and 28 days. The range of the compressive strength for concrete was between 111 N/mm² and 133 N/mm² based on 4 cubes, for the SFRC between 130 N/mm² and 143 N/mm².

Reinforcement. Table 3 shows the main properties of the reinforcement obtained from standard tensile tests. A company produced (as close as possible to the mesh produced by the robot) meshes 1000 × 1000 mm in size that were cut by hand to the required shape.

Table 3. Material properties of the reinforcement

	Ø [mm]	Steel type [–]	Yield stress f_y [N/mm²]	Ultimate stress f_u [N/mm²]	Strain at peak load ε_u [‰]
Mesh	6	B500A	513	540	29
	5	B500A	550	570	25
Strengthening	6	B500B	476	550	96

Properties of the Welding. The setup to measure the weld strength is shown in Fig. 5 (a). In this setup, cross-shaped samples consisting of two rebars (Ø5.0 and Ø6.0 mm) welded perpendicular to each other were tested by pulling the longer rebar. The specimens were randomly cut of the mesh and the vertical and the horizontal bars were carefully removed by hand.

The welding resistance obtained from a total of 6 samples exhibited a huge scatter in the weld strength. Two samples achieved a resistance of less than 1 kN (see Fig. 5 (b); specimens 5 and 6); these two samples were not taken into account for the determination of the mean value, as they had likely been damaged during preparation. This resulted in an average weld strength of 4.0 kN, much weaker than the meshes produced by the Mesh Mould tool-head which yields a weld strength of about 7.5 kN with a much reduced scatter.

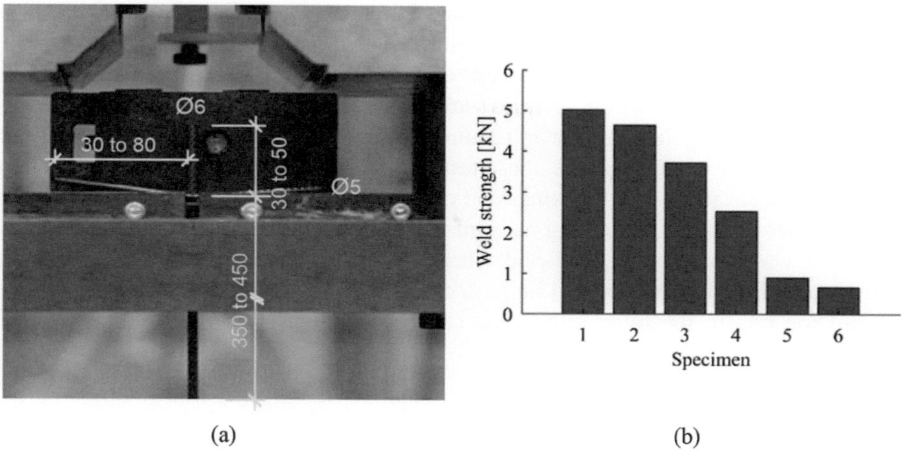

Fig. 5. (a) Setup of the welding test (dimensions in [mm]) and (b) Results of the welding tests sorted by strength

3 Results and Discussion

3.1 Predicted Behaviour of the Mesh

According to the results of the tension tests of the rebars and the weld strength, the welds are the weakest among the three possible failure mechanisms (see Fig. 3(b)). The mean weld strength is only capable of activating 35% of the ultimate strength of the discontinuous direction and 22% of the shear strength in the perpendicular direction. Hence, the capacity is strongly limited by the welds and most of the reinforcement strength will remain unused. Moreover, the high scatter observed in the weld strength could lead to significant differences in the results of the bending tests for identical configurations.

The bending capacity of the samples without fibres reinforced with the welded mesh (C+M) can be predicted by means of cross-sectional analysis neglecting the concrete tensile strength, considering the reinforcement capacity limited by the mean welding shear strength and the concrete at maximum capacity in compression. The predicted bending strength (0.70 kNm) is lower than the bending capacity of concrete (0.97 kNm), consequently, the tested meshes do not fulfil the requirements for minimum reinforcement in these samples. If the welds were made by the robot of the Mesh Mould project, then a bending strength of 1.24 kNm would have been possible, this would have fulfilled the requirement for the minimum reinforcement.

3.2 Four-Point Bending Test

Figure 6 shows the results of deflection versus bending moment for the different tests. In this plot, the horizontal lines show: (i) the mean peak bending capacity of plain concrete, (ii) the mean peak bending capacity of the SFRC, (iii) the predicted mean capacity of the samples without fibres reinforced with the mesh and (iv) the independent combination of the peak strength of the fibres and the reinforcing mesh.

The bending strength of the samples without fibres reinforced with the mesh (C+M) did not fulfil the requirements for minimum reinforcement in bending for the 80 mm deep elements as expected. Hence, these tests showed a softening branch after cracking with strain localisation. The cracking load of these elements corresponds to the results of the concrete samples containing no reinforcement at all. A post-peak plateau around 0.70 kNm was observed as predicted when neglecting the concrete tensile strength. Only the continuous rebar located in the localized crack showed a visible plastic shear deformation due to the force transfer. It can be concluded that the welding strength limited the bending capacity.

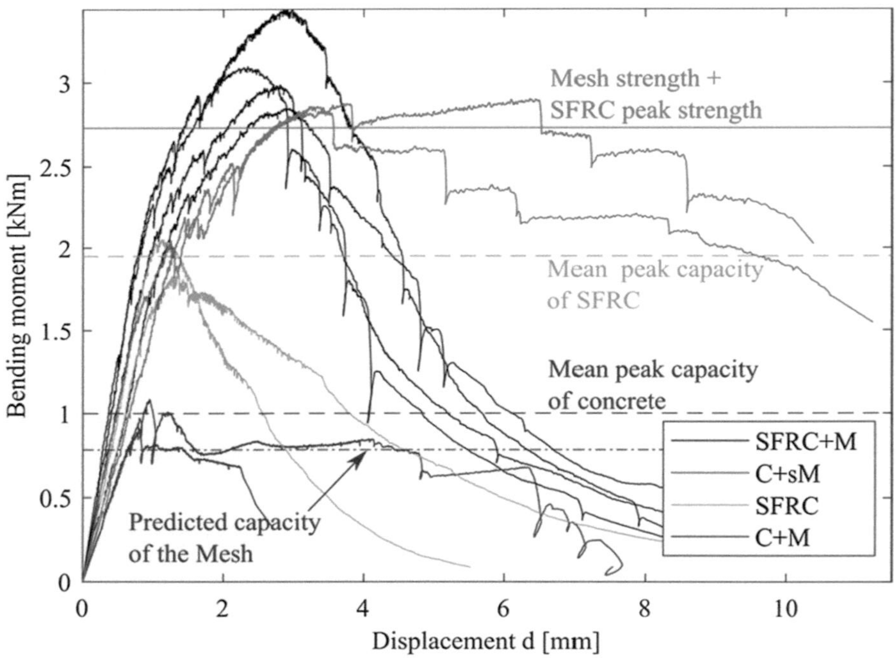

Fig. 6. Results of the four-point bending test of all samples

The configuration with fibres in addition to the mesh (SFRC+M) reached the desired deflection hardening and generated more cracks than the specimens without fibres. The ultimate bending strength increased significantly because of the addition of steel fibres. The capacity is 16% higher than the independent combination of the strengths of the SFRC (1.95 kNm) and the mesh (0.70 kNm) which is the residual strength of the C+M samples and was exceeded by all 4 tested specimens. This result shows the beneficial effect of SFRC when adding reinforcement because a more distributed crack pattern is produced, avoiding premature local failure in the weakest section. The softening branch in the post-peak behaviour is a combination of the smooth fibre pull-out in the localized crack and sudden failures of welding spots.

The two specimens with two additional rebars (C+sM) reached a mean bending capacity of 2.89 kNm higher than the predicted value around 2.50 kNm based on a cross-sectional analysis. Compared to specimens SFRC+M, the samples C+sM show a lower peak bending capacity but a higher ductility. These samples exhibited 3 to 4 cracks between the load application points at peak load, all of them extending over more than 50% of the depth. In the post-peak range, the drops in the bending moment are due to subsequent failures of welds and finally a rupture of the strengthening rebars.

The beneficial effect of crack distribution on the performance of fibre reinforced concrete can be clearly seen in Fig. 7, which shows the crack pattern near the peak load of two identical samples including fibres in addition to the mesh (SFRC+M). The specimen with more distributed and thinner cracks (SFRC+M#2, Fig. 7(b)) reached a significantly higher capacity than the other specimens, which all showed a more pronounced strain location in a single crack (as SFRC+M#1, Fig. 7(a)). In spite of this observation, the insufficient weld strength did not allow developing the envisaged crack spacing of 30 mm, with a crack at every transverse rebar. Hence, the observed behaviour of the mesh in this study is not representative of Mesh Mould structures, in which the stronger welding could lead to the envisaged distributed cracking and consequently a much better performance.

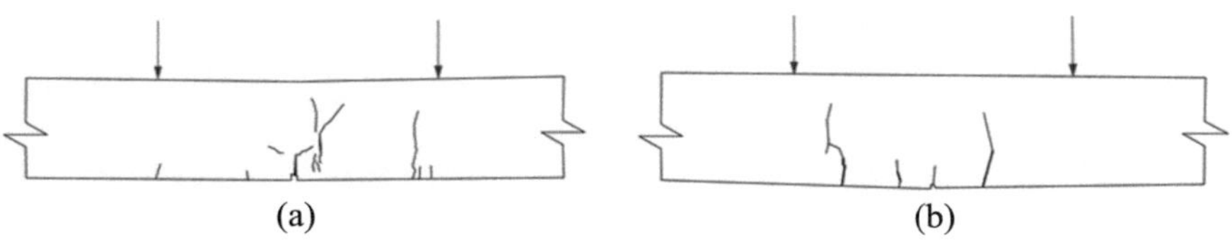

Fig. 7. Crack pattern near peak load: (a) SFRC+M#1, M_u = 3.1 kNm and (b) SFRC+M#2, M_u = 3.4 kNm

4 Conclusions and Outlook

This research has shown the beneficial effect of steel fibres to improve the mechanical behaviour of Mesh Mould structures, in particular in its weaker horizontal direction. The use of steel fibre leads to a significant increase in strength and ductility. Moreover, it has been observed that the capacity of steel fibre reinforced Mesh Mould elements is higher than the combination of the individual capacity provided by the fibres and the Mesh Mould meshes. This effect is attributed to the crack width control produced by the steel meshes, which allows a better performance of the SFRC.

However, careful attention must be paid to this main conclusion. In this preliminary study, only one fibre type in combination with a tailor-made reinforcement mesh was tested in a four-point bending test. The findings might not be transferable to all types of fibre or volumetric contents and the topic remains complex.

On a wider level, research is also needed to explore the possibilities of longer fibres for the application for Mesh Mould elements to more efficiently cause jamming during the filling process.

The use of Mesh Mould meshes with a higher weld strength would allow taking full advantage of the observed benefit of distributed cracking on the behaviour of SFRC, leading to a better performance than observed in this study.

Acknowledgments. This paper summarizes the results of a Master thesis carried out under research supported by the National Centre for Competence in Research in Digital Fabrication – Innovative Building Processes in Architecture (project number 51NF40_141853).

References

1. Hack, N., Lauer, W., Gramazio, F., Kohler, M., Blank, N.: Method of fabricating a 3-dimensional structure, mesh formwork element for fabricating a 3-dimensional structure, and method of fabricating the same. In: Google Patents (2015)
2. Kumar, N., Hack, N., Dörfler, K., Walzer, A., Gonzalo, J. Gramazio, F., Kohler, M., Buchli, J.: Design, development and experimental assessment of a robotic end-effector for non-standard concrete applications. In: 2017 IEEE International Conference on Robotics and Automation, Singapore, pp. 1707–1713 (2017)
3. Gebhard, L.: Mesh Mould: Model Material Filling Study. Zürich (2017, unpublished)
4. Hack, N., Wangler, T., Mata-Falcón, J., Dörfler, K., Kumar, N., Walzer, A., Graser, K., Reiter, L., Richner, H., Buchli, J., Kaufmann, W., Flatt, R., Gramazio, F., Kohler, M.: Mesh Mould: an on site, robotically fabricated, functional formwork. In: 11th High Performance and 2nd Concrete Innovation Conference, Tromsø (2017)
5. Agustí-Juan, I., Müller, F., Hack, N., Wangler, T., Habert, G.: Potential benefits of digital fabrication for complex structures: environmental assessment of a robotically fabricated concrete wall. J. Clean. Prod. **154**, 330–340 (2017)
6. DFAB HOUSE homepage. http://dfabhouse.ch/mesh_mould/. Accessed 5 Mar 2018
7. Mata-Falcón, J.: Mesh Mould structural tests. Zürich (2016, unpublished)
8. Pfändler, P.: Potential of steel fibre for Mesh Mould elements. Zürich (2017, unpublished)
9. DIN EN 14651: 2007 12: Prüfverfahren für Beton mit metallischen Fasern; Bestimmung der Biegezugfestigkeit (Proportionalitätsgrenze, residuelle Biegezugfestigkeit), Deutsche Fassung EN (2007)
10. Ruffray, J., Bernhard, M., Jipa, A., Meibodi, M., Montague de Taisne, N., Stutz, F., Wanglert, T., Flatt, R.J., Dillenburger, B.: Complex architectural elements from HPFRC and 3D printed sandstone. In: International Symposium on Ultra-High Performance Fibre-Reinforced Concrete UHPFRC 2017, Montpellier (2017)

Capillary Water Intake by 3D-Printed Concrete Visualised and Quantified by Neutron Radiography

Christof Schröfl(✉), Venkatesh Naidu Nerella, and Viktor Mechtcherine

Technische Universität Dresden, 01062 Dresden, Germany
Christof.Schroefl@tu-dresden.de

Abstract. Water uptake into two formulations of 3D-printed concrete via capillary suction was assessed by neutron radiography. The samples varied in their layer-to-layer deposition time intervals (TI) and the use of different binders. TI of two and 13 min were short enough to avoid preferential capillary suction at interlayer bonding areas in the fine-grained printable concretes containing supplementary cementitious materials. An increase in the time interval to 24 h gave rise to quick capillary suction through the layer-to-layer interfaces. However, moisture did not redistribute into the matrix regions from the interfaces. For mixture with Portland cement as sole binder and addition of a superabsorbent polymer (SAP), the short layer-to-layer deposition interval of two minutes resulted in tight interlayer bonds with quasi-null capillary suction. Intervals of 13 and 36 min, however, resulted in partially quick and intense absorption of water and immediate absorption by adjacent SAP particles.

Keywords: Extrusion-based additive manufacturing · Transport properties
Neutron radiography imaging · Superabsorbent polymers

1 Introduction

Producing civil engineering structures by concrete-3D-printing is a very promising approach to realise peculiar geometries and quick building progress [1, 2]. Many technical solutions to 3D-printing of cement-based building materials include layer-by-layer deposition of extruded fresh material. Upon release from the nozzle at the print head, the newly placed concrete has to bind to the previously generated layer. Such bonding zones, however, might be "weak links" inside an otherwise homogenous element. Their mineralogical composition can differ and their microstructure can be less dense in comparison to the interior of the layers [2]. Durability issues are thus a major concern because such weak interlayer regions may allow for intense capillary suction of aqueous liquids from the concrete surface deep into the structure. Visual impressions of several existing 3D-printed concrete structures have disclosed preferential water intake into the layer-to-layer bonding zones, see Fig. 1. Transport properties of 3D-printed concrete samples on the meso- and microscale have, however, not

T. Wangler and R. J. Flatt (Eds.): DC 2018, RILEM Bookseries 19, pp. 217–224, 2019.
https://doi.org/10.1007/978-3-319-99519-9_20

yet been scientifically investigated. The study at hand used neutron radiography imaging to shed light onto the open questions.

Fig. 1. Signs of local capillary suction at the interfaces between layers of printed concrete as well as at cracks, photo by V. Mechtcherine.

2 Experimental

Two fine-grained printable concrete mixtures were investigated in this study. Mixture A features a binder composite made of Portland cement, fly ash and silica fume. It is based on preceding investigations on interface strengths and microstructure analysis [2]. Nerella et al. [2] also investigated concretes with Portland cement as the sole binder showing its relatively weak performance in terms of mechanical interface strengths. In the study at hand, however, the mixture was modified by a superabsorbent polymer (SAP) for internal curing to improve its properties; it is denominated as Mixture G (Table 1).

Table 1. Materials and compositions of the 3D-printed concretes [kg/m^3].

Material	Mixture A	Mixture G
Portland cement (CEM I 52.5 R, OPTERRA, Karsdorf)	378.4	704.9
Silica fume (50 wt% aqu. suspension, EM-SAC 500 SE, Elkem)	206.4	0
Fly ash (class F, Steament H-4, STEAG)	206.4	0
Quartz fine sand 0.06 mm–0.2 mm	316.3	249.8
Quartz sand 0 mm–1 mm	278.0	249.8
Quartz sand 0 mm–2 mm	717.3	749.5

(*continued*)

Table 1. (*continued*)

Material	Mixture A	Mixture G
Tap water, $(w/c)_{eq} = 0.42$	133.7	296.0
High-range water-reducing admixture (SKY 593, BASF)	10.32	5.29
Superabsorbent polymer (0.4 wt% by weight of cement, an R&D sample provided by SNF Floerger, bulk-polimerised and crushed)	0	2.82

Production of 3D-printed concrete specimens followed an extrusion-based process that had been described earlier [3]. The time interval between depositing a subsequent layer was the single parameter of variation (Table 2) in the production process.

Table 2. Specimen nomenclature, layer-to-layer deposition intervals, and suction duration.

Mixture	Specimen	$1^{st}/2^{nd}$ layer	$2^{nd}/3^{rd}$ layer	$3^{rd}/4^{th}$ layer	$4^{th}/5^{th}$ layer	Suction time
A	A 4	13 min	13 min	13 min	13 min	26 h
A	A 19	2 min	2 min	2 min	24 h	26 h
G	G 1	13 min	13 min	13 min	13 min	54 h
G	G 16	2 min	2 min	2 min	36 min	26 h

Two minutes was the shortest time interval that was feasible on the 3D-printer at hand. 13 min represented a characteristic example in a range of ten to 20 min, which may be of practical relevance. Lastly, 24 h with Mixture A mimicked interruption of the building process and delayed restart of the work. For Mixture G, only 36 min were feasible for non-related technical reasons at the experimental facility. Each layer had a height of 20 mm and a width of 30 mm. The walls consisted of five layers and had a length of 120 cm. They were covered by a moist cloth for 24 h after production, prior to underwater curing. From the concrete age of seven days onwards, the walls were stored in a controlled climate (20 °C, 65% relative humidity) until being cut at the total age of 28 days. Square specimens were saw-cut from each wall, each of which contained four bonding zones and four layers. The samples were turned 90° anti-clockwise (Fig. 2). All lateral surfaces were tightly sealed by self-adhesive aluminium tape, which is almost transparent to neutrons, to avoid loss of moisture there.

Capillary suction experiments inside a neutron beam were performed in the NEUTRA beamline at Paul Scherrer Institute in Villigen/Aargau, Switzerland. Measurement position 3 was used. Primary data treatment, qualitative and (semi)quantitative image processing accorded to previous studies with a similar scope; see e.g. [4]. With the resolution of some few micrometers, areas with changing water contents as compared to the initial "dry" state disclosed themselves by time-resolved image referencing. Four stripes were laid over the entire width of a specimen's image to quantify the differential water contents in sufficient vertical resolution.

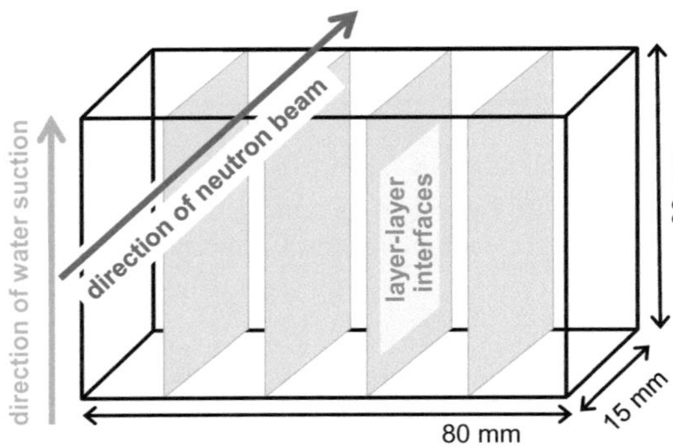

Fig. 2. Schematic of saw-cut 3D-printed concrete specimens bound for capillary suction in the neutron beam.

The figures display a selection of characteristic time steps during the imaging sequence. Each first picture and quantification chart indicate the first minute after the contact with water. Each final frame shows the state at the end of the experiments.

Three imperfections became clearly visible in the course of image evaluation, which are not directly linked to the 3D-printed concrete samples in their specific nature. They are just briefly mentioned here but will be discussed in a separate publication in-depth. Over time, the exact positions of the samples could slightly change and numerous images had to be re-localised for pixelwise matching. The success of these operations depended on the extent of the dislocalisation. Secondly, minor wrinkles in the self-adhesive aluminium tape at the edges of the specimens gave rise to capillary suction. The effect looks impressive, however, it is obvious that presence of water in the respective regions is not the consequence of capillary suction by concrete, but is due to the boundary conditions. As a third kind of imperfection, a non-zero moisture difference outside the specimens could be quantified. None of these three issues, however, limits the fundamental findings of this study related to the 3D-printed concretes.

3 Results and Discussion

3.1 Fine-Grained 3D-Printed Concrete (Mixture A)

Results obtained for the specimen A4 are displayed in Fig. 3 and those obtained for the specimen A19 in Fig. 4.

When printing with the practice-oriented layer-to-layer deposition interval of 13 min, no significant water uptake occurred throughout 26 h of capillary suction (Fig. 3); see also Sect. 3.2 and [4, 5] for comparison. The water content increased neither in the matrix regions nor at interlayer bonding zones to a measurable extent. The same interpretation applies to the two minute intervals in specimen A19 (Fig. 4). Contrarily, at the place of the bonding zone with deposition interval of 24 h a decisive intake of water was clearly visible in the neutron radiography images. The border of

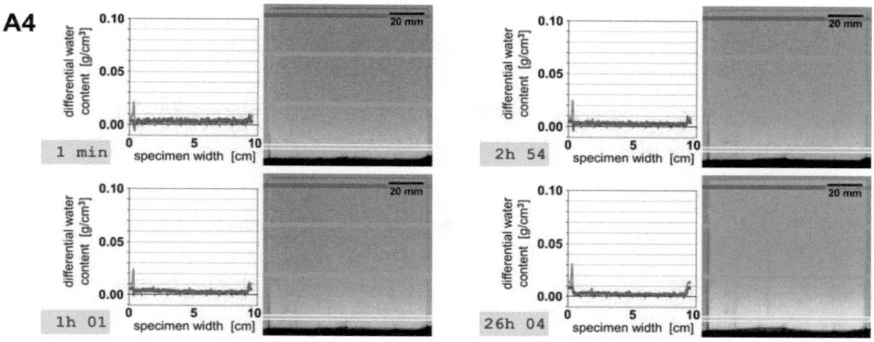

Fig. 3. Differential water content of specimen with 13 min layer-to-layer deposition interval (A4) at the four heights indicated in the images and neutron radiography referenced differential images.

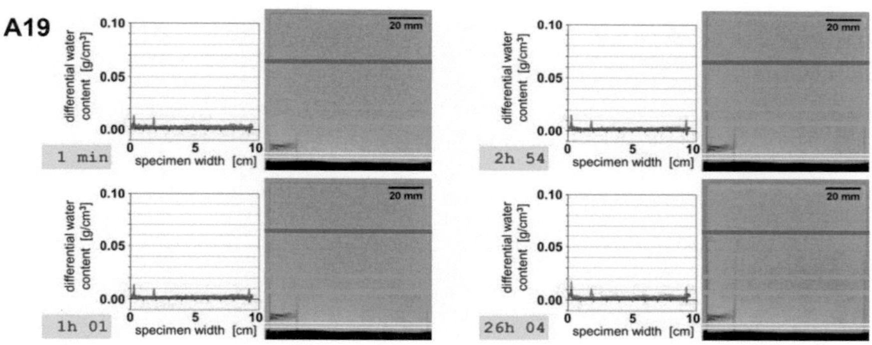

Fig. 4. Differential water content of specimen with one 24 h and three two minute layer-to-layer deposition intervals (A19) at the four heights indicated in the images and neutron radiography referenced differential images.

this water ingress is very sharp even on the long term and moisture is not redistributed by capillary suction into the interior regions of the deposited layers. The top-most layer, i.e. left one in the images, that was deposited after a delay of 24 h, was obviously cracked (probably due to shrinkage) and water followed this fine crack as the primary pathway of capillary suction. However, there as well, water was not redistributed into adjunct matrix regions within the layer.

To conclude, 24 h layer-to-layer deposition interval caused a "cold joint" through which water could easily ingress by capillary suction, whereas both two and 13 min intervals produced no absorbent bonding zones.

In comparison to typical capillary water uptake into both micro- and macrocracks in steel rebar reinforced and conventionally produced concrete (RC), the suction intensity of the specimens at hand is minor and can even be neglected [4, 5]. Furthermore, even micro-cracks in fibre-reinforced strain-hardening cement-based composites (SHCC) [4] or textile reinforced concrete (TRC) [6] have been shown to take up pronouncedly higher amounts very quickly than any parts of the present 3D-printed specimens at hand.

3.2 Portland Cement-Based 3D-Printed Concrete with Internal Curing by SAP (Mixture G)

Results of capillary suction by specimens prepared with Portland cement as the binder (without pozzolana) and enriched with a SAP are presented in Figs. 5 (G1) and 6 (G16).

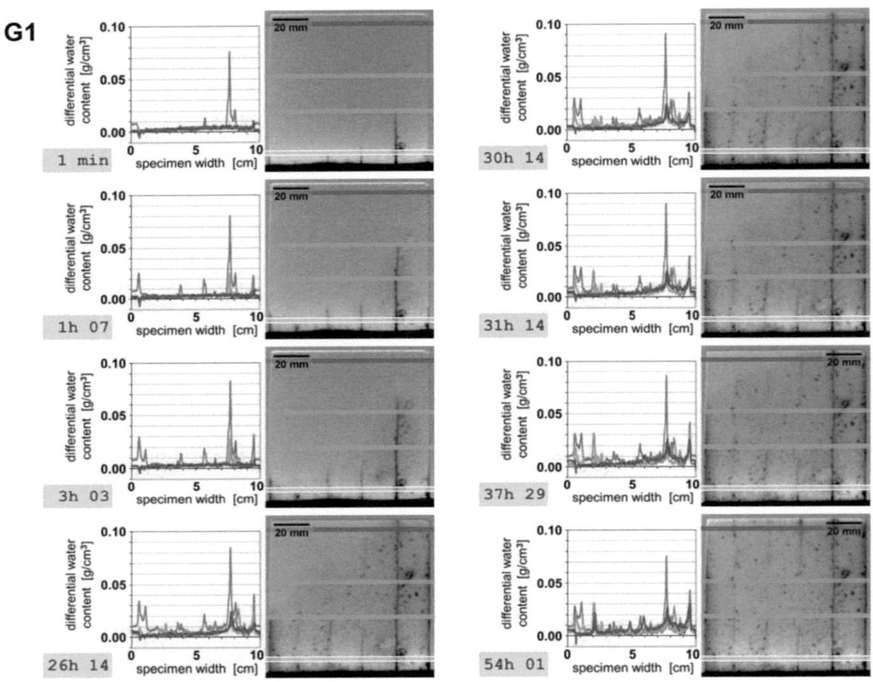

Fig. 5. Differential water content of specimen with 13 min layer-to-layer deposition interval (G1) at the four heights indicated in the images and neutron radiography referenced differential images.

Water was relatively quickly absorbed by capillary suction into the layer-to-layer bonding zones of specimen G1, which had been prepared with an interval of 13 min between the layers. The upper layers were slightly less active than the lower ones, but generally it is evident that capillary suction took place to a significant extent. Again, in comparison to RC, TRC or SHCC [4–6], the extent of water intake is very slow.

Most likely, the obvious but still quite minor increase in water content inside the matrix regions of G1 does not stem from the 3D-printed concrete itself. It is

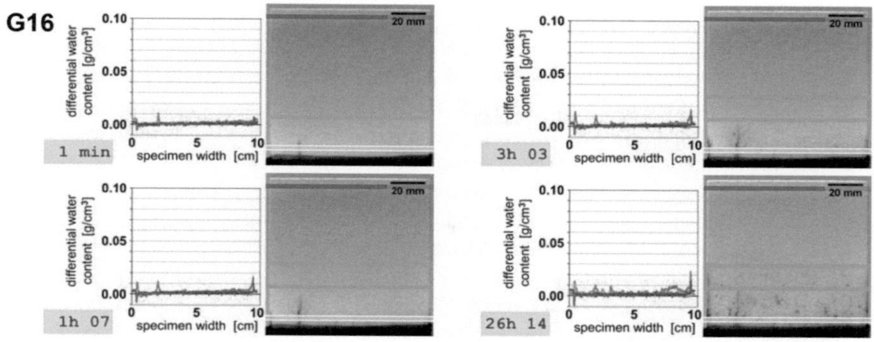

Fig. 6. Differential water content of specimen with one 36 and three two minute layer-to-layer deposition intervals (G16) at the four heights indicated in the images and neutron radiography referenced differential images.

presumably a result of debonding of the self-adhesive aluminium tape in the course of time, caused by swelling SAP particles at the respective interfaces, which opened up a tiny gap. Strikingly, any SAP particle that was reached by water, absorbed the arriving moisture quasi instantaneously. Assuming spherical hydrogels, i.e. the depth in the line of projection being equal to the visible diameter in the imaging plane, each SAP particle reached an equilibrium content of about 0.1 g water per gram dry SAP.

Finally, the results obtained from specimen G16 are discussed (Fig. 6).

The layer-to-layer deposition intervals between the last two layers (left in the images) was 36 min, whereas it was two minute at any other interface (three instances at 2 cm distances from the right rim of the specimen). The very short interval gave practically no rise to capillary suction at the respective bonding zones. However, the interfacial area at the 36 min delay lets ingress water by capillary suction. As compared to the specimen produced with deposition intervals of 13 min (cf. Fig. 5), the water intake speed was slower, the extent distinctly less and the final height only about two centimeters. Furthermore, the extent of aluminium tape debonding was clearly less than with G1. The SAP particles in G16 swelled upon supply with moisture, whereby they as well reached an equilibrium water content of about 0.1 g water per gram dry SAP.

To summarise the results obtained for Mixture G specimens, only a very short interval of two minute resulted in a tight bond between the extruded layers. Any longer time of 13 or 36 min caused a "weak joint" through which water could be taken in by capillary suction. Interestingly, the time intervals of 36 min resulted in less "active" joints than those obtained with the 13 min intervals. Further studies with respect to absorption and desorption kinetics of specific SAP are needed to clarify these differences. Presumably, the SAP had released a tiny but sufficient amount of internally stored liquid to the matrix after 36 min, which had not been freely available at 13 min.

References

1. Khoshnevis, B.: Automated construction by contour crafting—related robotics and information technologies. Autom. Constr. **13**(1), 5–19 (2004)
2. Nerella, V.N., Hempel, S., Mechtcherine, V.: Micro- and macroscopic investigations on the interface between layers of 3D-printed cementitious elements. In: Santhanam, M., Gettu, R., Pillai, R.G., Nayar, S.K. (eds.) International Conference on Advances in Construction Materials and Systems, vol. 3, pp. 557–565. RILEM Publications, Bagnieux (2017)
3. Nerella, V.N., Krause, M., Näther, M., Mechtcherine, V.: 3D printing technology for on-site construction. CPI Concr. Plant Int. **4**, 36–41 (2016)
4. Schröfl, C., Mechtcherine, V., Kaestner, A., Vontobel, P., Hovind, J., Lehmann, E.: Transport of water through strain-hardening cement-based composite (SHCC) applied on top of cracked reinforced concrete slabs with and without hydrophobization of cracks—investigation by neutron radiography. Constr. Build. Mater. **76**, 70–86 (2015)
5. Zhang, P., Wittmann, F.H., Zhao, T., Lehmann, E.H.: Neutron imaging of water penetration into cracked steel reinforced concrete. Phys. B **405**(7), 1866–1871 (2010)
6. Lieboldt, M., Mechtcherine, V.: Capillary transport of water through textile-reinforced concrete applied in repairing and/or strengthening cracked RC structures. Cem. Concr. Res. **52**, 53–62 (2013)

Corrosion Challenges and Opportunities in Digital Fabrication of Reinforced Concrete

M. Stefanoni[1(✉)], U. Angst[1], and B. Elsener[1,2]

[1] Institute for Building Materials, ETH Zurich, 8093 Zurich, Switzerland
matteost@ethz.ch
[2] Department of Chemical and Geological Science, University of Cagliari, 09100 Monserrato, CA, Italy

Abstract. This contribution addresses corrosion of steel in digitally fabricated concrete. In recent times the concrete processing for digital fabrication applications has been greatly advancing, rising the interest of research institutions, industrial partners, governments and public media. Nevertheless, for a broad large scale application, not just the technological feasibility, but also the long term durability needs to be ensured. This contribution presents a general overview of recently developed digital fabrication technologies and assesses them from the point of view of reinforcement corrosion risks. Experimental results are presented and a number of potential durability issues specific to digital fabrication are raised. On the other hand, we highlight opportunities for making more corrosion-resistant concrete structures by taking advantage of digital fabrication technology.

Keywords: Durability · Preferential attack · Admixtures

1 Introduction

Digital fabrication in the construction industry is increasingly receiving interest and bringing together different fields such as structural engineering, architecture, materials science, etc. It involves the application of digital technologies to the construction industry, promising to revolutionize it. For this reason it has been referred to as the "third industrial revolution" [1]. A particular challenge of digital fabrication of concrete is the introduction of reinforcement, which, in most applications, will be needed to ensure the tensile strength of the structural members or to counteract shrinkage cracking of the concrete. A range of different approaches can be pursued regarding the placement of reinforcement, either in the form of steel or carbon fibres, robot-assembled steel meshes, or manually placed reinforcing steel bars.

Digitally fabricated concrete structures will be exposed to the environment in the same way as traditionally fabricated structures. This includes exposure to CO_2 in the atmosphere with the risk of concrete carbonation, or exposure to chloride-bearing environments, both resulting in the risk of reinforcement corrosion. Corrosion of the reinforcement may significantly impair the safety and serviceability of digitally fabricated structures and also negatively affect their sustainability.

T. Wangler and R. J. Flatt (Eds.): DC 2018, RILEM Bookseries 19, pp. 225–233, 2019.
https://doi.org/10.1007/978-3-319-99519-9_21

There are a number of issues related to corrosion of steel in digitally fabricated concrete that differ from traditional construction, which may give rise to unexpected problems and lead to corrosion mechanisms differing from those experienced in conventional reinforced concrete. On the other hand, digital fabrication also offers a number of opportunities to improve the corrosion performance. As basis for the discussion we consider the review from Wangler et al. [2] as a comprehensive summary of the multitude of approaches in digital fabrication of concrete structures or structural members. We first focus on those techniques already able to implement reinforcement in a digital production context. Afterwards, the methodologies where reinforcing solutions are still under evaluation are taken into account, namely in layered extrusion processes, the most popular digital fabrication technique in concrete technology [2]. The starting point for an overall evaluation may be the differences between requirements for traditional concrete and for the specific application of concrete in digital fabrication. The acceptable operational window of the concrete mix, in terms of hydration kinetics, rheology and size limitations (i.e. coming for the use of nozzles and pumps), becomes much narrower. A very important issue is ageing and drying of the concrete after casting: due to the general absence of formwork in digital fabrication it becomes particularly challenging to provide adequate curing conditions. Finally, we present experimental data that shows the need for carefully evaluating the corrosion behavior of different approaches.

2 Corrosion Challenges in Digital Fabrication

At present, very little experience is available regarding the corrosion (or durability) performance of digitally fabricated concrete structures/elements. However, there are a number of distinct aspects inherent to digital fabrication processes that are expected to substantially increase the risk for corrosion. A first issue applying to all approaches in digital concrete fabrication is the drying-out of the surfaces of digitally produced concrete elements due to the absence of a conventional formwork, leading to a lower degree of hydration, higher porosity, and increased cracking tendency at the concrete surface. In principle, this leads to the need of rethinking the concept of concrete cover, as the protective effect of the cover for the reinforcing steel is impaired.

Further potential corrosion challenges are related to the type and method of placing reinforcing steel. According to Ref. [2] the technologies nowadays able to implement reinforcements in an automated manner (not manually placed) are:

- Steel welding (metal mesh mould approach [3]);
- Slipforming (i.e. Smart Dynamic Casting [4]).

These are discussed below in view of possible corrosion-related durability issues. Subsequently, more general comments will focus on the most widely used concrete digital fabrication approach, that is, the 3D printing by layer extrusion.

2.1 Corrosion Challenges in "Steel Welding" Technology

Robots are nowadays capable of cutting, shaping and welding steel bars. In the so-called mesh mould approach [3], this is utilized to create a reinforcing mesh that can also have the function of a formwork. This leads to a reinforcing steel mesh that differs in many regards from conventional reinforcing steel mats, which may ultimately affect the corrosion behavior. First, relatively short pieces of reinforcing steel are cut (of the order of a few centimeters) and welded to form a mesh. The frequent discontinuities lead to a comparatively high portion of "cut ends" being exposed. At these cut surfaces, a different steel microstructure and surface are exposed than the general surface of the reinforcing steel bars (that may also contain oxides such as mill-scale or native rust layers) [5]. Additionally, also the presence of welding points likely leads to a variation in composition and microstructure of the steel. This marked differences in steel surface conditions (rebar surface, cut sections, welding points) are expected to give rise to galvanic elements or to sites for preferential corrosion attacks. In general, welding is known to be a corrosion triggering process. Reasons are found in microstructural modifications, secondary phases precipitation, impurity inclusions, and morphological defects [6, 7].

The corrosion behavior of mesh mould samples (kindly provided by Alexander Walzer, ETH Zurich), each composed by 4 bars of 2 different diameters and steel types (6 mm B500B and 4.5 mm B500A) welded in 4 points, were tested for corrosion potential and chloride induced corrosion initiation (Fig. 1). A local electrochemical cell was used to perform local open circuit potential (or corrosion potential) measurements (Fig. 1b), in order to evaluate possible differences in the electrochemical behavior of the different surfaces found in mesh mould samples. The electrolyte during the localized testing was 0.1 M NaOH and 0.01 M NaCl. The detailed protocol for such measurements can be found in [8]. After local electrochemical characterization, the mesh mould samples were immersed in a solution containing 0.1 M NaOH and 0.1 M NaCl until corrosion started (ca. 5 h), to evaluate the location of pitting initiation. Finally, local open circuit potential measurements were performed again.

Figure 1a shows the results of the potential measurements before and after the immersion. Whereas only a slight decrease of the potential is observed in areas far from the welding points (on bar 1, 6 mm B500B, and 2, 4.5 mm B500A), a drop of more than 0.2 V was observed at the weldments, indicating the particular susceptibility of these points. This has been confirmed by visual inspection of the meshes after the immersion (Fig. 1c), showing a 100% occurrence of corrosion attack at the welding points but only sporadic localized attack on the bars far from the weldments. Other preferential points of corrosion initiation were also the cut edges of bar 1.

In a structural element that relies on thousands of welding points, the susceptibility to localized corrosion can become a serious issue, as it might lead in the first place to a detachment of the weld metal, already partially visible in Fig. 1c. The problem arises mainly in case of chloride induced localized corrosion. Therefore the application of steel welding technology must be carefully evaluated with respect to possible chloride exposure and transport properties of the concrete cover as the chloride threshold might be substantially lowered by the presence of weldments. Alternatively, the issue raised in Fig. 1 may be also approached from the standpoint of welding methodology [6], that is, to make welding points less susceptible to corrosion.

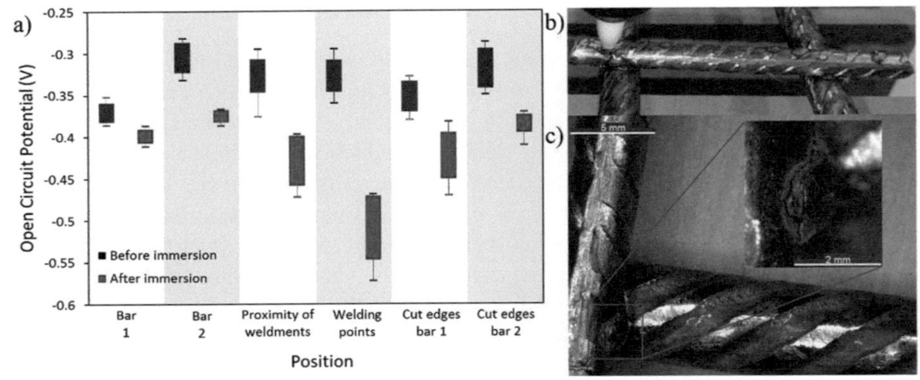

Fig. 1. Corrosion behavior assessment of mesh mould samples. (a) Open circuit potential of the different parts of the samples based on 20 point measurements per part, divided on 4 different samples (bar 1, 6 mm B500B, and bar 2, 4.5 mm B500A). Values showed as before and after immersion in a solution 0.1 M NaOH and 0.1 M NaCl; (b) example of point potential measurement by means of the EC pen; (c) example of preferential corrosion at the weldment with respect to the rest of the sample, with magnification of the corroded welding point.

2.2 Corrosion Challenges in "Slipforming" Technology

Robotically controlled slipforming production of concrete elements consists of a vertically movable, deformable formwork, which concrete enters in the fluid state and exits in a hardening state; this also allows for the direct incorporation of reinforcements [2]. A precise control of the cement hydration (by means of chemical admixture) in the moving formwork is essential [9, 10]. Such a need asks for the use of hydration accelerating admixtures and, in case modern and (presumably) environmentally friendly blended cements are used, an alkaline activator might also be needed. Among the most common admixtures belonging to these two categories we find sodium hydroxide (NaOH), as widespread alkaline activator, and calcium nitrate $Ca(NO_3)_2$ as most common accelerator (note that calcium chloride ($CaCl_2$ was banned decades ago due to corrosion problems). While on the one hand alkali activation has never been taken into account as a possible influencing parameter of reinforcement corrosion, on the other hand calcium nitrate has been found to be an effective inhibitor of chloride induced corrosion [11, 12], earning the title of "multifunctional admixture" for concrete [11].

Here the impact of these two types of admixtures has been studied in the frame of carbonation-induced corrosion, using the setup described in [13, 14]. After the specimens had been fully carbonated, the corrosion rate of embedded steel was measured, upon equilibration to different humidity exposures. The results detailed in Fig. 2 show a relevant effect of the admixtures on the corrosion behavior. The corrosion rate increases 2–3 times when 1 wt% NaOH is added to the mix. In the mix including also 1 wt% $Ca(NO_3)_2$ the corrosion rate increases by a factor 4 with respect to the mix containing only NaOH. The effect of NaOH might simply be due to a higher porosity, as it has been already noticed in the literature [15–17], leading to a higher exposed steel surface area as described in [18, 19]. The effect of $Ca(NO_3)_2$ – a well-known corrosion inhibitor – may initially seem very surprising, but once the electrochemical interaction of the compound with the steel is considered more closely, the reason of such behavior can easily be explained by sound science.

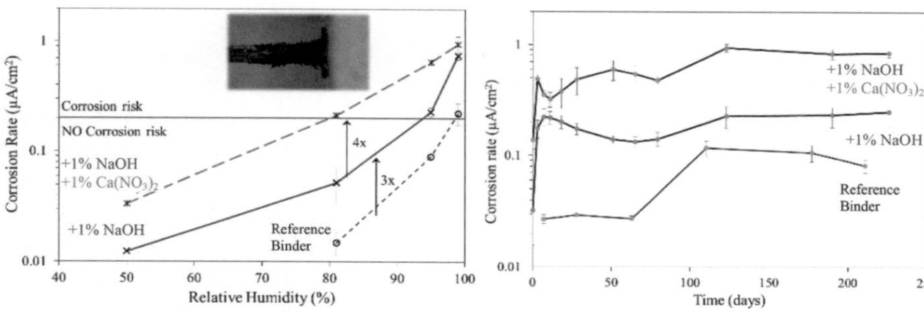

Fig. 2. Corrosion rate of steel in carbonated mortar with and without admixtures. (a) Behaviour at different relative humidity for the reference binder, with 1 wt% NaOH activator and with 1 wt % NaOH and 1 wt% of $Ca(NO_3)_2$ accelerator. The picture shows the formation of corrosion products sticking out of the mortar in case of use of both NaOH and $Ca(NO_3)_2$, never found in any other type of sample. (b) Corrosion rate of the three mixes over time at constant 95% RH (the increase around 100 days is due to replacement of the humidity sensor).

Steel reinforcement in alkaline concrete (with high pH) is protected by the spontaneous formation of a stable iron oxide film (the passive film). In such situation chlorides are usually the only threat, as they can locally break down the passive layer. Nitrate ions (NO_3^{2-}), released by the accelerator admixture, are powerful oxidizing agents, which are able to quickly re-oxidize the steel surface upon chloride induced breakdown, building back the stable protective layer; here explained the corrosion inhibiting nature of calcium nitrate [11, 12]. Moreover, nitrates compete with chloride (and hydroxyl) ions in migrating towards the corroding site, contributing to the repassivating effect.

In carbonated concrete, the pH is lowered to values where the protective passive film is no more stable [20]. Once the passive film is dissolved, the fast oxidation caused by the presence of nitrates is not leading to the formation of a protective layer anymore, but instead iron oxidation, thus corrosion, is enhanced by the presence of nitrates (see Fig. 2). The iron oxidation promoted by nitrate ions is not a new concept, application of such phenomenon has already been implemented in other fields, such as water purification from nitrate ions by means of iron powder [21, 22]. One may hypothesize that the negative impact of nitrates shown in Fig. 2 will only be active as long as nitrates are available. Thus, when they get consumed in the reaction, the corrosion rate may slow down. However, in this work, a decrease of the corrosion enhancing effect of the admixtures could not be noticed, at least not in the first 230 days from the specimens' carbonation.

In conclusion the use of activating and accelerating admixtures can lead to problems at longer ages or in case of a not sufficient concrete cover, leading to dangerous corrosion propagation rates even in case of modest relative humidity (81%, Fig. 2a).

2.3 Corrosion Challenges in "3D Printing by Layer Extrusion" Technology

As it was mentioned earlier, layer extrusion technology is by far the most used technique worldwide for digital fabrication with concrete. It is still not quite clear how to implement the presence of reinforcements in such processing approach [2]. At present different directions have been taken attempting to solve this issue, either by post-printing rebar placement (before concrete hardening) or by using fibre reinforced mixes. The processing needed for concrete pumping, common in 3D printing, can lead to a different microstructure that directly influences both initiation and propagation stages of reinforcement corrosion process; in particular the limited aggregate size tends to translate in higher porosity of the final product. Referring to the previous Sect. 2.2, also layer by layer extrusion requires a fine control of the hardening timing, possibly leading to similar concerns as already described.

However, the main durability issue related to this technology can be found in the so-called "cold joints", a problematic lack of intermixing between consecutive layers [23, 24]. Leaving aside the concerns related to structural properties as a consequence of an imperfect intermixing, the cold joints might represent areas of high porosity and thus "highways" for the ingress of aggressive species [25]. This would endanger the durability of any embedded steel causing the spread of localized corrosion at the printing joints.

According to [2], digital concrete can open a new field of application for fibre reinforced concrete, which could allow for infinite freedom of geometries. In this respect the material of the fibres would play a key role in defining possible corrosion phenomena (in case steel is used), as the concept of fibre reinforced concrete itself rules out the presence of any concrete cover, usually able to screen the steel from external aggressive species. Problems arising from this may be predominantly aesthetic, rather than structural, due to possible efflorescence at very early age of corrosion products from the fibres located in proximity of the concrete surface. Given that a high potential in digital concrete fabrication is seen in the context of architecture, rust stains at the surface may be a particular problem.

3 Opportunities for Corrosion Protection

So far, there is no experience supporting the opportunities for more durable reinforced concrete produced by digital fabrication. Despite some challenges presented above, we are convinced that digital fabrication also offers important possibilities for enhanced corrosion protection and durability monitoring. A selection is briefly outlined below:

Reduced Variability in Material Properties and Geometry: In most applications, reinforcement corrosion could be avoided by ensuring a certain minimum concrete cover and by controlling cover concrete quality (resistance against penetration of chlorides and carbonation). Compared to traditional engineering, the specific requirements in terms of mix design in digital fabrication substantially narrow the freedom in terms of concrete quality, due to the restrictions coming from the processing technique. This is perceived as an opportunity in terms of corrosion protection, because the significantly lower variability in concrete properties (compared to traditional reinforced concrete production) also makes it easier to stipulate clear guidelines (e.g. cover depth, etc.) to ensure the durability. In traditional reinforced concrete durability design, this is not always straightforward, which can be seen for instance from the somewhat confusing prescriptions made in different national standards or national appendices to standards. What is even more important in robotic fabrication, however, is that the variability in concrete properties achieved on site can be expected to be much lower compared to traditional concreting. Additionally, robotic placement of both concrete and reinforcing steel may also enhance the control of the cover depth on-site. In traditional work processes, it is an unfortunate fact that the prescribed concrete cover is not always achieved on-site.

Local Adjustments During Manufacturing: Furthermore, a digitally controlled process may allow for active intervention, during the production process, specifically on the "weak points". Locally controlling the material properties, such as automatically switching to stainless steel in those areas previously defined during the designing step as high corrosion risk or locally adjusting concrete properties may become possible with digital fabrication, while this is close to impossible in traditional processes. For example, the concrete placed around the reinforcing steel may be enriched with a corrosion inhibitor, while all the rest of the concrete is not (because there it is not needed). Another scenario may be to make the cover concrete more dense (resistant against penetration of aggressive species), while the internal parts of sections (not exposed to the environment) may be more porous.

Implementation of Sensors and Protection Instrumentation: Nowadays a great source of added expenses, when it comes to corrosion related maintenance, is the delay in condition assessment and intervention due to poorly organized or difficult monitoring. A great opportunity in digital fabrication lies in the possibility of more easily implementing sensors, such as for chloride and pH [26–29], or predisposed instrumentation for corrosion rate measurements and cathodic protection.

4 Conclusion

This contribution presented experimental results and raised potential steel corrosion problems specifically arising in digital reinforced concrete fabrication technologies. It was shown that the use of activators and accelerators, needed to enhance the rate of yield stress increase for enabling additive manufacturing such as slipforming or layered extrusion, may significantly increase the corrosion rate in the case of carbonation. Additionally, experimental results highlighted the risk of formation of galvanic elements in robotically assembled and welded mesh-type reinforcements, which also increases the corrosion risk. Thus, digital fabrication technology needs to be thought through with respect to durability properties, because potential sources of problems, characteristic for digital fabrication and different from traditional concreting, may show up unexpectedly.

Nevertheless, the highly controlled production process of digitally fabricated concrete elements might represent a great opportunity, for a relevant step ahead in the field of reinforcement corrosion prevention. At present, the main restrictions are the lack of awareness and/or knowledge. Thus, the already highly interdisciplinary field of digital fabrication of concrete (architects, concrete materials scientists, roboticists, structural engineers, etc.) may benefit from embracing established principles from corrosion science.

References

1. The third industrial revolution. Economist **403**(8781), 15 (2012)
2. Wangler, T., Lloret, E., Reiter, L., Hack, N., Gramazio, F., Kohler, M., Bernhard, M., Dillenburger, B., Buchli, J., Roussel, N., Flatt, R.: Digital concrete: opportunities and challenges. RILEM Tech. Lett. **1**, 67–75 (2016)
3. Hack, N., Wangler, T., Mata-Falcon, J., Dörfler, K., Kumar, N., Walzer, A.N., Graser, K., Reiter, L., Richner, H., Buchli, J., Kaufmann, W., Flatt, R.J., Gramazio, F., Kohler, M.: Mesh mould: an on site, robotically fabricated, functional formwork. In: Second Concrete Innovation Conference (2nd CIC), Paper no. 19, Tromsø (2017)
4. Lloret, E., Shahab, A.R., Linus, M., Flatt, R.J., Gramazio, F., Kohler, M., Langenberg, S.: Complex concrete structures: merging existing casting techniques with digital fabrication. Comput. Aided Des. **60**, 40–49 (2015)
5. Angst, U.M., Geiker, M.R., Michel, A., et al.: The steel–concrete interface. Mater. Struct. **50** (2), 143 (2017)
6. Davis, J.R.: Corrosion of Weldments. ASM International (2006)
7. Eid, N.M.: Localized corrosion at welds in structural steel under desalination plant conditions part I: effect of surface roughness and type of welding electrode. Desalination **73**, 397–406 (1989)
8. Stefanoni, M., Angst, U., Elsener, B.: Local electrochemistry of reinforcement steel–distribution of open circuit and pitting potentials on steels with different surface condition. Corros. Sci. **98**, 610–618 (2015)
9. Reiter, L., Palacios, M., Wangler, T., Flatt, R.J.: Putting concrete to sleep and waking it up with chemical admixtures. Spec. Publ. **302**, 145–154 (2015)
10. Reiter, L., Kaessmann, R., Wangler, T., Flatt, R.J.: Strategies to wake up sleeping concrete. In: Proceedings of International Conference on the Chemistry of Cement, Beijing (2015)
11. Justnes, H.: Calcium nitrate as a multi-functional concrete admixture. Concrete **44**(1), 34–36 (2010)
12. Myrdal, R.: Corrosion Inhibitors—State of the Art, COIN Project report no. 22, SINTEF (2010)
13. Stefanoni, M., Angst, U., Elsener, B.: Innovative sample design for corrosion rate measurement in carbonated concrete. In: 11th Annual International Concrete Sustainability Conference, Washington, DC (2016)
14. Stefanoni, M., Angst, U., Elsener, B.: A new setup for rapid durability screening of new blended cements. In: Concrete Innovation Conference, Trosmo (2017)

15. Shi, C.: Strength, pore structure and permeability of alkali-activated slag mortars. Cem. Concr. Res. **26**(12), 1789–1799 (1996)
16. Ravikumar, D., Peethamparan, S., Neithalath, N.: Structure and strength of NaOH activated concretes containing fly ash or GGBFS as the sole binder. Cem. Concr. Compos. **32**(6), 399–410 (2010)
17. Haha, M.B., Le Saout, G., Winnefeld, F., Lothenbach, B.: Influence of activator type on hydration kinetics, hydrate assemblage and microstructural development of alkali activated blast-furnace slags. Cem. Concr. Res. **41**(3), 301–310 (2011)
18. Stefanoni, M., Angst, U., Elsener, B.: Corrosion rate of carbon steel in carbonated concrete—a critical review. Cem. Concr. Res. **103**, 35–48 (2018)
19. Stefanoni, M., Angst, U., Elsener, B.: Electrochemistry and capillary condensation theory reveal the mechanism of corrosion in dense porous media. Sci. Rep-UK **8**(1), 7407 (2018)
20. Bertolini, L., Elsener, B., Pedeferri, P., Redaelli, E., Polder, R.B.: Corrosion of Steel in Concrete: Prevention, Diagnosis, Repair. Wiley (2013)
21. Xiaomeng, F.A.N., Xiaohong, G.U.A.N., Jun, M.A., Hengyu, A.I.: Kinetics and corrosion products of aqueous nitrate reduction by iron powder without reaction conditions control. J. Environ. Sci. **21**(8), 1028–1035 (2009)
22. Yang, G.C., Lee, H.L.: Chemical reduction of nitrate by nanosized iron: kinetics and pathways. Water Res. **39**(5), 884–894 (2005)
23. Roussel, N., Cussigh, F.: Distinct-layer casting of SCC: the mechanical consequences of thixotropy. Cem. Concr. Res. **38**, 624–632 (2008)
24. Nerella, V.N., Hempel, S., Mechtcherine, V.: Micro and macroscopic investigations on the interface between layers of 3D-printed cementitious elements. In: International Conference on Advances in Construction Materials and Systems, Chennai, 3–8 September 2017 (2017)
25. Keita, E., Bessaies-Bey, H., Zuo, W., Belin, P., Roussel, N.: Weak bond strength between successive layers in extrusion-based additive manufacturing: measurement and physical origin. In: Proceedings of 1st International Conference on Concrete and Digital Fabrication, Zurich (2018)
26. Femenias, Y.S., Angst, U., Elsener, B.: Monitoring pH in corrosion engineering by means of thermally- produced iridium oxide electrodes. Mater. Corros. **69**, 76–88 (2018)
27. Femenias, Y.S., Angst, U., Elsener, B.: PH-monitoring in mortar with thermally-oxidized iridium electrodes. RILEM Tech. Lett. **2**, 59–66 (2017)
28. Femenias, Y.S., Angst, U., Caruso, F., Elsener, B.: Ag/AgCl ion-selective electrodes in neutral and alkaline environments containing interfering ions. Mater. Struct. **49**, 2637–2651 (2016)
29. Angst, U., Elsener, B., Larsen, C.K., Vennesland, Ø.: Potentiometric determination of the chloride ion activity in cement based materials. J. Appl. Electrochem. **40**, 561–573 (2010)

The Effect of Print Parameters on the (Micro) structure of 3D Printed Cementitious Materials

J. Van Der Putten(✉), G. De Schutter, and K. Van Tittelboom

Ghent University, Ghent, Belgium
jolien.vanderputten@ugent.be

Abstract. The extrusion-based 3D printing method is one of the main additive manufacturing techniques worldwide in construction industry. This method is capable to produce large scale components with complex geometries without the use of an expensive formwork. The main advantages of this technique are encountered by the fact that the end result is a layered structure. Within these elements, voids can form between the filaments and also the time gap between the different layers will be of great importance. These factors will not only affect the mechanical performance but will also have an influence on the durability of the components. In this research, a custom-made 3D printing apparatus was used to simulate the printing process. Layered specimens with 0, 10 and 60 min delay time (i.e. the time between printing of subsequent layers) have been printed with two different printing speeds (1.7 cm/s and 3 cm/s). Mechanical properties including compressive and inter-layer bonding strength have been measured and the effect on the pore size and pore size distribution was taken into account by performing Mercury Intrusion Porosimetry (MIP) tests. First results showed that the mechanical performance of high speed printed specimens is lower for every time gap due to a decreased surface roughness and the formation of bigger voids. The porosity of the elements shows an increasing trend when enlarging the time gap and a higher printing speed will create bigger voids and pores inside the printed material.

Keywords: 3D printing · Mechanical properties · Microstructure Porosity · Pore size distribution

1 Introduction

In recent years, 3D printing of concrete has become one of the emerging technologies that can minimize the supply chain in the construction process by automatically producing building components with complex geometries layer by layer, directly from a digital model without human intervention [1, 2]. Hence to some extent, it could save material wastage, construction time and manpower. This new construction technique shows many advantages, but also has to challenge a couple of dualities compared to traditional casted concrete. First, the material has to be fluid to prevent any blocking or segregation during extrusion. On the other hand, printed layers must harden quickly to support the superposed layers. Another important inherent aspect of 3D printing is inter-layer bonding. Due to the lack of vibration or external force during deposition,

T. Wangler and R. J. Flatt (Eds.): DC 2018, RILEM Bookseries 19, pp. 234–244, 2019.
https://doi.org/10.1007/978-3-319-99519-9_22

layers must bond in an optimal way in order to create a homogeneous structure and provide adequate mechanical strength. The layered manufacturing process will also include voids in the specimen. These voids will create weak links in the interface behavior and will not only affect the mechanical behavior but also the durability as these voids are ideal ingress paths for chemical substances.

2 Materials and Methods

2.1 Materials and Mix Composition

Ordinary Portland cement (CEM I 52.5 N) was combined with standardized sand with a maximum particle size equal to 2 mm. To increase the flowability of the cementitious mixture, a polycarboxylic ether (PCE) with a molecular weight of approximately 4000 g/mol and 35% solids was used as a superplasticizer. The mix composition can be found in Table 1.

Table 1. Mix composition

Material [-]	Amount [g]
CEM I 52.5 N	675
Sand	1350
Water	246
PCE	0.26% [weight of cement]

The mix design aimed to meet 3D printing requirements, including extrudability, buildability and workability. To classify the composition as extrudable, two demands have to be fulfilled. First, the extrudability was tested by filling the 3D print apparatus and extruding one layer with a length of 300 mm. Once the mortar was expelled without blocking, segregation or bleeding, the mix composition met the first requirement. Second, the deformation of the layer after extrusion was considered. Deformations within a range of 10% were excepted. Buildability was obtained when different layers of material could be printed on top of each other and a general conclusion about the workability was made by performing Vicat tests (manual and automatic). These tests not only evaluate the yield stress of the material but also the buildup capacity as a function of time.

2.2 3D Printing Process

3D Printing Apparatus

A custom-made apparatus was used to simulate an extrusion-based 3D printing process (Fig. 1). The developed system is capable of printing up to 300 mm long concrete layers, at different speeds and different deposition rates. The nozzle of the print equipment has an elliptical shape (28 mm × 18 mm). The height of each layer is equal to 10 mm with an average layer width of 30 mm. For the purpose of this study, linear printing speeds of 1.7 cm/s and 3 cm/s were selected. The effect of a changing pressure inside the nozzle, accompanied with the different printing speeds, was not taken into account in this research.

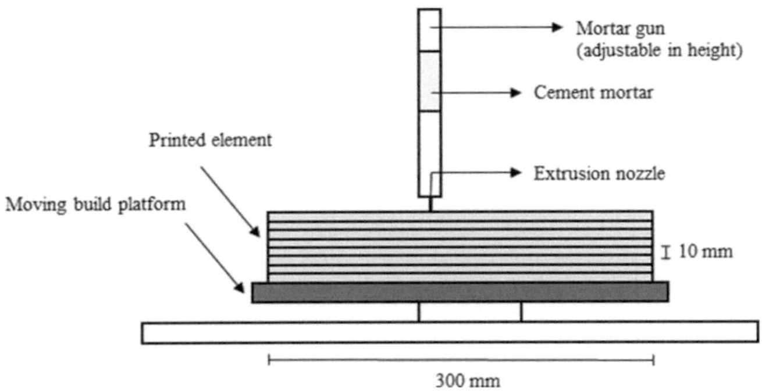

Fig. 1. Schematic illustration of the extrusion-based 3D printing process

Specimen Preparation

Sample preparation consists of filling the 3D printing apparatus and extruding material through the nozzle with a constant speed (1.7 cm/s and 3 cm/s). A single base layer, with a length of approximately 300 mm, was extruded for each specimen. After a predetermined time interval (0, 10 or 60 min), another layer was deposited on top of the previous one. In case of a zero time gap, the two layers were printed form the same batch of material. After 5 min, the material was difficult to print and therefore a fresh mortar mix was deposited on the first layer with 10 and 60 min time gap intervals. All the specimens were cured in a standardized environment (20 ± 3 °C) until the day of testing (i.e. 28 days after printing).

2.3 Surface Roughness

The surface roughness of the printed specimens was determined by the in house developed Automated Laser Measurement (ALM) technique, scanning the surface of the layers with a high precision beam, equipped with two stepping motors controlling the motion in X and Y direction. The profile measurements are used to calculate the center-line roughness (R_a) value of the printed specimens. This value is determined with an average line drawn through the profile and the center-line over a selected reference length (200 mm in Y-direction, 15 mm in X-direction), selected to include important roughness features, but exclude errors of form. Using the ALM, the R_a value can be derived with an accuracy of 7 µm. The surface roughness was measured every 5 and 50 mm in respectively X and Y direction (Fig. 2). Roughness measurements were performed for every printing speed.

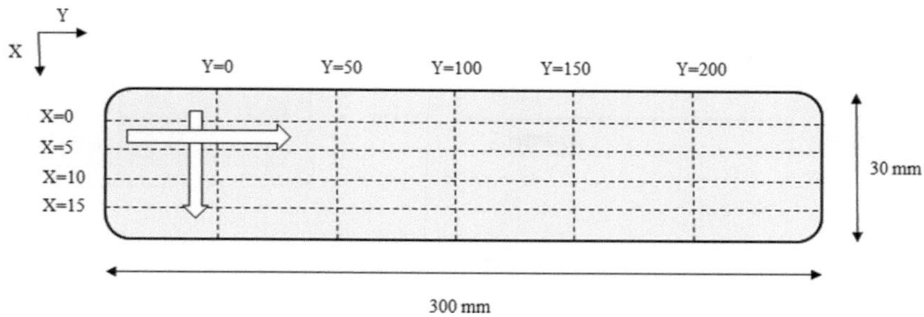

Fig. 2. Schematic representation of the surface roughness measurement (ALM)

2.4 Mechanical Properties Testing

Compressive Strength

The compressive strength of the specimens was tested in a universal testing machine under load control at a rate of 100 N/sec and is schematically illustrated in Figs. 3 and 4. For the compressive strength test, small cylinders were saw-cut from the printed elements with a height and diameter respectively equal to 20 and 25 mm. The specimens were loaded perpendicular to the print direction. At least 3 specimens were tested for every time gap and printing speed.

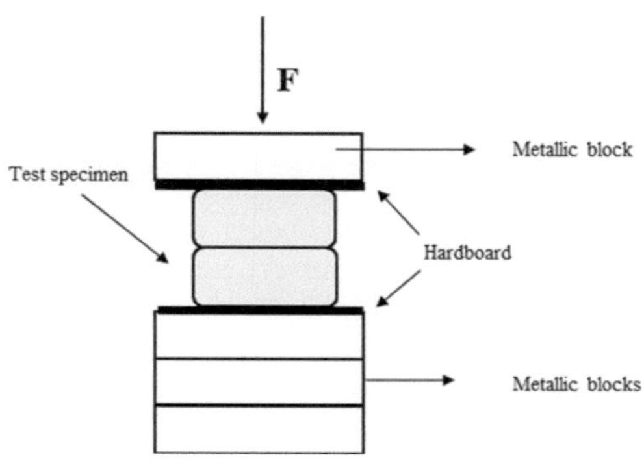

Fig. 3. Schematic illustration of the compressive strength test setup

Inter-layer Bonding Strength

The schematic illustration of the inter-layer bonding strength test setup is shown in Fig. 5. The test specimens were small cylinders, saw-cut from the original printed elements, with a height and diameter respectively equal to 20 and 25 mm. These specimens have analogue dimensions as the ones used for testing the compressive strength (Fig. 4). Two metallic brackets were epoxy glued on the top and bottom of the printed specimens. The inter-layer bonding strength test was conducted in an universal testing machine under displacement control at the rate of 50 N/s. Special attention was taken to align the specimens in order to avoid any eccentricity. At least 3 specimens were tested for every time gap and printing speed.

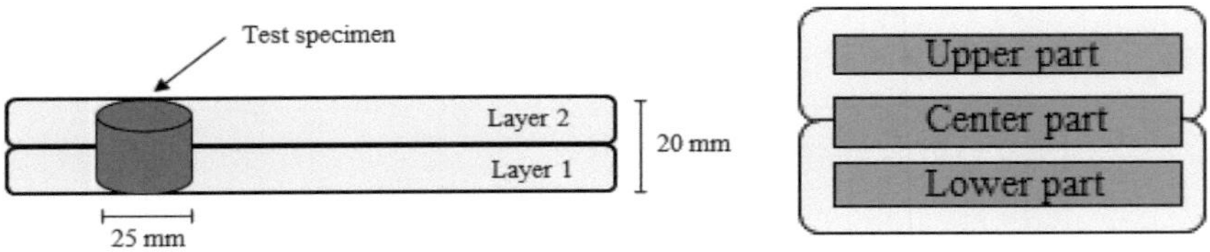

Fig. 4. Schematic illustration of sampling elements (left) and indication of the different zones in a printed element (right)

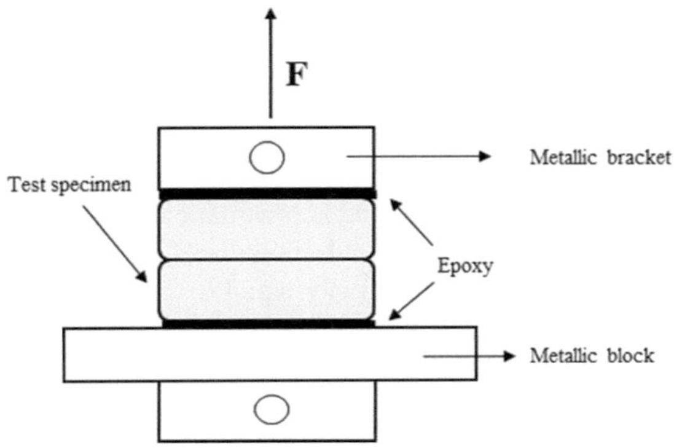

Fig. 5. Schematic illustration of the inter-layer bonding test setup

2.5 Porosity Measurements and Pore Size Distribution

Porosity tests were conducted on small cylinders, saw-cut from the original printed elements, with a height and diameter respectively equal to 20 and 14 mm (Fig. 4). The cylindrical elements were dried in an oven at 35 °C until constant weight. The weight of the samples at this stage is designated as dry weight and indicated as m_1 [g]. Thereafter, the samples were submerged in water with a temperature of 20 °C until complete saturation. The weight of the samples in soaked water was measured (m_2 [g]), followed by rolling the different sides on a damp cloth measured as m_3 [g]. From this, the apparent porosity (P_a) of the samples was calculated using Eq. (1).

$$P_a = (m_3 - m_1)/(m_3 - m_2) \cdot 100 \tag{1}$$

To measure the pore sizes and pore size distribution, Mercury Intrusion Porosimetry (MIP) was used. As a specific pressure corresponds to an aperture of a pore, and the amount of mercury intrusion approximates to the pores volume, the amount and size of the pores can be determined. MIP is performed for every series on samples taken from the upper, lower and center region after freeze-drying the samples for 7 days.

3 Results and Discussion

3.1 Surface Roughness

Table 1 gives an overview of the R_a-values of the printed specimens. The results in x-direction show the average values of 5 measurements, in y-direction 4 measurements on different positions were performed. One can see that the applied printing speed has a significant influence on the surface roughness. A lower speed introduces a higher roughness, which will have a positive effect on the mechanical properties, as can be seen in Sects. 3.2 and 3.3. In case of a lower printing speed, the surface roughness is more pronounced in X-direction. For the higher printing speed, the roughness values are more or less the same in both directions. This will also affect the anisotropic behavior of the printed elements, but these results are not included in this paper.

3.2 Compressive Strength

Figure 6 presents the compressive strength and associated standard deviation for a mixture printed with different speeds and different time gaps. One can see that the compressive strength decreases when increasing the time gap and the overall strength of the specimens is higher when printing on a lower speed. Table 2 also indicates the strength loss due to a layered manufacturing process. This loss is given as the ratio of ‘difference in compressive strength with certain time gap’ to the compressive strength of the basis one without time gap between the layers (Eq. (2)).

$$\Delta = \frac{\text{Strength}_0 - \text{Strength}_{\text{time gap}}}{\text{Strength}_0} \tag{2}$$

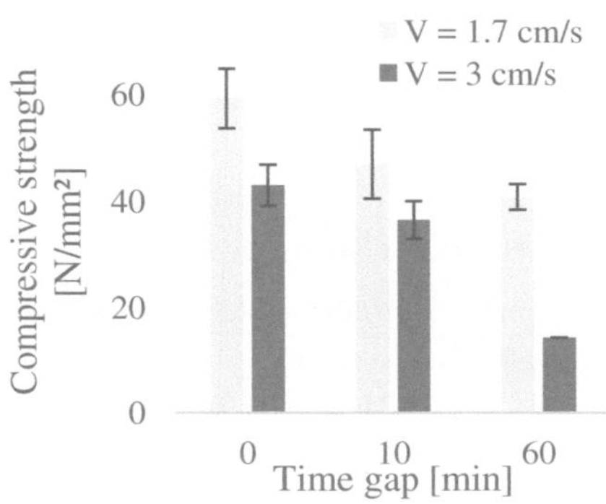

Fig. 6. Compressive strength

Table 2. Average roughness R_a of elements fabricated with different printing speeds

Printing speed [cm/s]	$R_{a,x}$ [mm]	stdev [-]	$R_{a,y}$ [mm]	stdev [-]
1.7	0.95	0.05	0.68	0.15
3	0.39	0.07	0.46	0.08

3.3 Inter-layer Bonding Strength

Figure 7 presents the inter-layer bonding strength and associated standard deviation for the different elements. For every printing speed, the interlayer bonding decreases when increasing the time gap and the overall strength of the printed specimens is higher in case of a lower speed. As indicated in literature [1, 3, 4], the inter-layer bonding strength between two different layers is related to the surface roughness. Due to the lower roughness in case of higher speed, the bonding between the layers decreases. Based on the results of Table 3, one can conclude that there is almost no bonding between the layers printed with a time gap of 60 min. The formation of this cold joint can be explained by a moisture exchange phenomenon that states, when the bottom layer becomes drier, it absorbs more water from the freshly deposited top layer and simultaneously, some air inside the bottom layer escapes out of it. This air stays entrapped at the interface and causes poor bonding [2]. Table 3 also indicates the strength loss due to a layered manufacturing process and this is again higher in case of a higher printing speed (Table 4).

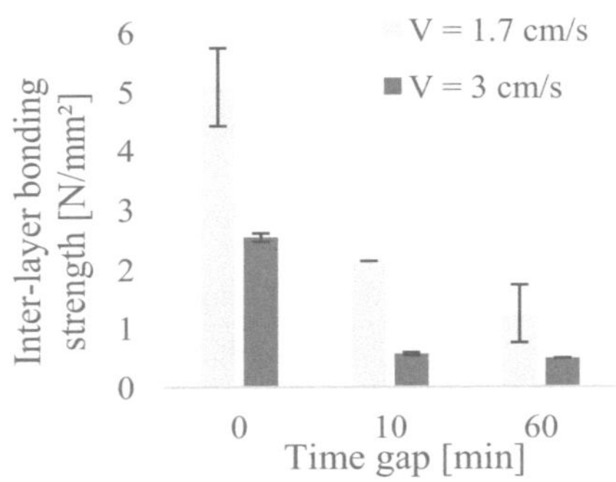

Fig. 7. Inter-layer bonding strength

Table 3. Compressive strength

Time gap [min]	F [N/mm^2]	stdev [-]	Loss Δ [%]
V = 1.7 cm/s			
0	59.21	5.58	-
10	46.92	6.45	20.77
60	40.79	2.40	31.12
V = 3 cm/s			
0	42.99	3.85	
10	36.49	3.50	15.12
60	14.29	-	66.75

Table 4. Inter-layer bonding strength

Time gap [min]	F [N/mm^2]	stdev [-]	Loss Δ [%]
V = 1.7 cm/s			
0	5.08	0.66	
10	2.14	-	57.51
60	1.24	0.49	75.43
V = 3 cm/s			
0	2.54	0.07	
10	0.56	0.02	88.87
60	0.49	-	96.53

3.4 Microstructure and Porosity

Based on Fig. 8A, it becomes clear that comparing different parts of a printed element with a time gap equal to zero, the pore size distribution is comparable for every region. This in contrast to what is obtained for elements with a 10 min time gap, where the amount of pores with a specific diameter is higher in the center and lower part of the layer (Fig. 8B). This probably indicates that printing a second layer will compact the layer underneath when printing without a time gap. In case of a time gap equal to 10 min, the first layer is already hardened to some extent and less affected by compaction due to printing the second layer. The same conclusion can be made for a time gap of 60 min (Fig. 8C). Figures 8D–F show a comparison between the different parts of a printed element, fabricated with different time gaps. Comparing the upper part, one can only see a shift towards bigger pores in case of a time gap equal to 10 or 60 min. The amount of pores is more or less the same. In case of the center and lower part of the specimens, there is not only a shift towards bigger pores but also the amount of pores increases significantly compared to a specimen fabricated without time gap. This can again be explained by a moisture exchange phenomenon where a drier bottom layer absorbs more water from the super positioned fresh layer and simultaneously, some air inside the bottom layer escapes out of it and will induce bigger pores in the lower and center part of the specimen. The same conclusions can be made for a higher printing speed. These graphs are not included in this paper.

Figure 9 shows a comparison between the pore size distribution of an upper part element, printed with a time gap equal to zero, for different printing speeds. One can see clear that printing on a higher speed will induce a higher amount of bigger pores in the system, what will have a negative effect on the durability of the material. Consequently, further research on this topic is mandatory.

Mercury Intrusion Porosimetry makes it possible to measure pores ranging from 0.005 μm to 10 μm. These pores are capillary pores and are formed during the hydration process of cement. The capillary pores will describe the microstructure of the element and are mainly affected by the amount of cement, the water/cement ratio and

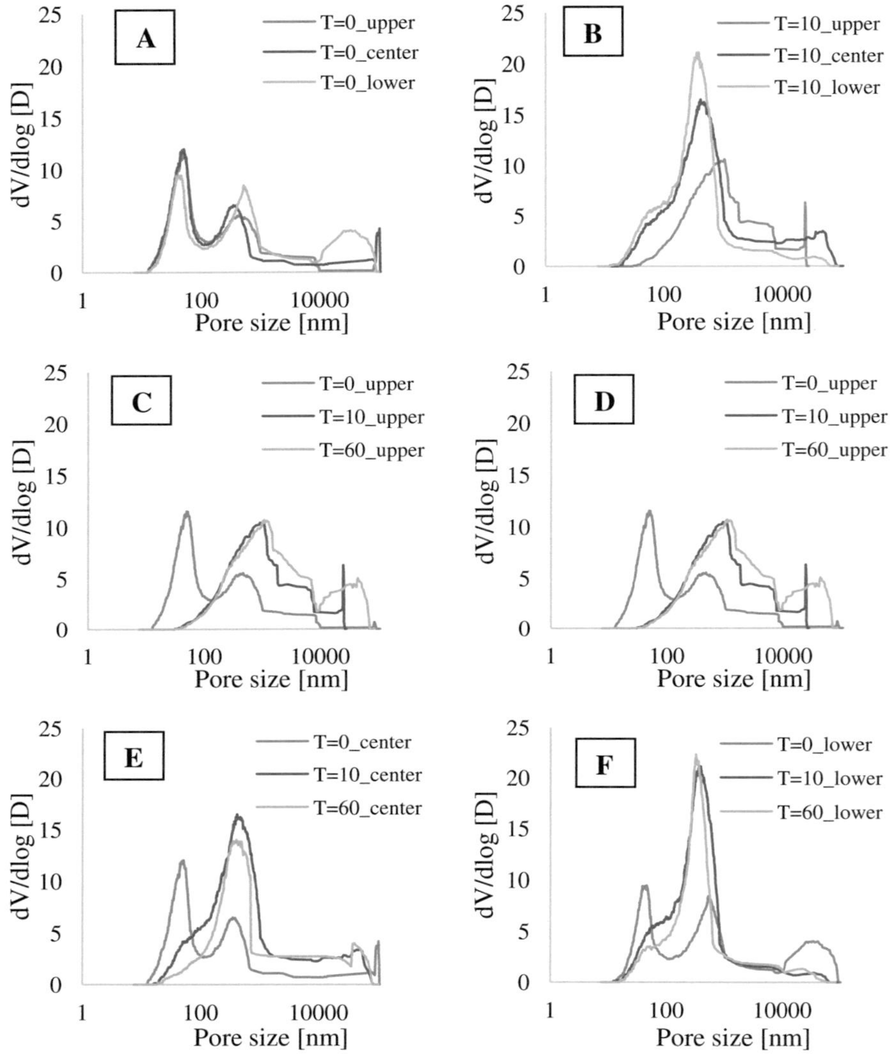

Fig. 8. Pore size distribution of elements with different time gaps and low printing speed

the degree of hydration [5]. Another type of porosity is the open porosity. These results are given by Fig. 10. Open porosity defines the pores on the outside of the printed elements. Here a decreasing trend is visible when increasing the time gap between the different layers. One can also see that in case of a high printing speed, the porosity is less affected.

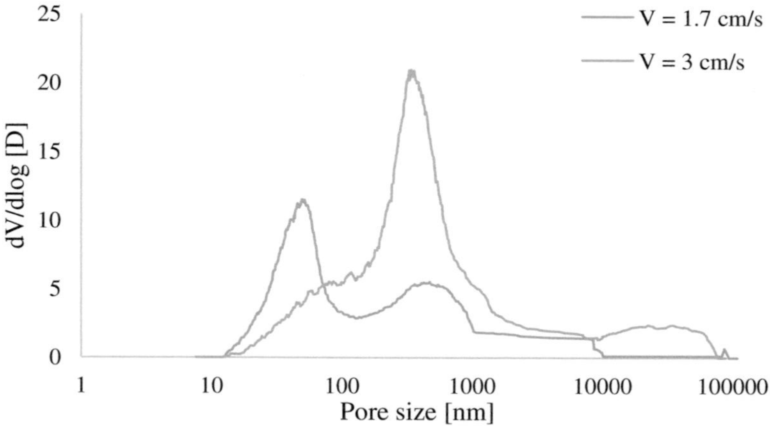

Fig. 9. Pore size distribution of the upper region in case of two different printing speeds

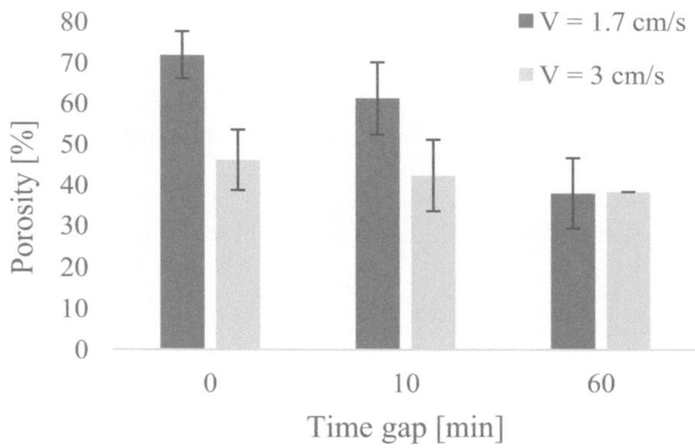

Fig. 10. Open porosity of printed elements fabricated with different speeds

4 Conclusions

The effects of different delay times and a changing print speed on the mechanical properties and the (micro)structure of 3D printed elements were investigated in this research. The mix composition was kept the same during all the experiments. The following conclusions can be drawn from this study:

1. The surface roughness of specimens fabricated with a higher printing speed was lower compared to the ones printed with a lower speed. There was a significant difference in roughness between the X- and Y- direction in case of a low printing speed. The difference in roughness was less pronounced for a higher printing speed.

2. The compressive strength and the inter-layer bonding strength of the elements was higher when fabricating the elements on a low speed and in case there was no time gap between the different layers.
3. Increasing the delay time induces a higher amount of bigger pores in the lower and center part of the printed specimens.
4. Increasing the printing speed introduces bigger pores. This not only affects the mechanical properties, but also the durability of the material.

Acknowledgements. The authors would like to acknowledge the support by EFRO for this C3PO-project with a total amount of € 668 320.41.

References

1. Zareiyan, B., Khoshnevis, B.: Interlayer adhesion and strength of structures in contour crafting: effects of aggregate size, extrusion rate, and layer thickness. Autom. Constr. **81**, 112–121 (2017)
2. Panda, B., et al.: Measurement of tensile bond strength of 3D printed geopolymer mortar. Measurement **113**, 108–116 (2018)
3. Zareiyan, B., Khoshnevis, B.: Effects of interlocking on interlayer adhesion and strength of structures in 3D printing of concrete. Autom. Constr. **83**, 212–221 (2017)
4. Eduardo Nuno Brito Santos, P.M.D., Eduardo, J.: Factors affecting bond between new and old concrete. Mater. J. **108**(4), 449–456 (2011)
5. Boel, V.: Microstructuur van zelfverdichtend beton in relatie met gaspermeabiliteit en duurzaamheidsaspecten, p. 320. Ph.D. dissertation, University of Ghent (2007)

Compressive Strength and Dimensional Accuracy of Portland Cement Mortar Made Using Powder-Based 3D Printing for Construction Applications

Ming Xia(✉), Behzad Nematollahi, and Jay Sanjayan

Centre for Sustainable Infrastructure, Faculty of Science, Engineering and Technology, Swinburne University of Technology, Hawthorn, VIC, Australia
{mxia, bnematollahi, jsanjayan}@swin.edu.au

Abstract. An innovative methodology has recently been developed by the authors of this study for geopolymer formulations for the requirements and demands of commercially available powder-based 3D printers. In this study, the formulation is extended to conventional Portland cement to expand the scope of printable materials that can be used in the commercially available powder-based 3D printers for construction applications. A Portland cement-based powder composed of Portland cement, amorphous calcium aluminate and fine silica sand was developed for powder-based 3D printing process. Effects of different printing parameters including binder saturation level (100%, 135% and 170%) and shell to core ratio (1:1 and 1:2) on dimensional accuracy and compressive strength of the green specimens have been investigated. A compressive strength of up to 8.4 MPa was achieved for the '*green*' 3D printed samples *before* any post-processing process. The results indicated that the increase in the binder saturation level and the change in the Shell/Core ratio from 1:1 to 1:2 significantly increased the compressive strength, but considerably reduced the linear dimensional accuracy of the green samples. The compressive strength and linear dimensional accuracy of the green samples exhibited an anisotropic behavior, depending on the testing direction.

Keywords: Powder-based 3D printing · Additive manufacturing
Portland cement · 3D concrete printing · Construction

1 Introduction

Three-dimensional (3D) printing, also known as additive manufacturing (AM) is a group of emerging techniques for fabricating a wide range of structures with complex geometries from digital models. The process involves of printing successive layers of materials that are formed on top of each other. Several industries including aerospace, automotive, biomedical have already explored the benefits of adopting this technology as an integral part of their product manufacturing process [1].

T. Wangler and R. J. Flatt (Eds.): DC 2018, RILEM Bookseries 19, pp. 245–254, 2019.
https://doi.org/10.1007/978-3-319-99519-9_23

In the recent years, the construction industry has attempted to adapt this technology for construction applications owing to its potential use for direct construction of buildings and other complex structures of virtually any shape without the use of expensive formwork. Elimination of the formwork would result in considerable cost savings, as the formwork represents 35–60% of the cost of construction of concrete structures [2]. In addition, the elimination of formwork also reduces the amount of wastages in construction, and therefore improves sustainability in construction, since the formwork represents a significant source of waste as all of it is discarded sooner or later.

To date, the construction industry has explored two different 3D concrete printing (3DCP) techniques. One of these techniques is extrusion-based 3DCP, which is analogous to fused deposition modeling (FDM) method that extrudes cementitious material from an extruder mounted on a gantry, crane or a robotic arm to print a concrete component layer-by-layer. Recently the authors of this study developed a 3D printable fly ash-based geopolymer suitable for extrusion-based 3DCP for construction applications [3, 4].

Another technique is powder-based 3DCP, which is capable of making complex structures with subtle details and intricate shapes. Figure 1(a) schematically demonstrates the powder-based 3DCP technique. A thin layer of powder is spread over the powder bed surface. Subsequently, binder droplets are selectively applied on the powder layer by a print-head, causing powder particles to bind to each other. Repeating the described steps, the built part is completed and removed after a certain drying time and unbound powder is removed by using an air blower. One of the advantages of this technique is that overhanging structures can be made without the necessity for using a supporting structure. This technique is an off-site process, which is highly suitable for manufacture of customized building components. Examples of technologies developed based on the powder-based 3DCP technique include D-shape [5] and Emerging Objects [6].

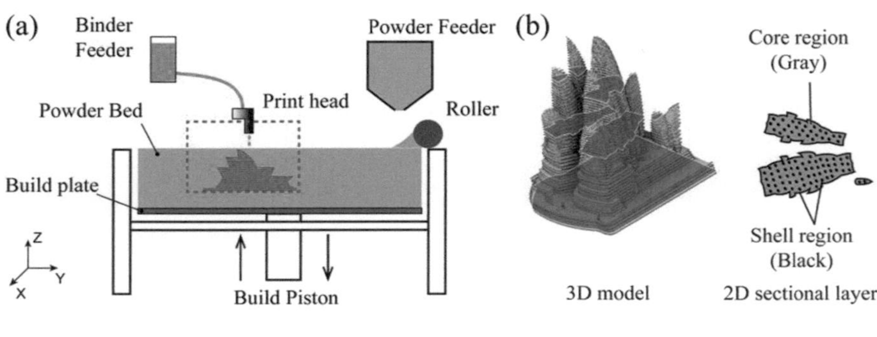

Fig. 1. (a) Schematic illustrations of powder-based 3DCP, (b) Schematic illustration of binder saturation.

Although the powder-based 3DCP technique can offer several advantages in the construction industry, there are a number of challenges to be overcome before the technique is fully utilized. One of the main challenges is that the proprietary printing materials that are typically used in the commercially available powder-based 3D printers are not suitable for the construction applications. To tackle this limitation, the authors of this study have been working on developing innovative methodologies for formulating geopolymer based materials which can be used in the commercially available powder-based 3D printers for construction applications. Geopolymer is a sustainable alternative to conventional Portland cement. It is synthesized by alkaline activation of fly ash and/or slag, being industrial by-products of coal power stations and iron manufacture, respectively [7]. The authors of this study have recently developed several slag/fly ash-based geopolymer formulations suitable for powder-based 3DCP for construction applications [2, 8–11].

Conventional Portland cement has been considered as the master construction material for its high strength and stability as well as its low cost for over 100 years, and will probably be produced and used for at least the next 100 years [12]. However, the setting characteristics of Portland cement limit its use for powder-based 3DCP. Few studies have reported the use of other types of cementitious materials. For instance, Gibbons et al. [13] conducted a preliminary study to investigate the feasibility of using a mixture of poly vinyl alcohol and rapid hardening Portland cement (RHPC) for powder-based 3DCP process for manufacture of biomedical implants. The printed specimens exhibited a maximum modulus of rupture of 2.4 MPa after 26-day water immersion at ambient temperature. The low strength of the developed RHPC powder limits its use for construction applications. Maier et al. [14] investigated a mixture of flash-setting calcium aluminate cement (CAC) for powder-based 3D printing to fabricate a bone regeneration scaffold. A compressive strength of up to 20 MPa was reported for the printed specimen after 3-day water immersion. The relatively high cost of the developed CAC powder (as compared to conventional Portland cement) may limit its use for large-scale construction applications. To the best of the authors' knowledge no study has yet been reported the use of conventional Portland cement based materials for powder-based 3D printing for construction applications. This study aims to fill this knowledge gap by developing a new printable Portland cement-based material to expand the severely limited scope of cementitious materials that can be used in the commercially available powder-based 3D printers for construction applications.

In the powder-based 3DCP process, binder saturation is a decisive parameter that directly affects the binder/powder interaction [11], which is somewhat similar to water-to-cement ratio in the conventional concrete casting process. The binder saturation is defined as the ratio between the volume of deposited liquid binder (V_{Binder}) and the volume of pores in the powder bed (V_{Pores}). As schematically illustrated in Fig. 1(b), two sub-variables of the binder saturation are employed in the powder-based 3D printers, namely shell and core. These two sub-variables are used to be able to print a sample in a short time, providing adequate stability and avoiding oversaturation which can lead to distortion [15, 16]. The shell refers to the region comprising the edges of the sample and parts of the interior area within the edges of the sample. The core refers to the remaining interior areas within the edges of the sample. The high binder saturation results in bleeding of the binder into the surrounding powder. However, if the binder saturation is too low, the printed sample would have extremely weak green strength owing to the poor bonding between particles of the powder [17]. It is therefore necessary to understand the quantitative influence of the binder saturation on the properties of powder-based 3D printed samples. This study investigates the quantitative effects of the binder saturation level and shell to core ratio on dimensional accuracy and compressive strength of 3D printed specimens using the novel Portland cement-based material.

2 Experimental Procedures

2.1 Materials

A conventional Portland cement conforming to the Australian Standard, AS 3972 general purpose (Type GP) cement was used in this study. The percentages of C_3S, C_2S, C_3A and C_4AF as the main constituents of Portland cement were 57.59%, 14.87%, 4.10% and 13.94%, respectively. A small amount of an accelerating additive was used to reduce the setting time of Portland cement, making it suitable for the powder-based 3DCP process. A high purity fine silica sand with an average particle size of 184 μm supplied by TGS Industrial Sand Ltd., Australia was also used in this study to function as an inert filler in the Portland cement-based powder to improve its printability. The raw materials were thoroughly dry mixed for 10 min in an Eirich mixer to achieve a homogenous mixture (visually assessed). The particle size distribution (PSD) of the developed Portland cement-based powder is given in Fig. 2. The PSD analysis result showed the average particle size, D_{10}, D_{50} and D_{90} values of the powder were 39.43 μm, 0.69 μm, 17.15 μm and 69.79 μm, respectively.

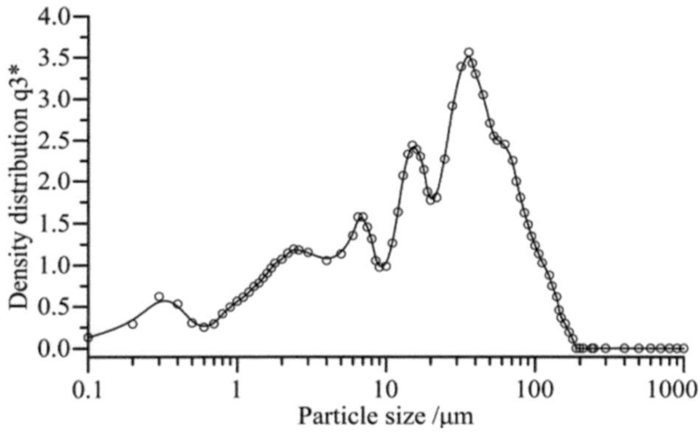

Fig. 2. Particle size distributions of the developed printable Portland cement-based powder.

2.2 Printing Process and Test Methods

3D Printing Process. A commercial powder-based 3D printer (Zprinter® 150) manufactured by Z-Corp, USA was used in this study. A commercial clear binder solution (Zb® 63) with a similar viscosity to pure water supplied by the printer's manufacture was used in this study. Details of the Zprinter® 150 and Zb® 63 binder can be found in the authors' previous work [8].

Figure 3 presents the 3D printed specimens using the developed Portland cement-based powder. A 20 mm cube specimen was printed to characterize dimensional accuracy and compressive strength of 'green' specimens. The green specimen refers to printed sample before any post-processing. An Opera House model with intricate shape was also designed and printed to demonstrate the feasibility of using the developed Portland cement-based powder for printing components with complex geometry. The powder layer thickness was set to 0.1016 mm. Three binder saturation levels of 100%, 135% and 170% and two shell to core ratios of 1:1 and 1:2 were selected in this study. The printed specimens were left within the powder bed for 2 h, and then de-powdering process was performed using a compressed air blower to remove the unbounded powder.

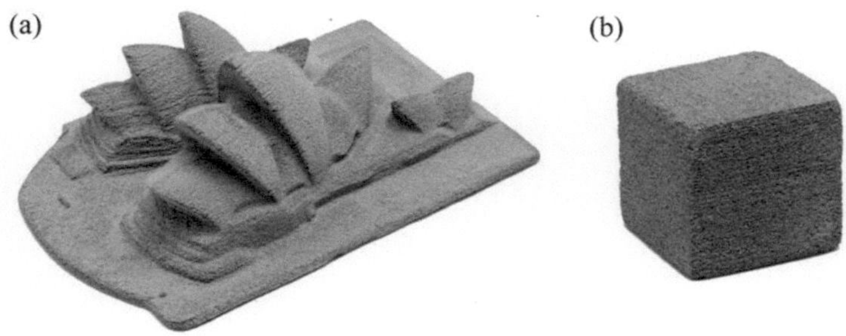

Fig. 3. 3D printed samples using the developed Portland cement-based powder (a) Opera House, (b) 20 mm Cubic structure.

Linear Dimensional Accuracy. A digital Vernier caliper with an accuracy of up to 0.01 mm was used to measure the dimensions of green cubic samples in three directions, namely X-direction (the direction of binder jetting), Y-direction (the direction of powder layer spreading) and Z-direction (the direction of layer stacking). The linear dimensional error was calculated based on the following equation:

$$Error = L_{\text{actual}} - L_{\text{nominal}} \tag{1}$$

Where L_{actual} is the measured length, whereas L_{nominal} is the length of the digital model. A population of 10 samples for each testing direction was used. For each sample, three measurements were taken for each testing direction and the mean error values were calculated to assess the linear dimensional accuracy.

Compressive Strength. The compressive strengths of green cubic samples in both X-direction and Z-direction were measured under load control at the rate of 0.33 MPa/s. A population of 10 samples for each testing direction was used.

3 Results and Discussions

3.1 Linear Dimensional Accuracy

The primary benefit offered by the powder-based 3DCP process is the capability of producing building components with intricate shapes without the use of expensive formwork. One of the important factors that define the printability of a newly developed powder system is the printing accuracy. Figure 4 presents the results of the linear dimensional accuracy of green cubic samples.

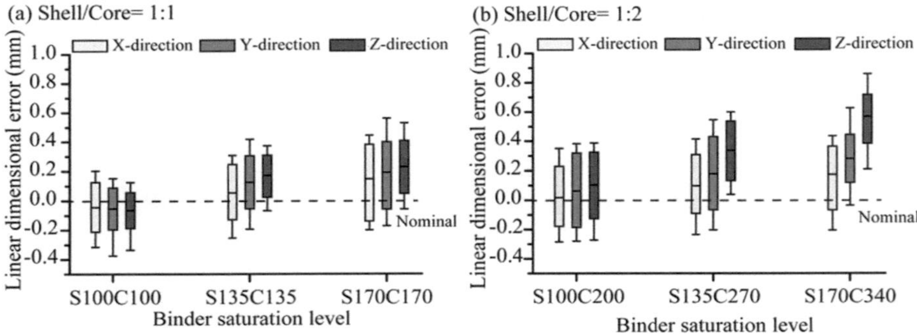

Fig. 4. Linear dimensional accuracy results of green cubic specimens printed with (a) Shell/Core = 1:1, (b) Shell/Core = 1:2. (Box: Mean ± standard deviation; Whisker: Minimum to Maximum).

According to Fig. 4, the mean error values of the samples printed with saturation levels of 135% and 170% in all directions were always greater than zero. This indicates that the measured dimensions of printed samples in all directions were more than those of the digital model. This pattern is true regardless of the shell to core ratio. With regards to the samples printed with a saturation level of 100%, the measured dimensions of the samples printed with Shell/Core of 1:1 were less than the digital model, whereas the measured dimensions of the samples printed with Shell/Core of 1:2 were more than the digital model.

It should be noted that an anisotropic phenomenon was observed regarding the linear dimensional accuracy of the green samples depending on the testing directions. For all binder saturation levels, the Z-direction had the highest mean error values. In other words, regardless of the binder saturation level, the Z-direction had always the lowest linear dimensional accuracy. This phenomenon is more pronounced in the samples printed with Shell/Core of 1:2. This anisotropic phenomenon might be associated with the different rates of binder penetrating in vertical (Z) direction and spreading in lateral (X and Y) directions [15, 18]. On the other hand, for all binder saturation levels, the X-direction had the lowest mean error values, thereby the highest linear dimensional accuracy. This may be because the X-direction (i.e. the binder jetting direction) is not affected by the powder spreading that takes place in Y-direction.

As can be seen in Fig. 4, the increase in the binder saturation level considerably increased the mean error values in all directions. For example, for the green samples printed with Shell/Core of 1:1 the mean error value in Z-direction significantly increased from −0.07 ± 0.08 mm in the case of S100C100 to 0.24 ± 0.17 mm in the case of S170C170. In other words, the increase in the binder saturation level significantly reduced the linear dimensional accuracy of the green samples in all directions.

According to Fig. 4, the change in the Shell/Core ratio from 1:1 to 1:2 significantly increased the mean error values in all directions. This is true regardless of the binder saturation level. For instance, for the samples printed with Shell/Core of 1:1 the mean error value in Z-direction significantly increased from 0.13 ± 0.23 mm in the case of S170C170 to 0.57 ± 0.18 mm in the case of S170C340 with a Shell/Core of 1:2.

In summary, it can be concluded that the increase in the binder saturation level and/or the shell to core ratio significantly reduced the linear dimensional accuracy of the green samples printed using the developed Portland cement-based powder. This might be explained by the bleeding mechanism [19], since at higher binder saturation level the excess binder spreads outside the edges of the printed sample, which results in a reduction of the linear dimensional accuracy.

3.2 Compressive Strength

Figure 5 shows the compressive strength of green samples printed using the developed Portland cement-based powder in both X-direction and Z-direction.

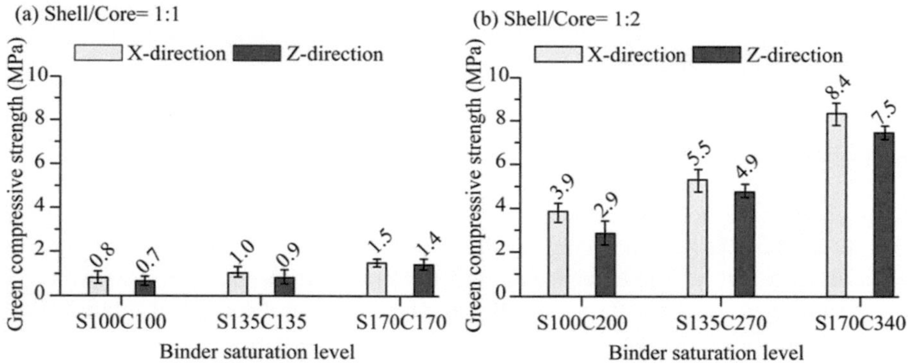

Fig. 5. Compressive strength of green cubic specimens printed with (a) Shell/Core = 1:1, (b) Shell/Core = 1:2.

As is shown in Fig. 5, an anisotropic phenomenon was observed regarding the compressive strength of the green samples depending on the loading directions. Regardless of the saturation level and shell to core ratio, the compressive strength was always higher in X-direction than in Z-direction. The anisotropy effect was more pronounced in the samples printed with Shell/Core of 1:2 as compared to Shell/Core of 1:1. For example, the compressive strength of the green samples printed with

S100C100 in X-direction was 14% higher than that in Z-direction. However, the compressive strength of the green samples printed with S100C200 in X-direction was 34% higher than that in Z-direction. This anisotropic phenomenon might be related to the preferential orientation of the powder particles during the powder spreading process [18].

According to Fig. 5, the change in the Shell/Core ratio from 1:1 to 1:2 significantly increased the compressive strength of the green samples. This is true regardless of the testing direction and binder saturation level. It is interesting to note that the rate of increase in the compressive strength of the green samples in X-direction was higher than in Z-direction. For instance, in X-direction the compressive strength of S100C100 samples was 4.9 times lower than that of the S100C200 samples. However, the corresponding value in Z-direction was 4.1 times. The green samples printed with Shell/Core of 1:1 exhibited relatively low compressive strength ranging from 0.7 to 1.5 MPa, depending on the testing direction and saturation level. However, it should be noted that the strength of all green samples printed with Shell/Core of 1:1 was already more than enough for the de-powdering process.

In both directions, the increase in the binder saturation level significantly increased the compressive strengths. However, the rate of increase in the compressive strength of the green samples with Shell/Core of 1:2 was higher than that with Shell/Core of 1:1. For instance, for Shell/Core of 1:1 and in X-direction the compressive strength of S170C170 samples was 1.9 times higher than that of S100C100 samples. However, the corresponding value for the Shell/Core of 1:2 is 2.15 times.

The inferior compressive strength of the green samples with lower binder saturation level and shell to core ratio is probably due to incomplete hydration process, which results in a weak bond between powder particles. Higher binder saturation level and core saturation use a higher volume of binder during the printing process, resulting in superior bonding between the powder particles.

4 Conclusions

A novel printable Portland cement-based powder was developed in this study which can be used in the commercially available powder-based 3D printers for the construction applications. The influence of binder saturation level and shell to core ratio on the linear dimensional accuracy and compressive strength of the green printed specimens was investigated. The following conclusions are drawn:

1. The increase in the binder saturation level and/or the core saturation significantly reduced the linear dimensional accuracy of the green samples. This is true regardless of the testing direction. This is probably because the excess binder at higher binder saturation spreads outside the edges of the printed sample, resulting in reduction of the linear dimensional accuracy.
2. The linear dimensional accuracy of the green samples exhibited an anisotropic behavior depending on the testing direction. The Z-direction always had the lowest linear dimensional accuracy, which may be associated with the different rates of binder penetrating in vertical (Z) direction and spreading in lateral (X and Y) directions. On the other hand, the X-direction always had the highest linear

dimensional accuracy, which may be because the X-direction (i.e., the binder jetting direction) is not affected by the powder spreading that takes place in Y-direction. This orthotropic behavior was true regardless of the binder saturation level, but more pronounced in the samples printed with Shell/Core of 1:2 than 1:1.
3. The compressive strength of the green samples also exhibited an orthotropic behavior depending on the testing direction. Regardless of the saturation level and shell to core ratio, the compressive strength in X-direction was always higher than in Z-direction, which may be related to the preferential orientation of the powder particles during the powder spreading process. This orthotropic behavior was more pronounced in the samples printed with Shell/Core of 1:2 than 1:1.
4. The change in the Shell/Core ratio from 1:1 to 1:2 significantly increased the compressive strength of the green samples. This is true regardless of the testing direction and binder saturation level.
5. In both X and Z directions, the increase in the binder saturation level significantly increased the compressive strengths. However, the rate of increase in the compressive strength of the green samples with Shell/Core of 1:2 was higher than that with Shell/Core of 1:1.
6. The inferior compressive strength of the green samples printed with lower binder saturation levels and shell to core ratios is probably due to incomplete hydration process due to insufficient amount of binder, which in turn results in a weak bond between the powder particles.

Future work will be focused on adjusting powder formulation and selecting effective post-processing method to improve the final characteristics of printed structures using the developed Portland cement-based powder.

References

1. Wohlers, T.: Wohlers Report 2016. Wohlers Associates Inc., Fort Collins (2016)
2. Nematollahi, B., Xia, M., Sanjayan, J.: In current progress of 3D concrete printing technologies. In: Proceedings of the International Symposium on Automation and Robotics in Construction (ISARC), Taipei (2017)
3. Nematollahi, B., Vijay, P., Sanjayan, J., Xia, M., Nerella, V.N., Mechtcherine, V.: Fresh and hardened properties of extrusion-based 3D printed geopolymer for construction applications. Materials (under review)
4. Nematollahi, B., Xia, M., Sanjayan, J.: Effect of type of fiber on inter-layer bond and flexural strengths of extrusion-based 3D printed geopolymer. In: Proceedings of the 2nd International Conference on Advanced Manufacturing and Materials (ICAMM), Tokyo (2018)
5. Cesaretti, G., Dini, E., De Kestelier, X., Colla, V., Pambaguian, L.: Building components for an outpost on the Lunar Soil by means of a novel 3D printing technology. Acta Astronaut. **93**, 430–450 (2014)
6. Rael, R., San Fratello, V.: Developing concrete polymer building components for 3D printing. In: Proceedings of the 31st Annual Conference of the Association for Computer Aided Design in Architecture (ACADIA), Banff (2011)
7. Nematollahi, B., Sanjayan, J., Shaikh, F.U.A.: Synthesis of heat and ambient cured one-part geopolymer mixes with different grades of sodium silicate. Ceram. Int. **41**, 5696–5704 (2015)

8. Xia, M., Sanjayan, J.: Method of formulating geopolymer for 3D printing for construction applications. Mater. Des. **110**, 382–390 (2016)
9. Xia, M., Sanjayan, J.: Post-processing methods for improving strength of geopolymer produced using 3D printing technique. In: International Conference on Advances in Construction Materials and Systems, ICACMS, Chennai (2017)
10. Xia, M., Nematollahi, B., Sanjayan, J.: Printability, accuracy and strength of fly ash/slag geopolymer made using powder-based 3D printing for construction applications. Autom. Constr. (under review)
11. Xia, M., Nematollahi, B., Sanjayan, J.: Influence of binder saturation level on compressive strength and dimensional accuracy of powder-based 3D printed geopolymer. In: Proceedings of the 2nd International Conference on Advanced Manufacturing and Materials (ICAMM), Tokyo (2018)
12. Biernacki, J.J., Bullard, J.W., Sant, G., Brown, K., Glasser, F.P., Jones, S., Ley, T., Livingston, R., Nicoleau, L., Olek, J., Sanchez, F., Shahsavari, R., Stutzman, P.E., Sobolev, K., Prater, T.: Cements in the 21st century: challenges, perspectives, and opportunities. J. Am. Ceram. Soc. **100**(7), 2746–2773 (2017)
13. Gibbons, G.J., Williams, R., Purnell, P., Farahi, E.: 3D printing of cement composites. Adv. Appl. Ceram. **109**(5), 287–290 (2010)
14. Maier, A.-K., Dezmirean, L., Will, J., Greil, P.: Three-dimensional printing of flash-setting calcium aluminate cement. J. Mater. Sci. **46**(9), 2947–2954 (2010)
15. Suwanprateeb, J., Sanngam, R., Panyathanmaporn, T.: Influence of raw powder preparation routes on properties of hydroxyapatite fabricated by 3D printing technique. Mater. Sci. Eng. C Mater. Biol. Appl. **30**(4), 610–617 (2010)
16. Castilho, M., Gouveia, B., Pires, I., Rodrigues, J., Pereira, M.: The role of shell/core saturation level on the accuracy and mechanical characteristics of porous calcium phosphate models produced by 3D printing. Rapid Prototyp. J. **21**(1), 43–55 (2015)
17. Fielding, G.A., Bandyopadhyay, A., Bose, S.: Effects of silica and zinc oxide doping on mechanical and biological properties of 3D printed tricalcium phosphate tissue engineering scaffolds. Dent. Mater. **28**(2), 113–122 (2012)
18. Shanjani, Y., Hu, Y., Pilliar, R.M., Toyserkani, E.: Mechanical characteristics of solid-freeform-fabricated porous calcium polyphosphate structures with oriented stacked layers. Acta Biomater. **7**(4), 1788–1796 (2011)
19. Stopp, S., Wolff, T., Irlinger, F., Lueth, T.: A new method for printer calibration and contour accuracy manufacturing with 3D-print technology. Rapid Prototyp. J. **14**(3), 167–172 (2008)

Impact of 3D Printing Direction on Mechanical Performance of Strain-Hardening Cementitious Composite (SHCC)

Jing Yu and Christopher K. Y. Leung(✉)

Hong Kong University of Science and Technology, Clear Water Bay, Hong Kong SAR, China
{ceyujing, ckleung}@ust.hk

Abstract. Automatically adding or even printing steel reinforcements into a 3D-printed concrete structure is antithetical to the design freedom as well as construction ease and efficiency. Strain-Hardening Cementitious Composite (SHCC) is a kind of short random fibre reinforced cementitious composites exhibiting robust tensile strain-hardening and multiple cracking, which has potentials to reduce or even eliminate the need for steel reinforcements in printed concrete structures. Since one of the main disadvantages of 3D-printed structures is the anisotropy, this study aims to evaluate the impact of 3D printing directions on the tensile and compressive performance of self-reinforced SHCC materials. Four series of SHCC specimens with the same mix proportion but different printing patterns (including Parallel, Perpendicular, Cross and Normal Casting) were prepared and tested under uniaxial tension and compression, and the single-crack fibre-bridging constitutive relations were micromechanically modelled to physically support the experimental results. The findings of this study can support the future design and manufacturing of 3D-printed concrete structures using fibre-reinforced materials.

Keywords: Strain-hardening cementitious composite · Printable concrete 3D printing · Additive manufacturing · Tensile performance · Fibre distribution

1 Introduction

The construction industry is anticipated to have a big revolution since construction automation is expected to intensely change traditional technologies. As an advanced manufacturing technology, 3D printing (also known as additive manufacturing) has been widely applied in many industries today. A recent application of this technology in construction industry shows the potential of advancing traditional building technologies while reducing construction time, cost and manpower [1–5].

Concrete is arguably the most important construction material in the modern world. Though many engineering challenges of the "printability" of concrete have been overcome, the intrinsic weaknesses of ordinary concrete and the involvements of these weaknesses in 3D-printed structures were thoughtless in most of the studies [6]. From its early days, plain concrete has been well known as a material that is strong in compression, but fails under tension in a brittle manner with little warning. While the

T. Wangler and R. J. Flatt (Eds.): DC 2018, RILEM Bookseries 19, pp. 255–265, 2019.
https://doi.org/10.1007/978-3-319-99519-9_24

inclusion of steel reinforcement has made concrete highly effective for practical constructions, automatically adding or even printing steel reinforcements into a 3D-printed structure is not so straight forward for the moment, as it is antithetical to the design freedom as well as construction ease and efficiency [1–6].

Fibre reinforcement, an effective way to control the cracking and introduce ductility into concrete, has led to the development of Fibre Reinforced Concrete (FRC) since 1960's. As a special kind of FRC with short random fibre reinforcements, Strain-Hardening Cementitious Composites (SHCCs) exhibit tensile strain-hardening and multiple cracking. At the ultimate state under uniaxial tension, the strain of SHCCs can reach 3–8% [7, 8] (Fig. 1), which is about 300 to 800 times the tensile strain capacity of ordinary concrete and fibre reinforced concrete (around 0.01%). In addition, the tensile strength of SHCCs can reach 10–20 MPa, and the compressive strength ranges from 20 MPa to 150 MPa depending on the matrix proportion [7, 8]. Moreover, the crack width can be typically controlled to less than 100 μm [7, 8] (Fig. 1). With these characteristics, SHCCs have potentials to be a kind of self-reinforced cementitious composite and reduce or even eliminate the need for steel reinforcements in 3D-printed structures [6], providing more efficiency and freedom for 3D printing processes.

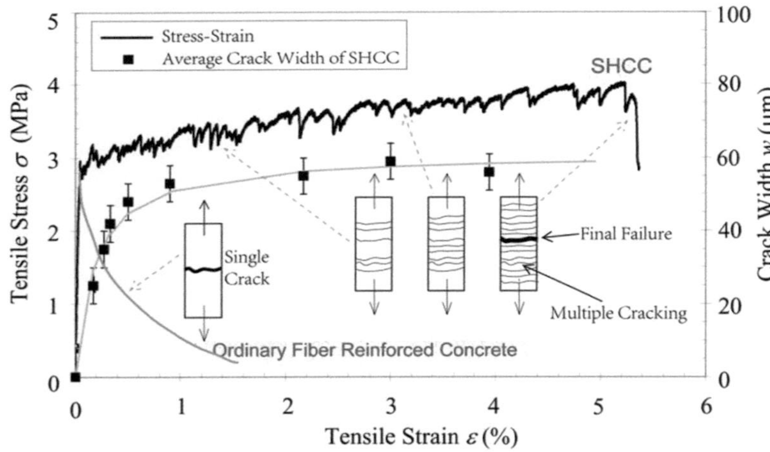

Fig. 1. Typical tensile stress-strain curve and crack width development of SHCCs (adapted from Li [7]).

One of the main disadvantages of 3D-printed structures is the anisotropy [2, 3]. Specifically, the printing directions can significantly affect the fibre distributions, and then the tensile and compressive performance of self-reinforced SHCCs when loaded in different directions. This study aims to evaluate the impact of 3D printing directions on the tensile and compressive performance of SHCCs. Four series of SHCC specimens with the same mix proportion but different printing directions were prepared and tested under uniaxial tension and compression, and the single-crack fibre-bridging constitutive relations were micromechanically modelled.

2 Materials and Methods

2.1 Experimental Program

The mix proportion of SHCC used in this study is listed in Table 1, and additional details of the ingredients and matrix design can be found in Yu et al. [9, 10]. Table 2 shows the physical properties of the discontinuous Kuraray™ K-II REC15 polyvinyl alcohol (PVA) fibres, which are commonly-used in SHCCs.

Table 1. Mix proportion of SHCC (by weight).

Material	Binder (B)		Sand/B	Water/B	SP/B	VMA/B	PVA Fibre
	OPC	FA					
Weight Ratio	0.2	0.8	0.2	0.3	0.0037	0.001	0.02 (by volume)

Note: OPC: Type I ordinary Portland cement 52.5 N; FA: Fly ash (ASTM C168 Class F); Sand: Silica sand with the nominal diameter from 120 μm to 212 μm; SP: Grace™ ADVA 189 polycarboxylate-based super-plasticizers in solution form (with 30% solid content); VMA: Viscosity modifying admixture Hydroxypropyl Methylcellulose (HPMC).

A Hobart™ HL200 mortar mixer was used to prepare the SHCC mixture. The binder and sand were first dry-mixed for 2 min. Water and super-plasticizers were then added and mixed for 3 min. After the liquefied fresh mortar matrix reached a consistent and uniform state, PVA fibers were added into the mixture and mixed for 4 min until fibers were evenly distributed. Finally, the VMA was added and mixed for 1 min. The mixing process as well as the fresh-state properties of cementitious materials have been identified as being crucial to their printability [6, 12–15]. The fresh property of SHCC mixtures was evaluated by a mini-slump flow test. A truncated cone mould with a diameter of 100 mm at the bottom and 70 mm at the top and a height of 60 mm was placed on a smooth plate, filled with mortar, and lifted upward. After that, the table was immediately dropped 25 times in 15 s [16]. The average of two orthogonal diameter measurements after 25 times dropping was 130 mm. The rheological properties are not the focus of this paper, and therefore they are not further discussed here.

Since no ready-made concrete printer was available in our lab, a small-scale manual extrusion technique was utilized as the extrusion-based printing approximation for the present study. After the SHCC mixture with a small workability loss was ready, the mixture was filled in a mechanically actuated caulk gun with a circular nozzle diameter of 10 mm for further manual extrusion, and the thickness of each printed strip was around 6–7 mm. A similar approach was also reported in Soltan and Li [6].

Cube specimens measuring 50 mm × 50 mm × 50 mm were manually printed in 8 layers with moulds for the compressive tests, and commonly-used dumbbell specimens with a cross section of 30 mm (width) × 13 mm (thickness) at the middle part (Fig. 2) were manually printed in 2 layers with moulds for the uniaxial tension test following the recommendation by the Japan Society of Civil Engineers [17].

Table 2. Physical properties of polyvinyl alcohol (PVA) fibre [11].

Length L_f (mm)	Diameter d_f (μm)	Aspect ratio	Modulus of elasticity E_f (GPa)	Tensile strength σ_f (MPa)	Density (g/cm^3)
12	39	308	16.9	1275	1.30

During sample preparations, four kinds of printing patterns including Parallel, Perpendicular, Cross and Normal Casting were used (Fig. 3). Since the length of fibres (12 mm) was larger than the diameter of the circular nozzle (10 mm), it is expected that the fibres are highly aligned with the extrusion direction due to the restricted strip dimension and flow direction associated with the extrusion process (Fig. 3a). Therefore, fibres were highly parallel to the loading direction for SHCC-Parallel (SHCC-PA for short), are almost perpendicular to the loading direction for SHCC-Perpendicular (SHCC-PE for short), are almost symmetrically 45° aligned to the loading direction for SHCC-Cross (SHCC-CR for short), and are randomly distributed for SHCC-Normal Casting (SHCC-NC for short), as shown in Fig. 3.

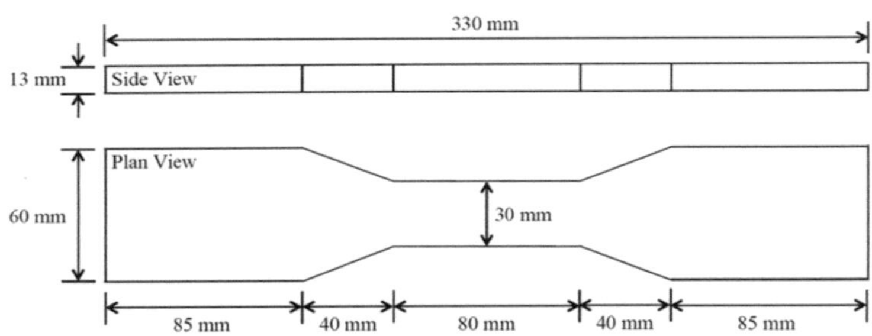

Fig. 2. Dimensions of dumbbell tensile specimens.

After printing/casting, the specimens were kept at room temperature for 24 h prior to demoulding, and then cured for another 6 days in a fog room at a temperature of 23 ± 2 °C and relative humidity of 95 ± 5%.

An automatic compression testing machine was used to perform the compressive tests, with a loading rate of 0.3 MPa/s. During uniaxial tension test, two external linear variable displacement transducers (LVDTs) were attached to both side of the middle part of the tensile specimen to measure the elongation over a gauge length of 80 mm, for calculating the tensile strain. The tensile test was performed in a 25-kN servo-hydraulic MTS™ 810 testing system, with a loading rate of 0.2 mm/min.

2.2 Micromechanical Modelling

For fibre-reinforced materials, the single-crack fibre-bridging σ-δ constitutive relation describes the relationship between fibre-bridging stress σ transferred across a crack and

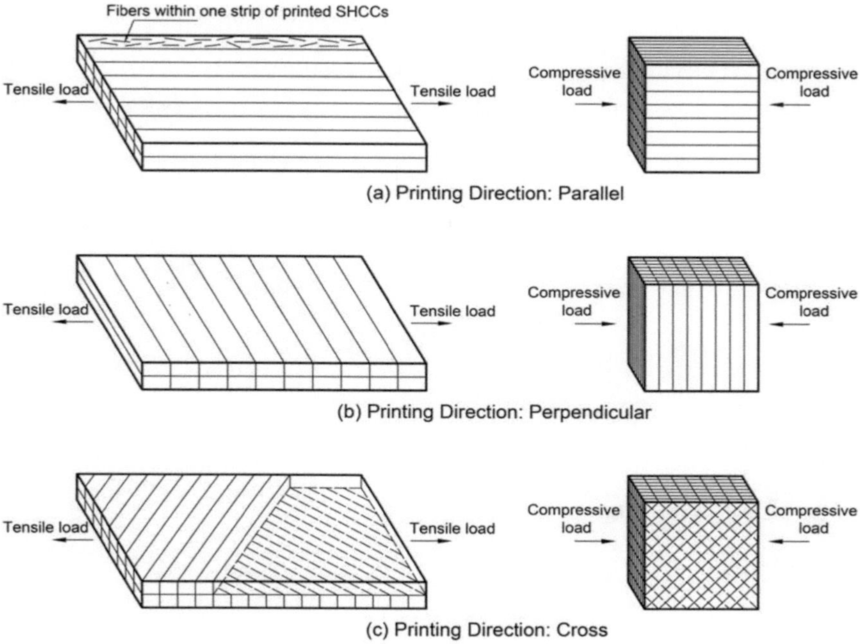

Fig. 3. Different printing patterns for SHCCs used in this study.

the corresponding crack opening δ, which relates the material parameters (micro-scale) to the composite performance (macro-scale). Therefore, control of the σ-δ constitutive relation by tailoring material parameters is the key to a successfully design of SHCC, in terms of ultimate tensile strength and strain, as well as steady-state crack width in particular [18]. In this study, the impact of 3D printing directions on the σ-δ relations of SHCCs was theoretically investigated using a micromechanical model proposed in Yang et al. [18], where the fibre rupture, fiber two-way debonding and pull-out, matrix micro-spalling and Cook-Gordon effects were considered.

3 Results and Discussion

3.1 Experimental Results on Mechanical Properties

The uniaxial tensile stress-strain curves of SHCCs with different printing directions are shown in Fig. 4, and the compressive strength as well as the major tensile characteristics for tensile performance are summarized in Table 3. For the compressive strength (Table 3), SHCC-PA shows the lowest value, while the other three cases have very similar values about 19 MPa. With a further observation of the tested SHCC-PA samples, it was found that most of final failures were resulted from the separation between printing layers. Reduced compressive strength of 3D-printed cement paste and fibre reinforced cementitious composites resulting from separate printing layers is also reported in other studies [19, 20]. To obtain reliable engineering performance in construction applications, further studies on improving the inter-layer bonding strength of 3D-printed cementitious composites should be conducted.

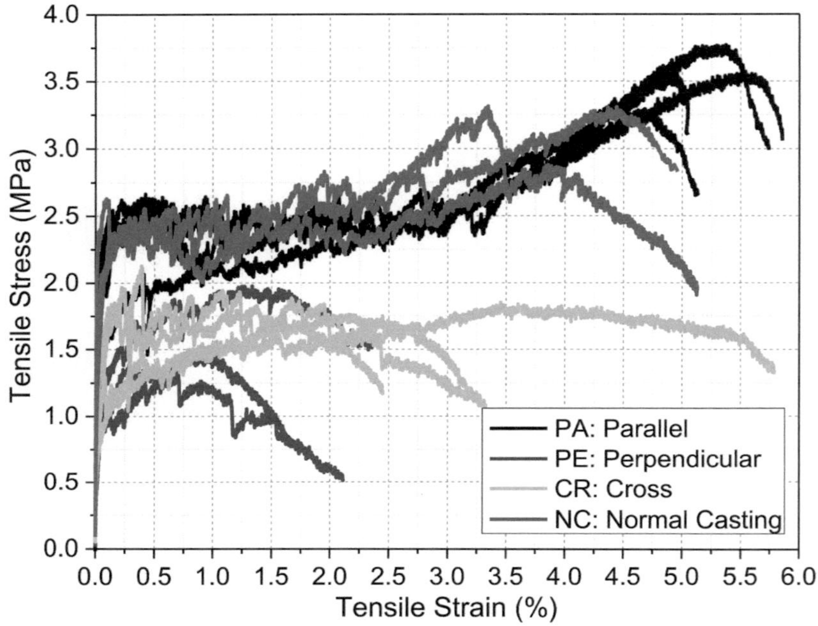

Fig. 4. Uniaxial tensile stress-strain curves of SHCCs with different printing directions

Table 3. Summary of main experimental results for 7-day-old SHCC

SHCC	Printing pattern	Compressive strength (MPa)	First cracking strength (MPa)	Ultimate tensile strength (MPa)	Ultimate tensile strain (%)
PA	Parallel	16.45 ± 0.32	2.20 ± 0.12	3.54 ± 0.20	5.13 ± 0.39
PE	Perpendicular	18.68 ± 0.48	1.02 ± 0.19	1.60 ± 0.30	1.00 ± 0.59
CR	Cross	19.23 ± 0.23	1.35 ± 0.47	1.76 ± 0.09	2.41 ± 0.86
NC	Casting	18.77 ± 0.46	2.17 ± 0.16	3.16 ± 0.27	4.01 ± 0.48

For the uniaxial tensile properties, SHCC-PA with the highly aligned fibres parallel to the loading direction performs the best, in terms of both the ultimate tensile strain and strength (Fig. 4 and Table 3). Specially, compared to SHCC-NC, SHCC-PA shows 12% and 28% improvements in ultimate tensile strain and strength, respectively. Similar results showing the advantages of parallel printing (fibres were highly parallel to the tensile loading direction) over normal casting for fibre-reinforced cementitious composites under tension or bending are also reported in the literature [6, 21]. On the contrary, SHCC-PE shows the poorest performance (Fig. 4 and Table 3). The first-cracking strength of SHCC-PE is only 1.02 MPa, which to some extent reflects the inter-layer bonding strength of 3D-printed SHCC in this study. In addition, the tensile strength of SHCC-PE is only 1.60 MPa, which is mainly attributed to severe fiber rupture and this point is further discussed in Sect. 3.2. The significant differences of tensile performance between SHCC-PE and SHCC-NC indicate that a high-performance fibre-reinforced material can perform quite poor if it is not well prepared. Finally, compared to SHCC-NC, SHCC-CR shows a 44% reduction in ultimate tensile strength and a 40% reduction in ultimate tensile strain (Table 3).

Theoretically speaking, for tensile crack bridging, the highly aligned fibres are more effective when the fibres are parallel to the tensile loading direction and are least effective when the fibres are perpendicular to the tensile loading direction, which is further theoretically discussed in Sect. 3.2. In addition, it should be pointed out that the degree of fibre alignment is likely to be lower with the increasing strip size for process scaling, but the alignment is still anticipated due to the extrusion process [6].

3.2 Modelling of Fibre-Bridging Constitutive Relations of SHCCs

The micromechanical parameters used as model input for all the SHCCs (Table 3) are listed in Tables 2 and 4. Fibre parameters in Table 2 were obtained from a single fibre tensile test. Interface parameters (Table 4) including chemical bond, frictional bond and slip-hardening coefficient, were obtained from a single fibre pull-out test. More details on the single fibre tensile test and the single fibre pull-out test can be found in Lin et al. [22].

Table 4. Micromechanical parameters used as modelling input for fibre-bridging relations

Matrix parameters	Modulus of elasticity E_m (GPa)	11.8 (calculated from Fig. 4)
	Tensile strength σ_m (MPa)	2.17 (obtained from Table 3)
	Fracture toughness K_m (MPa$\sqrt{\text{m}}$)	0.30 (test result)
	Poisson's ratio ν	0.2 (assumed as in [11])
	Matrix-spalling parameter k	500 (assumed as in [18])
Interface parameters	Interfacial chemical bond G_d (J/m^2)	0.66 (test result)
	Interfacial frictional bond τ_0 (MPa)	0.636 (test result)
	Slip-hardening coefficient β	0.60 (test result)
	Snubbing coefficient f	0.5 (assumed as in [11])
	Fibre strength reduction factor f'	0.5 (assumed as in [11])
	Cook-Gordon effect parameter α	78 (assumed as in [18])

To reflect the impact of 3D printing directions, the only difference in the modelling was the inclination angle of fibres. Since the length of fibres was 12 mm and the diameter of the circular nozzle was 10 mm, it was assumed that the inclination angle of a fibre θ ranges from 0° to 30° within one strip of printed material (Fig. 5), by considering the restriction of fibre distribution near the border of the strip caused by the extrusion process. In other words, taking the tensile loading direction as a reference, the inclination angles of fibres range from 0° to 30° for SHCC-PA, from 60° to 90° for SHCC-PE and from 15° to 75° for SHCC-CR.

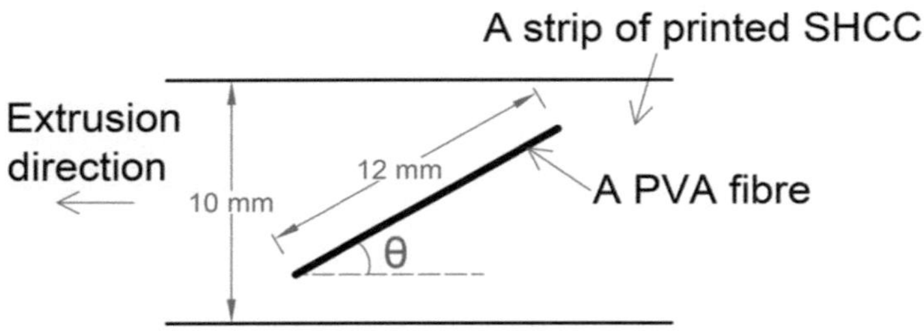

Fig. 5. Inclined fibres within one strip of printed material

The modelling results of the single-crack fibre-bridging σ-δ constitutive relations of SHCCs with different printing directions are shown in Fig. 6. The modelled peak fibre-bridging stresses (ultimate tensile strength) are much higher than those from experiment as shown in Fig. 4 and Table 3, which may be caused by the accuracy of the micromechanical tests (especially the single-fibre pull-out tests) as well as the model itself. Further consideration on the modelling is currently undertaken by the authors.

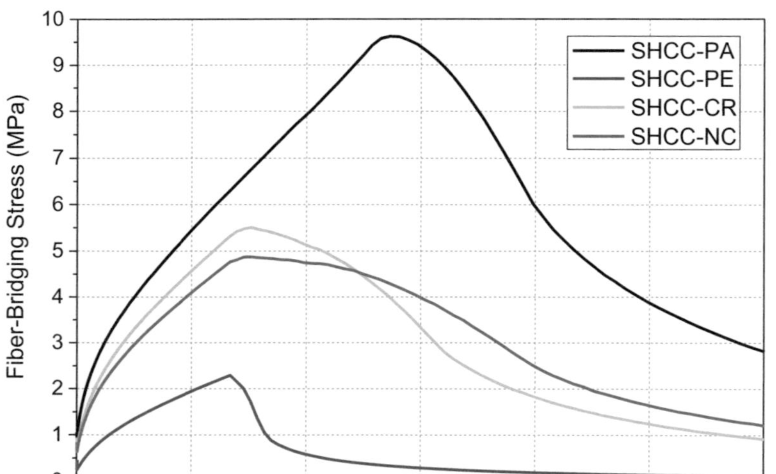

Fig. 6. Modelled fibre-bridging relations of SHCCs with different printing directions

However, the modelling results of the σ-δ constitutive relations still theoretically reveal the impact of 3D printing directions on the tensile performance of SHCCs. As shown in Fig. 6, SHCC-PA is significantly better than SHCC-CR and SHCC-NC, while SHCC-PE shows very poor fibre-bridging performance. These trends physically support the test results for SHCC-PA, SHCC-PE and SHCC-NC shown in Fig. 4 and Table 3. As discussed in Soltan and Li [6], the control of fibre alignment to be parallel to the tensile loading direction provides solutions to (1) reinforcing tension-critical areas of concrete structures, (2) mesoscale structuring of materials for functional grading, and (3) optimizing mechanical performance for a given structure part.

For SHCC-CR, the modelling results indicate that it can show comparable or even better performance to SHCC-NC, which is not the case in experiment (Fig. 4). The dissatisfactory test results for SHCC-CR come from the asymmetry of the tensile specimen, since only two layers of materials were printed in this study (Fig. 3c). Further experiments using modified tensile specimens with more printing layers to ensure the symmetry (different from Fig. 2) should be conducted to accurately evaluate the tensile performance of SHCC-CR.

4 Conclusions

This study evaluated the impact of 3D printing directions on the tensile and compressive performance of SHCCs. Four series of SHCC specimens with the same mix proportion but different printing patterns (including Parallel, Perpendicular, Cross and Normal Casting) were prepared and tested under uniaxial tension and compression, and the fibre-bridging constitutive relations were micromechanically modelled. Based on the materials used and the test results of this study, the following conclusions can be drawn, and these findings can support the future design and manufacturing of 3D-printed concrete structures using fibre-reinforced materials:

(1) 3D-printed SHCCs are highly anisotropic. SHCCs with parallel printing direction (fibres were highly parallel to the loading direction) show the best tensile performance and the lowest compressive strength, while SHCCs with perpendicular printing direction (fibres were highly perpendicular to the loading direction) show the worst tensile performance.
(2) The impact of 3D printing directions on the tensile performance of SHCCs was physically supported by the micromechanical modelling.
(3) Both the experimental and modelling results reveal the advantage of parallel printing (fibres were highly parallel to the loading direction) over normal casting for fibre-reinforced cementitious composites under tension.

References

1. Wu, P., Wang, J., Wang, X.: A critical review of the use of 3-D printing in the construction industry. Autom. Constr. **68**, 21–31 (2016)
2. Bos, F., Wolfs, R., Ahmed, Z., Salet, T.: Additive manufacturing of concrete in construction: potentials and challenges of 3D concrete printing. Virtual Phys. Prototyp. **11**(3), 209–225 (2016)
3. Tay, Y.W.D., Panda, B., Paul, S.C., Noor Mohamed, N.A., Tan, M.J., Leong, K.F.: 3D printing trends in building and construction industry: a review. Virtual Phys. Prototyp. **12**(3), 261–276 (2017)
4. Ma, G.W., Wang, L., Ju, Y.: State-of-the-art of 3D printing technology of cementitious material—an emerging technique for construction. Sci. China Technol. Sci. (2017). https://doi.org/10.1007/s11431-016-9077-7
5. Delgado Camacho, D., Clayton, P., O'Brien, W.J., Seepersad, C., Juenger, M., Ferron, R., Salamone, S.: Applications of additive manufacturing in the construction industry—a forward-looking review. Autom. Constr. **89**, 110–119 (2018)
6. Soltan, D.G., Li, V.C.: A self-reinforced cementitious composite for building-scale 3D printing. Cem. Concr. Compos. **90**, 1–13 (2018)
7. Li, V.C.: Engineered cementitious composites (ECC): material, structural, and durability performance. In: Nawy, E. (ed.) Concrete Construction Engineering Handbook. CRC Press, Boca Raton (2008)
8. Yu, J.: Multi-scale study on strain-hardening cementitious composites with hybrid fibers. Ph.D. Thesis, Hong Kong University of Science and Technology, Hong Kong (2017)
9. Yu, J., Li, H., Leung, C.K.Y., Lin, X., Lam, J.Y.K., Sham, I.M.L., Shih, K.: Matrix design for waterproof engineered cementitious composites (ECCs). Constr. Build. Mater. **139**, 438–446 (2017)
10. Yu, J., Leung, C.K.Y.: Strength improvement of strain-hardening cementitious composites with ultrahigh-volume fly ash. J. Mater. Civ. Eng. **29**(9), 05017003 (2017)
11. Yu, J., Yao, J., Lin, X., Li, H., Lam, J.Y.K., Leung, C.K.Y., Sham, I.M.L., Shih, K.: Tensile performance of sustainable Strain-Hardening Cementitious Composites with hybrid PVA and recycled PET fibers. Cem. Concr. Res. **107**, 110–123 (2018)
12. Le, T.T., Austin, S.A., Lim, S., Buswell, R.A., Gibb, A.G.F., Thorpe, T.: Mix design and fresh properties for high-performance printing concrete. Mater. Struct. **45**(8), 1221–1232 (2012)
13. Ma, G., Wang, L.: A critical review of preparation design and workability measurement of concrete material for largescale 3D printing. Front. Struct. Civ. Eng. (2017). https://doi.org/10.1007/s11709-017-0430-x
14. Paul, S.C., Tay, Y.W.D., Panda, B., Tan, M.J.: Fresh and hardened properties of 3D printable cementitious materials for building and construction. Arch. Civ. Mech. Eng. **18**(1), 311–319 (2018)
15. Weng, Y., Li, M., Tan, M.J., Qian, S.: Design 3D printing cementitious materials via Fuller Thompson theory and Marson-Percy model. Constr. Build. Mater. **163**, 600–610 (2018)
16. ASTM: Standard Test Method for Flow of Hydraulic Cement Mortar C1437. ASTM International, West Conshohocken (2007)
17. JSCE: Recommendations for design and construction of high performance fiber reinforced cement composites with multiple fine cracks (HPFRCC), Tokyo (2008). https://www.jsce.or.jp/committee/concrete/e/hpfrcc_JSCE.pdf
18. Yang, E.-H., Wang, S., Yang, Y., Li, V.C.: Fiber-bridging constitutive law of engineered cementitious composites. J. Adv. Concr. Technol. **6**(1), 181–193 (2008)

19. Le, T.T., Austin, S.A., Lim, S., Buswell, R.A., Law, R., Gibb, A.G.F., Thorpe, T.: Hardened properties of high-performance printing concrete. Cem. Concr. Res. **42**(3), 558–566 (2012)
20. Hambach, M., Volkmer, D.: Properties of 3D-printed fiber-reinforced Portland cement paste. Cem. Concr. Compos. **79**, 62–70 (2017)
21. Panda, B., Chandra Paul, S., Jen Tan, M.: Anisotropic mechanical performance of 3D printed fiber reinforced sustainable construction material. Mater. Lett. **209**, 146–149 (2017)
22. Lin, X., Yu, J., Li, H., Lam, J.Y.K., Shih, K., Sham, I.M.L., Leung, C.K.Y.: Recycling polyethylene terephthalate wastes as short fibers in strain-hardening cementitious composites (SHCC). J. Hazard. Mater. **357**, 40–52 (2018)

Applications and More

Feasibility of Using Low CO_2 Concrete Alternatives in Extrusion-Based 3D Concrete Printing

Yu Chen(✉), Fred Veer, Oguzhan Copuroglu, and Erik Schlangen

Delft University of Technology, 2600 AA Delft, The Netherlands
Y.Chen-6@tudelft.nl

Abstract. In conventional concrete, replacing high-volume (more than 45%) of ordinary Portland cement (OPC) by supplementary cementitious materials (SCMs) is not a novel CO_2 reduction method, whereas rarely in 3D printable concrete. This study attempts to explore the feasibility of using SCMs in 3D printable concrete. Initially, the existing binder mixes, required fresh properties and a research method of 3D printable concrete are investigated by reviewing the relevant papers. Additionally, the constraints and opportunities of using SCMs in 3D printable concrete are illustrated and summarized. Finally, it has been found that up to 45% of cement can be replaced by a blend of fly ash and silica fume. The essential fresh properties of 3D printable concrete include extrudability, workability, open time, buildability and structural build-up, which are influenced by the binder mix, particle size distribution, water to binder ratio, binder to aggregate ratio, admixture addition, the dosage of reinforced-fibers, etc. On the other hand, there are many limitations to develop SCMs-based 3D printable concrete, such as few relevant studies, a lack of the certificated standard, massive related-parameters and the shortage of common SCMs. For the first three problems, it can be solved with the development of 3D printable concrete. For the last one, calcined clay is one potential alternative for developing sustainable 3D printable concrete in the areas where are in short supply of fly ash and silica fume.

Keywords: 3D printable concrete · Low CO_2
Supplementary cementitious materials · Fresh properties

1 Introduction

In the recent periods, extrusion-based 3D concrete printing (3DCP) as a novel concrete construction method has been significantly developed by many research institutions and enterprises throughout the world. 3DCP can be defined as a fabrication method that employs an additive, layer-based manufacturing technique to make concrete components without formwork [1]. The potential advantages of 3DCP include increasing flexibility in architecture [2], reducing labor usage, as well as saving in-situ construction time and the building costs [3]. As a future construction trend, 3DCP may be a potential low CO_2 approach [2] since no formwork is needed. Moreover, decreased

T. Wangler and R. J. Flatt (Eds.): DC 2018, RILEM Bookseries 19, pp. 269–276, 2019.
https://doi.org/10.1007/978-3-319-99519-9_25

amounts of ordinary Portland cement (OPC) might be consumed by using 3DCP. Lim et al. [4] identified that the 3D model could be optimized for strength before concrete printing and thus the final print only requires the minimum amount of concrete. Besides, it has been found that small amounts of supplementary cementitious materials (SCMs) can be blended with OPC to improve the fresh properties of printing concrete. In conventional concrete, replacing the high-volume of OPC by SCMs is a CO_2 reduction strategy, especially when SCMs are sourced from industrial by-products like slag and fly ash [5]. However, there is no exploration for using the high-volume of cementitious alternatives in printable concrete. The experiences of using SCMs in conventional concrete cannot be directly referenced in 3D printable concrete since the different manufacturing processes are employed. Thus, this study initially aims to investigate existing binder mixes and the required fresh properties of 3D printable concrete by reviewing the relevant publications over the last 20 years. Furthermore, the constraints and opportunities of using those low CO_2 concrete alternatives in extrusion-based 3DCP are illustrated and discussed.

2 Binder Mix and Fresh Property of 3D Printable Concrete

2.1 Literature Survey of 3D Printable Binder Mix

As shown in Table 1, fly ash, and silica fume, as well as limestone, have been mixed into the binder of printing concrete in different research groups. The total amount of those cement alternatives in binders is around 10–45% by weight. OPC still possesses the highest content of binder mix in the existing printing concrete proposals. The primary objective of blending SCMs and limestone in the binder mix may be to achieve the required rheological requirements for printable concrete. Kazemian et al. [6] stated that adding silica fume in printing concrete could improve its cohesion property at the fresh state as well as the mechanical performance and impermeability when hardened. Adding ultra-fine fly ash is beneficial to the workability of 3D printable concrete by reducing yield stress and viscosity [11] at the early stage. Limestone as a kind of inert filler is commonly used to improve the workability of self-compacting concrete [12].

Table 1. Cementitious binder content of 3D printable concrete.

Source:	Cementitious binder content (by weight)
[6]	OPC (90%) and silica fume (10%)
[7, 8]	OPC (70%), fly ash (20%) and silica fume (10%)
[9]	OPC (55%), fly ash (22%) and silica fume (23%)
[10]	OPC (60–67%), limestone filler (17–20%) and silica fume (17–20%)

2.2 Fresh Property of 3D Printable Concrete

Compared with conventional concrete, 3D printable concrete has many required properties especially in the fresh state, such as no slump and fast setting, due to the absence of formwork [2]. Utilizing different types and amounts of SCMs in the binder mix may significantly affect the fresh properties of 3D printable concrete. The required fresh properties and dominant parameters are demonstrated in Fig. 1.

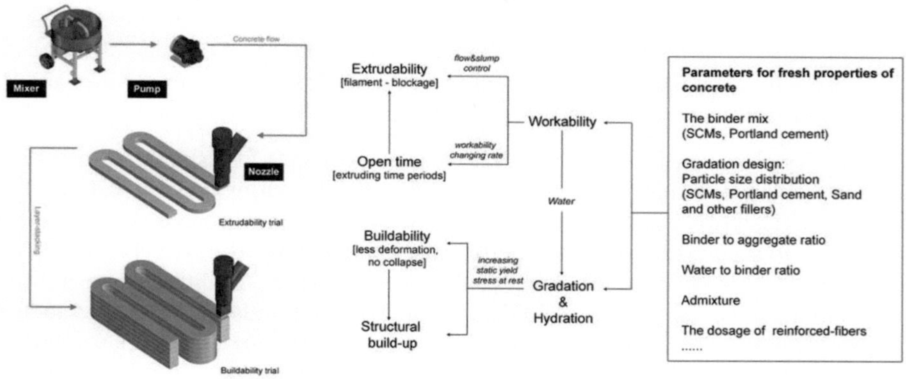

Fig. 1. Fresh properties of 3D printable concrete and dominant parameters

Extrudability. It is used to describe the property of a material that could be quickly and reliably delivered out from the transmission system [4, 7, 8]. Extrudability can be determined by using the visual inspection method. The comparable extrudability is evaluated by the continuity and conformity of the extruded filaments [6, 8]. According to Ma, Li, and Wang [8], the extrudability is primarily affected by the amount and distribution of the dry components in the blend. Besides, Le et al. [7] pointed out that the particle size distribution, binder to aggregate ratio, the dosage of superplasticizer and fibers influence the extrudability of printable concrete as well.

Workability. The conventional evaluation methods include slump, flow, and compact tests, which are inadequate for the printable concrete research. The workability of fresh printable concrete is feasible to be determined by conducting a rheological test [7]. In the study of Paul et al. [3], a Viskomat Rheometer is utilized to examine the workability of fluid concrete. By using a calibration coefficient [13], the viskomat values can be transferred to the parameters of plastic viscosity and yield stress which are expressed in the Bingham model for non-Newtonian flow [3]. The proper amount of superplasticizer is added to printable concrete to achieve the appropriate workability of fresh concrete with the lower water to binder ratio (0.2–0.3) [8, 9]. Additionally, the workability of fresh printable concrete may be influenced by using different types of cementitious alternatives as stated in Sect. 2.1.

Open Time. It should be defined as the time period for printing fluid concrete with proper workability [8]. It starts with extruding stable and consistent filaments and ends up with hardly printing the filament with standard quality. Open time is closely related to the changes of workability which can be determined by measuring shear strength of concrete with time by using a shear vane apparatus [7]. The decrease of workability with time is mainly due to the loss of water in fresh concrete. Both hydration and evaporation processes contribute to the water consumption at this stage. For a specific environment condition (temperature, humidity, and wind), the length of open time is directly decided by the decreasing rate of workability which may depend on the hydration rate of printable concrete. Apart from the environmental factors, the parameters which can influence the hydration rate of concrete, like the water content, types of SCMs and admixture also affect the open time of 3D printable concrete. Besides, the impact of physical operations is non-negligible. According to Le et al. [7], the agitated fresh concrete shows longer open time than the non-agitated.

Buildability. It is considered as the ability of fresh concrete to resist the deformation and avoid collapse during the layer-based additive manufacturing process [4, 6, 8]. For adequate buildability, it is necessary that the first layer of printed concrete has sufficient yield stress to sustain the weight from itself and upper deposited layers [14]. Based on the study of Kazemian et al. [6], layer settlement and cylinder stability tests are utilized for determining the buildability of fluid concrete. The authors also illustrate that adding the proper dosage of rheology (or viscosity) modifier, silica fume or nano-clay will help to achieve the required shape stability of fresh printable concrete. On the other hand, Weng et al. [15] point out that the continuous gradation of particles in concrete will benefit to get the high yield stress of the printable mixture for better buildability. However, only the sand particle gradation was applied in their research. It is necessary to implement continuous gradation design for all constituents including cement, SCMs, and other fillers.

Besides, the time interval between two layers is also a critical parameter in this sector. If the time interval is not long enough, four filaments may form a void due to the deformation property of fresh concrete (Fig. 2). The higher porosity will affect the mechanical performance and durability of printed concrete. The longer time interval, the better shape stability of deposited layers can be got. Whereas, the longer time interval will also weaken the bond strength between two layers. Therefore, buildability and layer adhesiveness need to be considered together in further research.

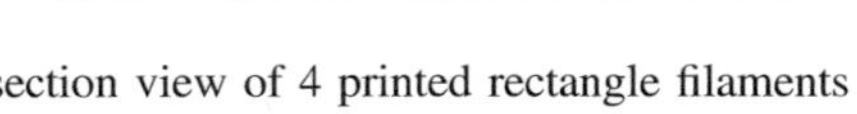

Fig. 2. A section view of 4 printed rectangle filaments (deformation)

Structural Build-up. It is a similar concept to buildability. The structural build-up is defined as the fact that the stiffness of fluid concrete increases with time due to hydration and physical operations [8, 16]. It is required to achieve a high structural build-up rate in 3D printable concrete. However, the higher structural build-up rate leads to lower bond strength between layers. In contrast to thixotropy, the concept of structural build-up is applicable in both revisable and irreversible processes of fluid cement-based materials [16]. The penetration resistance method is utilized for measuring the structural build-up rates of printed concrete at different rest times by Ma, Li, and Wang [8]. In the study of Yuan et al. [16], replacing the partial amount of Portland cement by SCMs can affect the structural build-up of fresh concrete. To what extent the structural build-up behavior will be influenced should depend on the specific physical and chemical characteristics of SCMs.

A Trial & Error Process. Overall, based on the roles of different fresh properties of printable concrete, a testing method to explore the printability of mix designs is generated (Fig. 3). It is practicable to develop SCMs-based printable concrete by using this method.

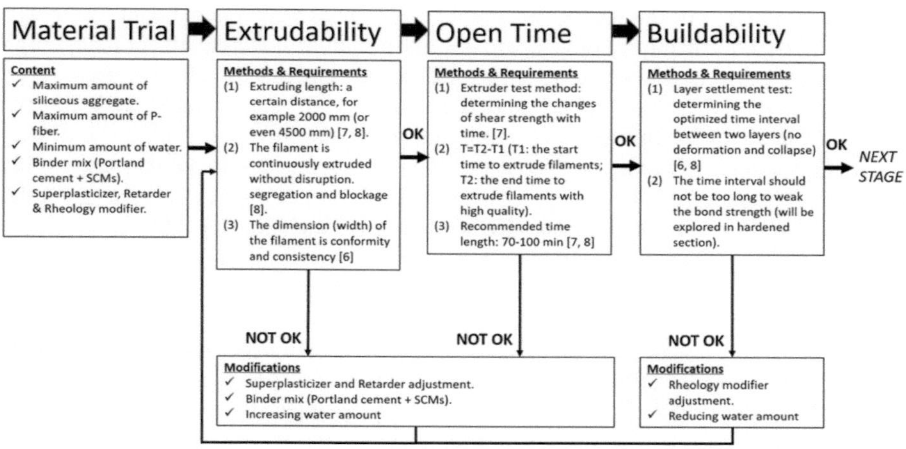

Fig. 3. A trial & error process for exploring the printability of SCMs-based concrete

3 Constraint and Opportunity to Develop SCMs-Based Printable Concrete

Utilizing the high-volume of SCMs or the mixture of limestone and SCMs to replace more than 45% of clinker is not a novel CO_2 reduction method in the conventional concrete industry. However, according to Table 1, this low CO_2 method is not widely adopted in 3DCP at present. There are four constraints to develop SCMs-based 3D printable concrete as follow. First, few studies have attempted to explore low CO_2 binder mixes of 3D printable concrete. The effects on fresh and hardened properties of different types and amounts of SCMs substitution in 3D printable concrete are unknown. Second, 3DCP is a novel technique. There is no certified standard for 3D printable concrete at present [17]. Current research efforts are built on the experiments from only a limited number of academic institutions. More specific and efficient test methods for determining the fresh and hardened properties of 3D printable concrete need to be developed and evaluated further. Third, besides the binder mix, there are many other parameters that affect the fresh properties of 3D printable concrete, such as the particle size distribution, water to binder ratio, binder to aggregate ratio, admixtures addition, and dosage of fiber-reinforcement. Thus, to develop the low CO_2 printable cement by using large amounts of SCMs becomes more complicated and difficult. Fourth, replacing the clinker by the common SCMs (fly ash, slag, and silica fume) might not be a proper way for the long-term development of 3D printable concrete. The world production of silica fume is about 0.5–1.0 million tons per year which is quite limited compared to other SCMs [18]. According to Scrivener [19], the total amount of slag is only 5% of clinker, and the fly ash which is unavailable in many countries is around 30% of clinker worldwide. Therefore, it is necessary to seek the new and widely available source of SCMs.

The first, second and third problems may be solved with the development of 3D printable concrete. For the fourth constraint, in countries with an abundant resource of fly ash, it is worthwhile to develop low CO_2 printable concrete by utilizing high-volumes of fly ash. However, for countries lacking a supply of fly ash, it is necessary to use alternative SCMs which are abundant locally. In the conventional concrete industry, calcined clay has attracted more and more attention from researchers. Kaolinitic clays abundantly exist in the crust of the earth. After the dehydroxylation of the kaolinitic clay under a calcining process between 600 and 800 °C, metakaolin which shows comparable pozzolanic properties will be generated [20, 21]. Most properties of concrete can be enhanced by adding limestone and calcined clay [22]. The characteristics of the ternary blend have been illustrated by Antoni et al. [20], and Avet et al. [23]. The mortars which contain about 45% of metakaolin and limestone with a 2:1 proportion in the binder mix demonstrates better mechanical performance than the mortars with 100% of OPC at 7 and 28 days [20]. However, the price of the pure metakaolin in the study of Antoni et al. [20] is about three times of OPC. The high-grade kaolinitic clay and metakaolin usually are used by other industries, for example, ceramics, and paper [24]. Utilizing lower grade kaolinitic clays which are widely available and much cheaper to substitute clinker may be an ideal solution. According to the study of Avet et al. [23], replacing partial clinker (even 50% in the LC3 blend: 15% of limestone, 30% of calcined clay, and 5% of gypsum) by the lower grade calcined clay which contains at least 40% of calcined kaolinitic clays can achieve the same compressive strength after seven days. Thus, it is feasible to use lower grade calcined kaolinitic clays in the concrete industry. However, no one attempted to implement calcined clay cement or limestone calcined clay cement in 3DCP currently.

4 Conclusion

Overall, through reviewing the relevant literature published over the past 20 years, it is found that SCMs like fly ash, silica fume, and limestone have been applied for making printable concrete. Up to 45% of OPC can be substituted by the blend of fly ash and silica fume in the binder mix of 3D printable concrete. Additionally, this study reports the required fresh properties of 3D printable concrete, including extrudability, workability, open time, buildability and structural build-up. Those properties are significantly affected by the binder mix, particle size distribution, water to binder ratio, binder to aggregate ratio, admixture addition, the dosage of reinforced-fibers, etc. Based on those fresh properties and parameters, a trial & error process method for testing the printability of SCMs-based concrete is generated.

However, many constraints still exist for using SCMs as low CO_2 cementitious alternatives in extrusion-based 3DCP. Only a few studies have attempted to explore the feasibility of SCMs as OPC replacement in 3D printable concrete, especially using high-volume of SCMs as substitutions. No certified standard of printable concrete is currently available. The fresh properties of 3D printable concrete depend on not only the binder mix but other material conditions. Increasing the amount of SCMs in 3D printable concrete should also consider other parameters, which would make the experimental process difficult and complicated. The geographical distribution of fly ash is uneven worldwide. Silica fume and slag are in limited supply and cannot satisfy a long-term global demand.

After a series of analysis, the opportunities of developing low CO_2 printable concrete by using SCMs are summarized as follow. In the place with abundant sources of fly ash, it is worth to explore the printability of high-volume of fly ash-based blends. Calcined kaolinitic clays as a widely available SCM has been investigated and applied in the conventional concrete industry. Using lower-grade calcined clay or the blend of limestone and calcined clay as low CO_2 alternatives is one potential direction for making 3D printable concrete in the future.

References

1. Lim, S., Le, T., Webster, J., Buswell, R., Austin, A., Gibb, A., Thorpe, T.: Fabricating construction components using layered manufacturing technology. In: Global Innovation in Construction Conference, pp. 512–520. Loughborough University, Civil and Building Engineering, Loughborough (2009)
2. Bos, F., Wolfs, R., Ahmed, Z., Salet, T.: Additive manufacturing of concrete in construction: potentials and challenges of 3D concrete printing. Virtual Phys. Prototyp. **11**(3), 209–225 (2016)
3. Paul, S.C., Tay, Y.W.D., Panda, B., Tan, M.J.: Fresh and hardened properties of 3D printable cementitious materials for building and construction. Arch. Civ. Mech. Eng. **18**(1), 311–319 (2018)
4. Lim, S., Buswell, R.A., Le, T.T., Austin, S.A., Gibb, A.G., Thorpe, T.: Developments in construction-scale additive manufacturing processes. Autom. Constr. **21**, 262–268 (2012)
5. Meyer, C.: The greening of the concrete industry. Cem. Concr. Compos. **31**(8), 601–605 (2009)
6. Kazemian, A., Yuan, X., Cochran, E., Khoshnevis, B.: Cementitious materials for construction-scale 3D printing: laboratory testing of fresh printing mixture. Constr. Build. Mater. **145**, 639–647 (2017)
7. Le, T.T., Austin, S.A., Lim, S., Buswell, R.A., Gibb, A.G., Thorpe, T.: Mix design and fresh properties for high-performance printing concrete. Mater. Struct. **45**(8), 1221–1232 (2012)
8. Ma, G., Li, Z., Wang, L.: Printable properties of cementitious material containing copper tailings for extrusion based 3D printing. Constr. Build. Mater. **162**, 613–627 (2018)

9. Nerella, V.N., Krause, M., Näther, M., Mechtcherine, V.: Studying printability of fresh concrete for formwork free Concrete on-site 3D Printing technology (CONPrint3D). In: Proceeding of the 25th Conference on Rheology of Building Materials. Tredition GmbH, Hamburg (2016)
10. Gosselin, C., Duballet, R., Roux, P., Gaudillière, N., Dirrenberger, J., Morel, P.: Large-scale 3D printing of ultra-high performance concrete—a new processing route for architects and builders. Mater. Des. **100**, 102–109 (2016)
11. Weng, Y., Lu, B., Tan, M.J., Qian, S.: Rheology and printability of engineered cementitious composites—a literature review. In: Proceedings of the 2nd International Conference on Progress in Additive Manufacturing (Pro-AM 2016), pp. 427–432. Research Publishing, Singapore (2016)
12. Mahoutian, M., Shekarchi, M.: Effect of inert and pozzolanic materials on flow and mechanical properties of self-compacting concrete. J. Mater. **2015**, 1–11 (2015)
13. Flatt, R.J., Larosa, D., Roussel, N.: Linking yield stress measurements: spread test versus Viskomat. Cem. Concr. Res. **36**(1), 99–109 (2006)
14. Perrot, A., Rangeard, D., Pierre, A.: Structural built-up of cement-based materials used for 3D-printing extrusion techniques. Mater. Struct. **49**(4), 1213–1220 (2016)
15. Weng, Y., Li, M., Tan, M.J., Qian, S.: Design 3D printing cementitious materials via Fuller Thompson theory and Marson-Percy model. Constr. Build. Mater. **163**, 600–610 (2018)
16. Yuan, Q., Zhou, D., Li, B., Huang, H., Shi, C.: Effect of mineral admixtures on the structural build-up of cement paste. Constr. Build. Mater. **160**, 117–126 (2018)
17. Panda, B., Paul, S.C., Hui, L.J., Tay, Y.W.D., Tan, M.J.: Additive manufacturing of geopolymer for sustainable built environment. J. Clean. Prod. **167**, 281–288 (2017)
18. Glavind, M.: Sustainability of cement, concrete and cement replacement materials in construction. In: Khatib, J.M. (ed.) Sustainability of Construction Materials, pp. 120–147. Wood Head Publishing in Materials, Cambridge (2009)
19. Scrivener, K.L.: Options for the future of cement. Indian Concr. J. **88**(7), 11–21 (2014)
20. Antoni, M., Rossen, J., Martirena, F., Scrivener, K.: Cement substitution by a combination of metakaolin and limestone. Cem. Concr. Res. **42**(12), 1579–1589 (2012)
21. Tironi, A., Scian, A.N., Irassar, E.F.: Ternary blended cement with limestone filler and kaolinitic calcined clay. In: Editor, S., Editor, F. (eds.) Calcined Clays for Sustainable Concrete 2015, vol. 10, pp. 195–201. Springer, Dordrecht (2015)
22. Berriel, S.S., Favier, A., Domínguez, E.R., Machado, I.S., Heierli, U., Scrivener, K., Hernandez, F.M., Habert, G.: Assessing the environmental and economic potential of Limestone Calcined Clay Cement in Cuba. J. Clean. Prod. **124**, 361–369 (2016)
23. Avet, F., Snellings, R., Diaz, A.A., Haha, M.B., Scrivener, K.: Development of a new rapid, relevant and reliable (R3) test method to evaluate the pozzolanic reactivity of calcined kaolinitic clays. Cem. Concr. Res. **85**, 1–11 (2016)
24. Avet, F.H.: Investigation of the grade of calcined clays used as clinker substitute in Limestone Calcined Clay Cement (LC3). Ph.D. Thesis, EPFL (2017)

Experimental Investigation on the Mechanical Strength and Thermal Conductivity of Extrudable Foamed Concrete and Preliminary Views on Its Potential Application in 3D Printed Multilayer Insulating Panels

Devid Falliano[1(✉)], Ernesto Gugliandolo[2], Dario De Domenico[1], and Giuseppe Ricciardi[1]

[1] University of Messina, 98166 Messina, Italy
dfalliano@unime.it
[2] G. Gugliandolo s.r.l., Via Galileo Galilei, 98100 Messina, Italy

Abstract. This contribution is focused on the properties of a particular type of foamed concrete, the extrudable foamed concrete, which is characterized by the dimensional stability in the green state, that is the ability to maintain its shape in the fresh state (green strength). In particular, after an overview of both the compressive and the indirect tensile strength, the effect of density on thermal conductivity values is presented. Interestingly, the thermal conductivity of this particular kind of lightweight cementitious material is lower compared to both classical foamed concrete and aerated autoclaved concrete (AAC) at comparable density. Moreover, the remarkable inherent green strength makes this material potentially suitable for in situ 3D printing applications in co-extruded elements with both thermal insulation and structural purposes.

Keywords: Extrudable foamed concrete · Concrete 3D printing
Thermal conductivity

1 Introduction

Foamed concrete belongs to the wider category of lightweight concrete. It is characterized by a density which is generally ranging from 200 kg/m^3 to 2000 kg/m^3 and typically comprises cement, water, preformed foam, fine sand and, eventually, fine particles such as fly ash or silica fume and, if necessary other additives. The foam component mixed with the cement paste give rise to the development of a system of air-voids in the cementitious matrix. The peculiar characteristics of foamed concrete elements are summarized in the following aspects: (1) lightness, which allows reducing the dimensions of the resisting frame structure and is advantageous in the scope of refurbishment or seismic retrofitting in seismically vulnerable areas; (2) thermal insulating properties, especially in the case of low densities; (3) good resistance against

T. Wangler and R. J. Flatt (Eds.): DC 2018, RILEM Bookseries 19, pp. 277–286, 2019.
https://doi.org/10.1007/978-3-319-99519-9_26

fire as compared to ordinary concrete; (4) improved workability; (5) cost-effectiveness due to the usage of quite simple constituting elements that are easily available locally.

Foamed concrete elements characterized by medium to low densities, because of their poor mechanical strength, are commonly employed in the non-structural field in order to exploit their properties specified above; in this regard, several studies in the relevant literature have focused on way to develop the mechanical properties of these lightened cementitious materials, especially when low densities are employed [1]. However, only in the case of densities higher than 1500 kg/m^3 they are usually employed for structural purposes.

2 Materials and Methods

The experimental campaign was focused on the evaluation of compressive and indirect tensile strength, qualitative "green strength" and thermal conductivity of a new kind of foamed concrete: the extrudable foamed concrete. In particular, cubes of 5 cm side to evaluate the compressive strength (according to ASTM C109 standard), prismatic specimens of 4 $\times$ 4 $\times$ 16 cm sides to investigate the indirect tensile strength from three-point-bending tests (according to UNI EN 196-1) and slabs with dimensions of 50 $\times$ 50 $\times$ 3 cm for the evaluation of the thermal conductivity values by means of heat flow meter method (with cold plate temperature of 15 °C and hot plate temperature of 40 °C) were used.

The specimens were prepared with cement CEM II A-L 42,5 R (in accordance with UNI EN 197-1 standard) and a protein-based foaming agent called Foamin C® (trademark name), commonly employed in the field of foamed concrete [4], was used to generate a foam with density of 85 $\pm$ 5 g/l through an opportune foam generator (foaming agent concentration equal to 3% in volume and air pressure equal to 3 bar). The preformed foam was then introduced into the mix of cement, water and additive; the latter is an appropriate chemical agent, belonging to the category of Viscosity Enhancing Agents (VEA) [2], which significantly improves the cohesion and the viscosity of the fresh lightweight cementitious paste. The mixing was carried out using a vertical mixer with a speed of 3200 rpm for at least 2 min and, anyway, until a homogeneous paste was reached.

Three different target dry densities, namely 400, 600 and 800 kg/m^3, and respectively target fresh densities (that is the density at the end of the mixing phase) equal to 650, 790 and 1040 kg/m^3, with a tolerance of $\pm$50 kg/m^3, were investigated. The amount of cement, for the three different target densities, was equal respectively to 350, 470 and 650 kg/m^3, while the correspondingly quantity of foam was equal to 130, 145 and 125 kg/m^3. It is interesting to report that the foam to cement ratio (namely 0.37, 0.31 and 0.19 respectively for the three target densities analyzed) decreased with increasing the density as expected. The water to cement ratio was fixed equal to 0.3.

With regard to compressive and indirect tensile strength investigations, three different curing conditions were explored: in air at environmental temperature of 20 °C and relative humidity (RH) of 75%, wrapped in cellophane at the same external environmental conditions and in water at controlled temperature of 30 °C. Instead, the slabs for the determination of the thermal conductivity values were cured only in air at environmental temperature of 20 °C and relative humidity (RH) of 75%. All tests were performed after 28 days of maturation in the curing conditions specified above.

The testing equipment for the evaluation of the mechanical resistance consists in a CONTROLS test frame model 65-L1301/FR with 250 kN load capacity for the compressive strength tests and 15 kN load capacity for the indirect tensile strength tests.

3 Results and Discussion

3.1 Extrusion Process

Classical foamed concrete can be only poured into molds or cast in situ and so it cannot be used for 3D printing applications as the dimensional stability in the green state is not permitted: so this special type of lightweight concrete is not extrudable. Furthermore, viscosity enhancing agents (VEA) used to extrude traditional concrete [2], in the case of foamed concrete cause the collapsing of air bubbles, so that it is impossible to extrude a lightweight product. To overcome this drawback an appropriate additive was employed in this experimental campaign: in particular, it modifies the rheology of foamed concrete fresh paste by increasing cohesion and viscosity without worsening the workability and, at the same time, stabilizes the bubbles into the cementitious matrix, in order to allow the dimensional stability in the green state and, consequently, the extrusion of a lightweight product.

From the comparison in Fig. 1, it is possible to notice not only the remarkable differences between the fresh state properties of extrudable and classical foamed concrete, but also that this extrudable lightweight cementitious material is characterized by air bubbles of smaller dimensions than those pertinent to the classical one; besides, the bubbles are more homogeneously distributed within the cementitious matrix.

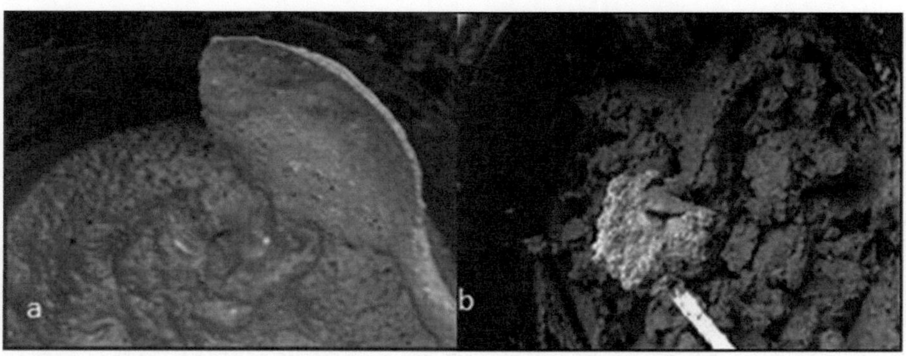

Fig. 1. Fresh state differences between classical (a) and extrudable (b) foamed concrete.

Moreover, in the sequence of photographs shown in Fig. 2, it is possible to notice an "home-made" extrusion process: after filling the mold and shaving the upper surface, the steel formwork was pushed down in such a way that the hardened cubes pushed up the extrudable foamed concrete.

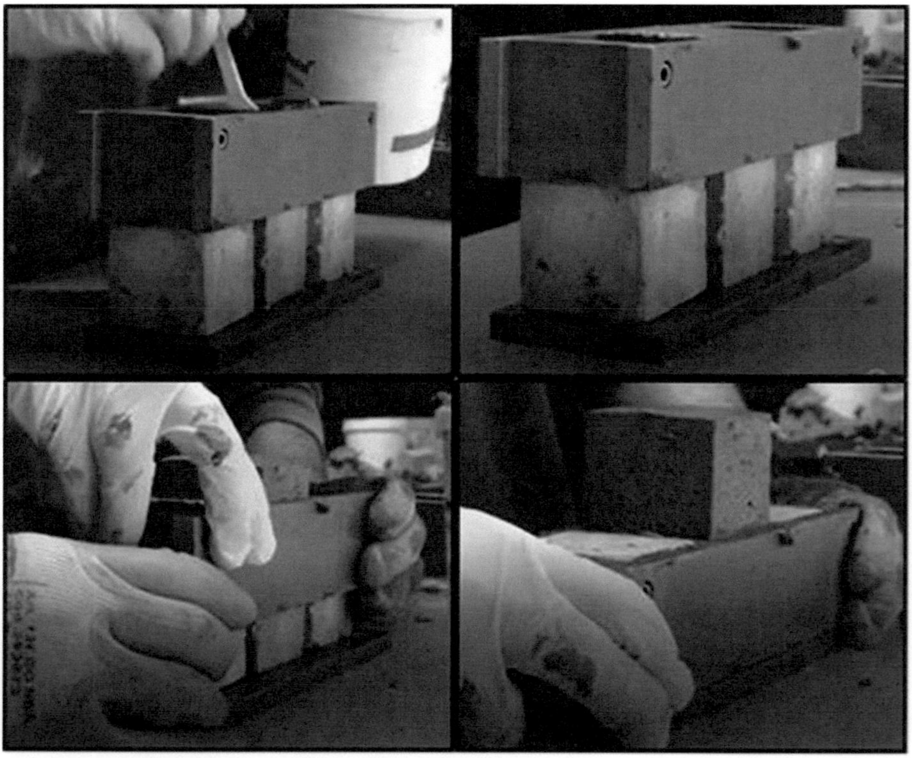

Fig. 2. "Extrusion process" and green strength of extrudable foamed concrete sample.

It should be emphasised the dimensional stability of the sample in the "green state", that is its ability to maintain its shape in the fresh state: this property is known as "green strength".

Quantitative analysis of rheological changes as well as alteration in microstructural properties performed through the use of the additive is under investigation and will be presented in a forthcoming study.

3.2 Compressive Strength

A first experimental campaign [3] focused on the mechanical characterization of this new kind of lightweight cementitious material showed that the compressive strength values are slightly higher than those related to the classical foamed concrete [4] evaluated in the same conditions of dry density, foaming agent, curing conditions, cement type and water to cement ratio. This is confirmed in the experimental campaign reported in this study.

In particular, in Fig. 3 the effects of both the target dry density and the three different curing conditions investigated on the achievement of the compressive strength values of extrudable foamed concrete cubes are reported. The highest strength values are associated with cellophane curing conditions for all the target dry densities investigated. Moreover, in Table 1 the experimental compressive strength as well as the porosity values are listed for the different specimens analysed. The porosity (ε) of the specimens was evaluated as reported in [4].

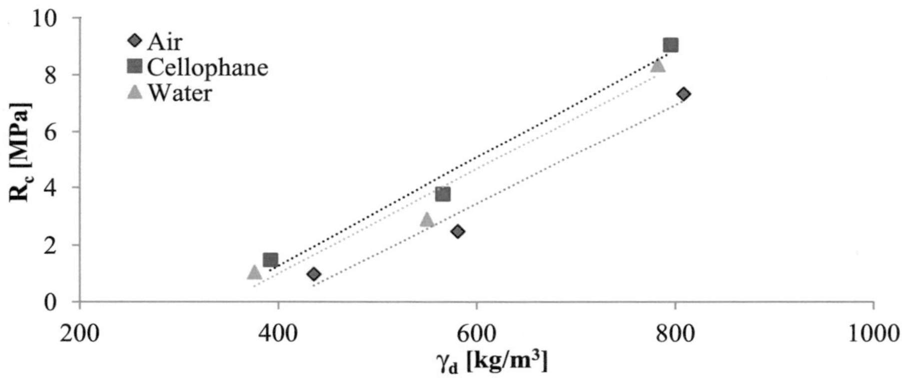

Fig. 3. Effect of the curing conditions and dry density γ_d on the compressive strength R_c of extrudable foamed concrete specimens (CEM II A-L 42.5 R, $w/c = 0.3$).

It is possible to notice that the increase of the compressive strength with the density is almost linear in the density range investigated; this is highlighted in linear regression curves reported in Fig. 3: the coefficient of determination R^2 is equal to 0.97 for air curing condition, 0.98 for cellophane curing condition and 0.96 for water curing condition.

The percentage increase of the compressive strength values in the dry density transition between 400 kg/m^3 to 600 kg/m^3 is 158%, 157% and 181%, while between 600 kg/m^3 to 800 kg/m^3 is 196%, 140% and 188% for air, cellophane and water curing conditions respectively.

3.3 Indirect Tensile Strength

The effect of both the curing conditions and target dry density on the achievement of the indirect tensile strength values of extrudable foamed concrete specimens with CEM II A-L 42.5 R cement type is reported in Fig. 4.

It is possible to realize that, differently from the compressive strength, for extrudable foamed concrete specimens with CEM II A-L 42.5 R cement type, it is more important the arrangement of the microstructure in the collapse zone rather than the curing

Table 1. Compressive strength values and porosity of extrudable foamed concrete specimens prepared with CEM II A-L 42.5 R and *w/c* = 0.3.

Dry density [kg/m^3]	Curing conditions	Compressive strength [Mpa]	Porosity [%]
436	Air	0.96	76
581	Air	2.48	68
808	Air	7.34	55
392	Cellophane	1.47	78
566	Cellophane	3.78	69
795	Cellophane	9.07	56
376	Water	1.03	79
550	Water	2.90	69
782	Water	8.36	57

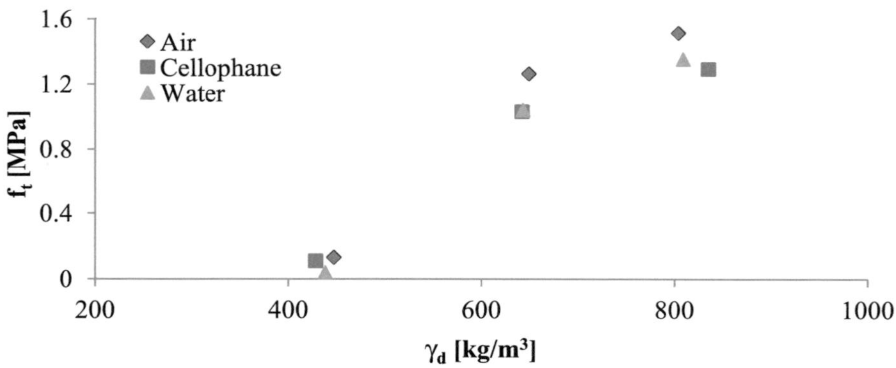

Fig. 4. Effect of the curing conditions and dry density γ_d on the indirect tensile strength f_t of extrudable foamed concrete specimens (CEM II A-L 42.5 R, *w/c* = 0.3).

conditions: indeed, the highest indirect tensile strength values are associated with the worst curing condition (i.e. air). Furthermore, the increase in the indirect tensile strength is much more pronounced in the range between 400 kg/m^3 to 600 kg/m^3 rather than 600 kg/m^3 to 800 kg/m^3. In Table 2 the experimental indirect tensile strength as well as the porosity values are listed for the different specimens analyzed.

3.4 Thermal Conductivity

The experimental campaign to evaluate the thermal conductivity has comprised a set of three extrudable foamed concrete specimens. These specimens are slabs with dimensions of 50 × 50 × 3 cm. Besides, in order to compare the results with those relating to the aerated autoclaved concrete, even an AAC slab, with a wet density (that is the density at environmental conditions after at least 28 days of air curing conditions) of about 550 kg/m^3, was subjected to the same experimental determination.

Table 2. Indirect tensile strength values and porosity of extrudable foamed concrete specimens prepared with CEM II A-L 42.5 R and *w*/*c* = 0.3.

Dry density [kg/m^3]	Curing conditions	Indirect Tensile strength [Mpa]	Porosity [%]
447	Air	0.13	75
649	Air	1.26	64
804	Air	1.51	55
428	Cellophane	0.11	76
642	Cellophane	1.03	64
835	Cellophane	1.29	54
438	Water	0.04	76
643	Water	1.04	64
809	Water	1.35	55

To determine the thermal conductivity values, the heat flow meter method was used [5, 6] with cold plate temperature of 15 °C and hot plate temperature of 40 °C. The thermal conductivity was evaluated at ambient temperature of 25 °C on the three specimens in natural moisture conditions and wet density of 316 kg/m^3, 551 kg/m^3 and 828 kg/m^3, respectively. The dry density of the specimens, evaluated after the thermal conductivity test, was equal to 260 kg/m^3, 450 kg/m^3 and 672 kg/m^3, respectively.

As expected, the thermal conductivity values significantly decrease with decreasing density of the lightweight cementitious material and the experimental findings are perfectly described by the exponential regression curve (R^2 = 1) reported in Fig. 5.

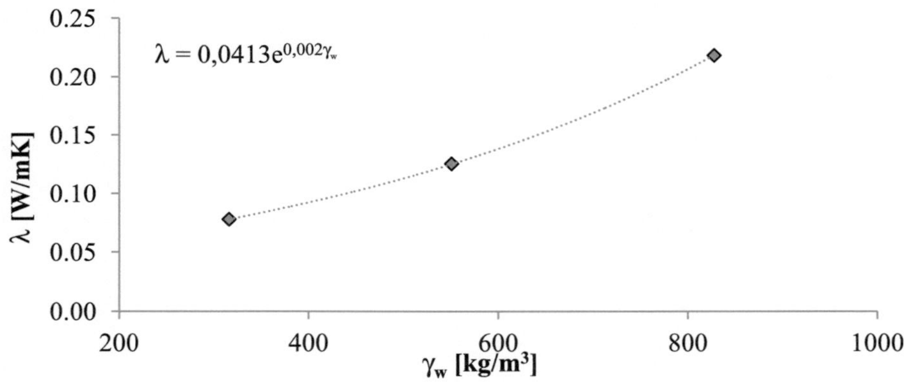

Fig. 5. Effect of wet density γ_w on the thermal conductivity values λ of Extrudable Foamed Concrete (CEM II A-L 42.5 R, *w*/*c* = 0.3).

It is interesting to note that, due to the smaller size of the air bubbles of extrudable foamed concrete compared to those characterizing classical one (cf. Fig. 1), the experimental evidences show that the thermal conductivity values associated with the new type of foamed concrete are notably lower than that of the classical one, at the same density, reported in the literature [7]. For the same reason, the thermal conductivity of the extrudable foamed concrete slab is 18% lower than that of the aerated autoclaved concrete one at comparable wet density, as shown in Fig. 6. The substantial differences between the dimensions and the distributions of the voids in the EFC and in the AAC are qualitatively well highlighted in the microscopic images shown in Fig. 7.

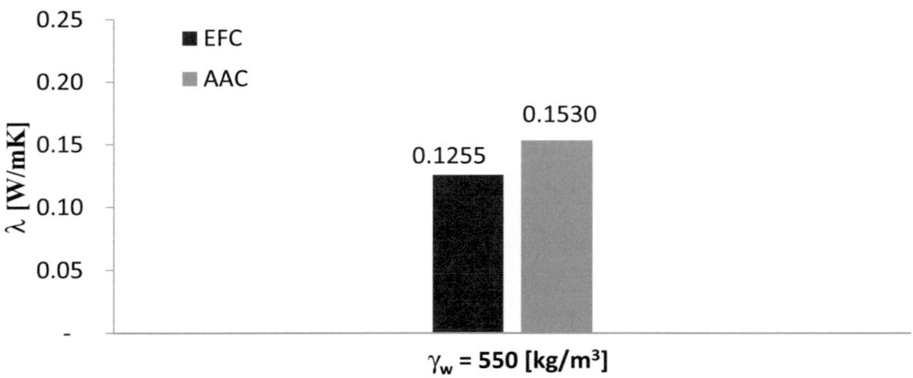

Fig. 6. Comparison of the thermal conductivity values λ of Extrudable Foamed Concrete (EFC) and Autoclaved Aerated Concrete (AAC) slabs at the same wet density γ_w.

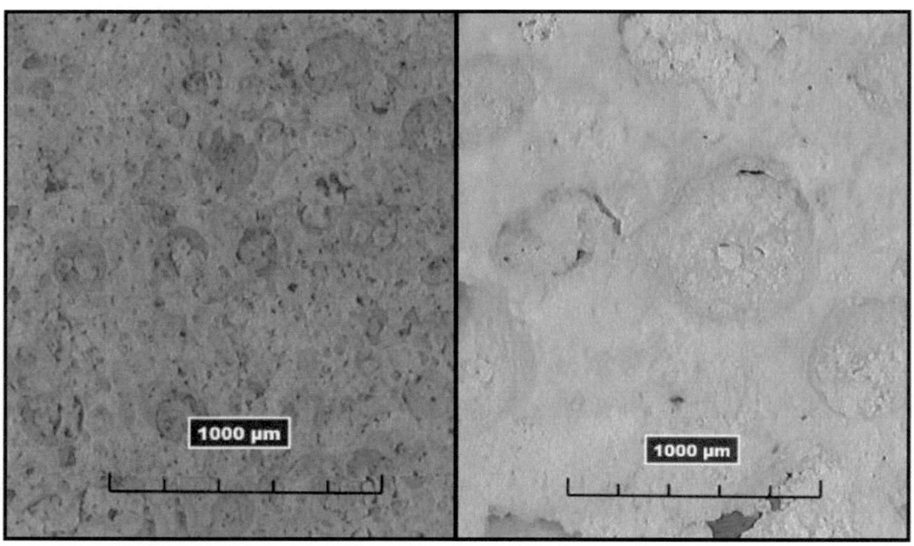

Fig. 7. Microscopic images of EFC (left) and AAC (right) at the same wet density γ_w of 550 kg/m^3.

4 Outlook: Multilayer Insulating Panels

The extrudability makes this new kind of foamed concrete suitable for a simple, effective, efficient and innovative, in this field, production process characterized by: no formwork, increased flexibility both of the product and the production process and suitability for both non-structural and structural purposes via the possibility to obtain extruded elements with density ranging from 200 kg/m^3 to 2000 kg/m^3. But, beyond the possibility to obtain precast lightweight concrete elements, due to the chance of 3D printing this extrudable foamed concrete, an interesting application is the formation of multilayer insulating panels directly in situ. The number of layers, their density and thickness can be designed according to the needs. Besides, the circumstance that it is possible to insert fiber or netting reinforcements both longitudinally, at the layer interface, and transversely, at the subsequent material stratifications, allows to obtain an original and innovative constructive element with both structural and thermal insulation properties.

Obviously, more research is clearly needed in this interesting area especially about the quantitative evaluation of the rheology of fresh extrudable foamed concrete, the evaluation of the interface properties between two co-extruded layers at different densities and the possible impact of cold joints on both mechanical and thermal properties.

Finally, future efforts will converge also towards the design of appropriate matrices for the simultaneous extrusion of layers of extrudable foamed concrete at different densities. The perspective is to move these matrices by means of opportune robotic arms or, more hopefully, with other remotely operated equipment specially designed in order to overcome some limits related to intrinsic weight and excessive complexity of the existing solutions.

5 Conclusion

In this work, some properties of a new kind of foamed concrete, characterized by an enhanced viscosity and cohesion of the fresh cementitious paste, which allows the material to be processed with the extrusion process, have been highlighted. In particular, the compressive strength is actively affected by the curing conditions and for CEM II A-L 42.5 R cement type and water to cement ratio equal to 0.3 the highest compressive strength values are associated with cellophane curing conditions. On the contrary, it seems that the indirect tensile strength values are more strongly influenced by the arrangement of the microstructure in the rupture zone rather than the different curing conditions. As regards thermal conductivity, the experimental campaign showed that, as expected, it is strongly influenced by the density but, more interestingly, because of the smaller size of the air bubbles qualitatively observed, the thermal conductivity values of extrudable foamed concrete are lower than those of both classical foamed concrete and aerated autoclaved concrete at the same density. Forthcoming studies on the microstructure of this material will give more indications in this regard. Finally, the particular properties of this material suggest its use not only in an efficient and economical industrialization process in a precast plant perspective, but also in the field of 3D prints directly in situ in an automated and robotized way, transforming the construction site in a smart factory and several studies are currently underway for this purpose.

References

1. Just, A., Middendorf, B.: Microstructure of high-strength foam concrete. Mater. Charact. **60**(7), 741–748 (2009)
2. Micaelli, F.: Proprietà reologiche di paste di cemento estrudibili ad elevato contenuto di ceneri di carbone. Ph.D. Thesis, University of Pisa (2008)
3. Falliano, D., De Domenico, D., Ricciardi, G., Gugliandolo, E.: Mechanical characterization of extrudable foamed concrete: an experimental study. World Acad. Sci. Eng. Technol. Int. J. Civ. Environ. Eng. **12**(3), 228–232 (2018)
4. Falliano, D., De Domenico, D., Ricciardi, G., Gugliandolo, E.: Experimental investigation on the compressive strength of foamed concrete: effect of curing conditions, cement type, foaming agent and dry density. Constr. Build. Mater. **165**, 735–749 (2018)
5. CTI: Thermal performance of building materials and products—determination of thermal resistance by means of guarded hot plate and heat flow meter methods—products of medium thermal resistance. UNI EN 12667
6. CTI: Thermal performance of building materials and products—determination of thermal resistance by means of guarded hot plate and heat flow meter methods—dry and moist products of medium and low thermal resistance. UNI EN 12664
7. Mydin, A.O.: Effective thermal conductivity of foamcrete of different densities. Concr. Res. Lett. **2**(1), 181–189 (2011)

Development of a Shotcrete 3D-Printing (SC3DP) Technology for Additive Manufacturing of Reinforced Freeform Concrete Structures

H. Lindemann[1(✉)], R. Gerbers[1], S. Ibrahim[2], F. Dietrich[1], E. Herrmann[1], K. Dröder[1], A. Raatz[2], and H. Kloft[1]

[1] Technical University Braunschweig, 38106 Brunswick, Germany
h.lindemann@tu-braunschweig.de
[2] Leibniz University Hannover, 30167 Hannover, Germany

Abstract. In this paper, a novel Additive Manufacturing (AM) technique for robot-based fabrication of large-scale freeform reinforced concrete elements is presented. The AM technology, called Shotcrete 3D Printing (SC3DP), is based on an automated shotcreteing process and offers the ability to integrate structural reinforcement in both principal directions and enables printing of horizontal cantilevers onto vertical surfaces. Moreover, the SC3DP technique effectively addresses the problem of cold joints that is inherent to other 3D printing techniques. However, as controlling the process parameters of the SC3DP technique is significantly more complex than it is for conventional 3D concrete printing processes, several closed-loop online control routines were developed and integrated. The resulting gain of control for this adaptive fabrication process is demonstrated through a case study for the production of a complexes reinforced concrete component. Moreover, its conceptual implications are discussed and an outlook for future work is given.

Keywords: Robotic fabrication · Online path-planning
Concrete reinforcement · 3D printing · Shotcrete 3D printing

1 Introduction

In 2016 the Universities of Braunschweig, Clausthal and Hannover started an interdisciplinary research project on Additive Manufacturing (AM) with sprayed concrete. Six different institutes are working on the development of a new AM method, which has the potential to produce large scale, reinforced concrete elements with a high surface accuracy, based on the principals of shotcrete technology. The research activities are consolidated in a novel large scale a robotic research facility the "Digital Building Fabrication Laboratory" (DBFL) for additive and subtractive manufacturing of reinforced freeform concrete structures [1].

This paper presents the ongoing research on Shotcrete 3D Printing (SC3DP), which is currently being carried out on the DBFL. The SC3DP technology has the potential of producing structural freeform concrete elements with large overhangs and a high

T. Wangler and R. J. Flatt (Eds.): DC 2018, RILEM Bookseries 19, pp. 287–298, 2019.
https://doi.org/10.1007/978-3-319-99519-9_27

surface quality. The paper focuses on offline path planning based on experimental knowledge and the development of control algorithms for the AM process. In particular, a senor guided online path planning technology for SC3DP is presented since model-based offline planning alone could not achieve high component accuracies.

1.1 Additive Manufacturing with Concrete

Although several AM technologies are used in the manufacturing industry today [2], the research for construction scale 3D printing is focusing on two different technologies: Firstly, powder bed printing [3] and secondly extrusion based methods. The extrusion based AM-methods are the most common in research and are about to be used in industrial applications. The first viable approach of large-scale concrete printing was made by Behrokh Khoshnevis with "Contour Crafting" (CC) [4]. Based on the extrusion of viscous-like materials layer-by-layer to build-u a 3D object. Two trowels attached to the printing nozzle are used to achieve a smooth and accurate planar surface. Another extrusion based process is the "Concrete Printing" technology developed at the Loughborough University [5]. Unlike CC, no trowel is used, therefore the resolution needs to be smaller to achieve the required accuracy. Both projects show the great potential of AM technologies in the building industry. Another AM approach has been made at the ETH in Zurich with the project "mesh mold" [6]. Instead of AM manufacturing the concrete element, a rebar formwork, is manufactured by a robot and gets filled with concrete afterwards, which allows the production of non-standard elements with a fully functional structural reinforcement. Nevertheless, there are still major challenges for AM of structural reinforced concrete, such as implementation of reinforcement, surface quality and the cold joint problem [7]. Therefore, further research on AM with concrete is still necessary.

1.2 Potential of Shotcrete 3D Printing as an Additive Manufacturing Technique

The shotcrete technology has been around for more than 100 years and is mainly used in tunnel construction and mining industry [8]. It is fast and easy to apply on all kind of surfaces and allows the integration of reinforcement even with nonstandard geometries. The manufacturable complexity can be seen in projects like the Taichung opera house of Toyo Ito, which has been completed in 2016. The quality and accuracy of the process is highly depending on the skills and experience of the operator. The nozzle operator can adjust the strategy of material application according to the environment parameters in order to fill holes, avoid nesting or to avoid slipping of the applied material.

Compared to the established extrusion-based material application processes [6], spraying concrete offers several different advantages. It is possible to spray the concrete not only as horizontal layer (Fig. 1a), the material can also be placed laminar (Fig. 1b) on surfaces with altering orientations (Fig. 1c). By introducing additional accelerator, the time of solidification can be adjusted. This not only allows to print overhangs up to 90° (Fig. 1d), it reveals new possibilities to manufacture complex 3-dimensional concrete components. Using different amounts of the accelerator the curing process of the material can be adjusted to offer support for upcoming layers or to be workable for additional post processing (Fig. 1e). Therefore, the material for the application process is being further developed at the TU Braunschweig together with industrial Partners. Additionally, the flexibility of the spraying process enables the integration of reinforcements into the production process, since it is possible to inject them during manufacturing of the component (Fig. 1f). Another unique feature is the reduction of cold joints, due to the impact energy of the shotcrete and the resulting geometrical interlocking of the layers. In conclusion, SC3DP offers the potential to address several disadvantages of existing AM technologies for manufacturing of freeform concrete elements.

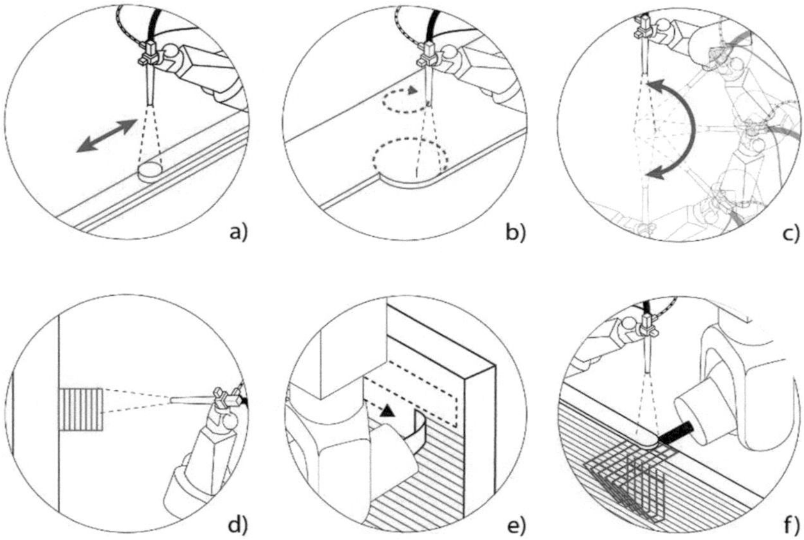

Fig. 1. Specific features of the SC3DP process: layer-by-layer application (a), laminar application (b), Variation of the application angle from 0° to 180° (c), overhang spraying (d), several post processing options (e), implementations of reinforcement and build-in-parts (f)

1.3 Challenges of Shotcrete 3D Printing

The application of sprayed concrete for AM of concrete components is a major challenge, as a large number of process parameters have to be kept in a constant ratio in order to achieve a homogeneous concrete application and therefore an acceptable part accuracy. The accuracy depends on parameters such as the concrete volume flow rate, the delivery pressure, the air volume flow rate, the air pressure, the concrete accelerator dosage and parameters related to the path planning such as the nozzle distance, velocities, layer spacing, application angle or times between layer applications. The variation of these parameters has a great influence on the associated target values, such as early strength, layer thickness, concrete quality (e.g. compressive strength, tensile strength) and rebound. In particular, the early strength and the layer thickness pose a major challenge for additive production. The early strength results in a limited time window in which a next layer of concrete can be applied. If this time window is missed, the concrete is either too soft or already too hard. In the first case, material displacement occurs due to the high air pressure from the nozzle. In the second case, the concrete quality suffers from a lack of adhesion between the layers. A constant width of the application layers is also important for AM, so that the planned paths lead to the desired component. In conclusion, a good accuracy of the component is difficult to achieve in Shotcrete 3D printing due to various disturbances such as uneven concrete mixtures, fluctuating pressures in the supply hoses or suboptimal path parameters.

2 Offline Process Control Based on Experimental Knowledge

Parameter studies were conducted in order to identify appropriate spraying parameters for offline process planning. Therefore, it was mandatory to keep as many parameters as steady as possible to be able to find the adequate process parameters and to understand its influence on the outcome. Environmental conditions (e.g. temperature and humidity) cannot be controlled in an appropriate manner, but still have a great influence on the concrete behavior. Examinations on the flow spread were used to adapt the mixture to ensure a repeatable material behavior. The conveyance line of the concrete air can be an influencing parameter as well. Hose length, diameter and curvature have to stay the same for a consistent result. While the length and the diameter are easy to fix, the curvature is constantly changing due to the moving robot. Closed-loop controls for the material supply were used to compensate such disturbances (see Sect. 3.3). The main influencing parameters, which have been tested, are the nozzle distance, material deposition rate, flow of the compressed air and application speed. The results of the parameter studies have shown that the deposition rate and the application speed have nearly linear dependencies on the layer thickness. The spraying distance on the other hand has a linear dependency on the layer width. However, the air flow also has a major influence on the material distribution and therefore the layer geometry and the resulting rebound. Since this influence is not linear it is therefore considered as a constant parameter for offline process planning. Next to the parameter studies, several strategies, for manufacturing of concrete elements have been investigated. These experiments were used to develop algorithms for offline path planning to generate the robot motion for the production of the components (Fig. 2) [14]. The challenge here is to synchronize the material behavior and the results of the parameter studies with the robot movement. This challenge will be explained in more detail using the example of building a thin, curved concrete wall.

The first printed wall elements were single line elements, in which the shape was created by spraying layer on layer back and forth. The first results turned out to be promising. The used concrete, allowed the production of elements up to 100 cm height without the introduction of additional accelerator. However, one major problem turned out to be sagging on unsupported edges. On 90° corners the sagging is negligible (Fig. 2a), on a 180° turn (Fig. 2b) the sagging of the material increased with every single layer which made the production of precise components impossible (Fig. 3a).

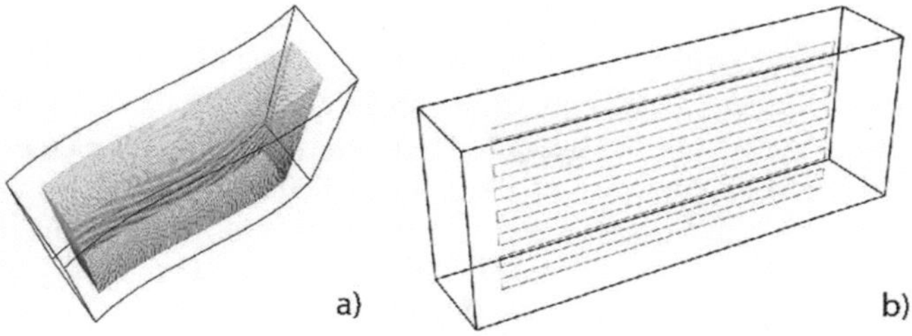

Fig. 2. Path planning for continues application (>90° corners) (a) and single line application (b)

Fig. 3. Sacking on unsupported edges (180° turns)

Avoiding 90-degree turns, would lead to extremely limited application strategies, and concrete elements could only be manufactured with a minimum thickness of at least twice the layer width. This would pose a major disadvantage for SC3DP since a single layer width is already wider than established application techniques. In order to build up a wall with the thinnest width possible, it is necessary to overcome the limitations of the "edge-sacking". Experiments have shown, that decreasing the application speed towards the edges counteracts upcoming sagging. Additional material needed, can be applied in order to compensate for a deviation from the initial programmed (offline) target geometry. However, this behavior is highly dependent on the individual component geometry and cannot be predicted without complex simulations. Such experiences have led to the insight that adaptive processes based on measurement technology for the detection of geometric inaccuracies are essential in order to develop a reliable AM process.

3 Online Measurement and Process Control for SC3DP

The results of the presented experiments demonstrate the need for online measurement and control of the component geometry. The adjustment of the robot velocity as well as the distance between the surface and the nozzle has been identified as a suitable method to avoid inaccuracies of the component geometry such as edge-sacking or material displacement. An online velocity control, which regulates the layer thickness, can also counterbalance disturbances of the material supply e.g. failures in the conveyer system or can be used as a controlled process for introducing additional elements, like reinforcement or other in-build-parts. To be able to realize such a closed-loop control system, a measurement technique is required that can be used for online measurement of the deviations from the desired shape. Therefore, the State-of-the-Art for control systems for layer-based Additive manufacturing serves as a basis for the selection of suitable measuring and control methods.

3.1 Measurement and Control Concepts for Layer-Based Additive Manufacturing

The lack of quality assurance in AM is an important technological barrier that prevents AM to be used for industrial production. [13] In all AM technologies, disturbances (e.g. material inconsistency) or environmental influences (e.g. variations in temperatures) lead to inconsistent component qualities. In contrast to the established subtractive processes, where in-process monitoring is common today, additive technologies have not yet established standardized online process monitoring to detect discontinuities during manufacturing. However, research is currently focusing increasingly on approaches that enable the online detection of inaccuracies and their compensation using adaptive processes. In particular, adaptive technologies for layer-based additive processes are of interest for the transferability to additive production with sprayed concrete. Heralić et al. [9] developed an in-situ monitoring technology for laser deposition welding based on a 3D laser scanner. The team was able to prove that 3D sensor data can be used for in-process detection of disturbances and their compensation. For this purpose, a control algorithm was developed using an iterative learning approach that incorporates the 3D data to adjust the layer height during the process in order to avoid geometric inaccuracies. Xiong et al. [11] developed a vision sensor system and a process controller for AM with gas metal arc welding (GMAW). The system uses a CCD camera to monitor the distance between the nozzle and the surface. A control system was implemented, that uses the nozzle distance as input to adjust the deposition rate to control the layer height in closed-loop. Holzmond et al. [12] developed a system based on 3D digital image correlation as in-situ measurement technology for monitoring the surface geometry in fused deposition modeling (FDM). The system uses stereoscopy to captures 3D point cloud data of the printed geometry. The 3D point cloud data is then compared with the digital CAD model to detect manufacturing errors. Wolfs et al. developed a height measurement and feedback system for 3D concrete printing [10]. The system utilizes a 1D Time-of-Flight (TOF) distance sensor to measure the distance between the nozzle and the surface for closed-loop control of the nozzle position. The results from the State-of-the-Art indicate that in-situ measurement can be utilized to increase the component quality in layer-based AM. Nevertheless, the results also emphasizes the advantages and disadvantages of the different measuring technologies. For example control approaches based on 2D cameras are limited for measuring the distance between nozzle and surface, concepts based on 3D measurement and point cloud data require constant ambient conditions and laser triangulation is limited to measuring 2D profiles. The development of a control concept for additive concrete spraying therefore requires a proper evaluation of the different measuring technologies.

3.2 Online Geometry Measuring for Shotcrete 3D Printing

In order to select a measuring technology for closed-loop control of the component geometry, laser triangulation was compared with the TOF. In-process measurements of both principles were carried out and compared with each other. The results show that the disturbing influences of the spraying process, especially the rebound, results in insufficient measuring stability for the TOF camera. The results of the laser sensor, on the other hand, exceeded the expectations. Here it was possible to get appropriate measuring results even with the spray-jet between the CCD camera and the laser line. Laser triangulation may have the disadvantage of only being able to update a 2D section of the component and not an entire area. However, this disadvantage is compensated by the higher precision and the robust measuring process. Therefore, laser triangulation was chosen for implementation of the control concept in this project (Fig. 4).

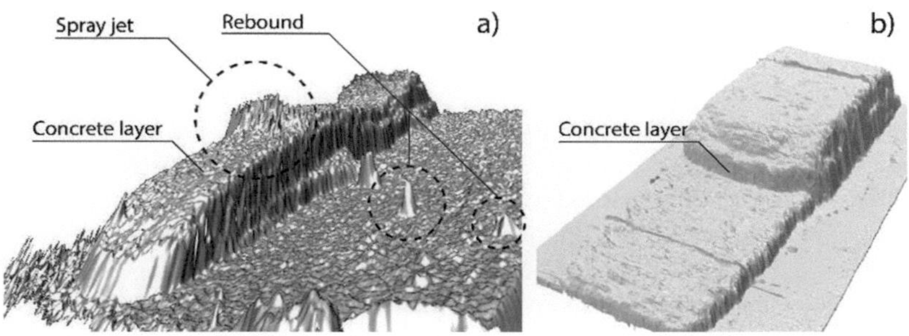

Fig. 4. Online measuring for Shotcrete 3D Printing (a) Time-of-Flight and (b) laser triangulation

3.3 Control Concept for Shotcrete 3D Printing and Experimental Implementation

Compared to conventional extrusion technique for additive manufacturing with concrete the SC3DP method is influenced by a larger number of process parameters, which have to be kept in a constant relation to produce high quality components. In order to allow comprehensible results of the parameter studies, it was decided, to keep as many parameters as steady as possible. Therefore, all process parameters for the material supply, which are directly measurable (e.g. flow rate of the compressed air) have been included in separate close-loop controls (see Fig. 5) to compensate for disturbances "D". This multi-loop control approach was chosen due to its reasonable implementation effort. It turned out that the influence of the process parameters among each other are neglectable and therefore it can be expected that a multi-variable control approach would not yield significant improvements.

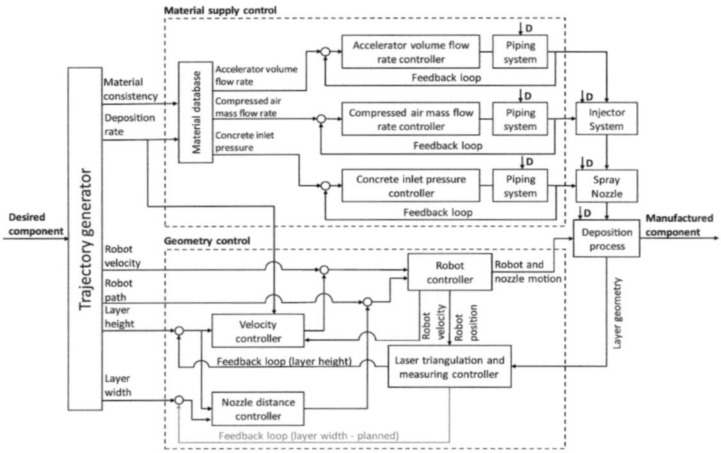

Fig. 5. Implemented control concept for Shotcrete 3D Printing (left) and online measuring using a laser triangulation sensor (right)

All controllers are implemented using a Beckhoff control system. The current control concept uses a "Trajectory generator" that generates the time-dependent and offline planned variables for the different controllers. This includes the desired material consistency and deposition rate, the robot path and velocity and the geometry of the deposition layers. To reduce the offline planning effort for the material supply a "Material database" with the experimental knowledge of the parameter studies is used define the set points for the closed-loop controller. The path of the robot is generated from CAD data and is then continuously send to the "Robot controller" of the DBFL that is run by a Siemens Sinumerik 840D. A Stemmer C5 laser triangulation sensor is used to measure the component geometry while the concrete is applied. The sensor data is processed and synchronized with the current robot position to enable measuring in six dimensions using the "Laser triangulation and measuring controller".

The geometry of the deposition layer (width & height) has been included in two additional control algorithms to compensate for geometric inaccuracies (e.g. material displacement). For the development of these controllers, the approaches presented in the state of the art were adapted for Shotcrete 3D printing (see Sect. 3.1). The first control algorithm uses the deviation between the planned and the actual layer height as well as the planned deposition rate to calculate a velocity offset for the robot. The "Velocity controller" not only ensures a constant layer height of the deposition process, it also enables an adaptive material deposition process to fill gaps or holes and to avoid material accumulations. The second controller leverages the distance between the nozzle and the surface to control the layer width. The "Nozzle distance controller" therefore ensures a constant distance between the nozzle and surface depending on the desired layer width. The controller utilizes the deviation between the planned and the actual layer height to calculate a correction vector for the robot path to control the nozzle distance. However, this only allows indirect control of the layer width.

Therefore, it is planned to measure the layer width additionally in order to enable direct closed-loop control of the layer width in the future.

The control concept is currently being examined in the DBFL. Figure 6 shows the first results of an experimental investigation of the control algorithms to compensate for irregularities. Therefore, a hole was inserted manually during the production of a straight wall and the system's reaction to this geometric inaccuracy was observed. It can be seen that the algorithm is able to compensate the geometric error in a few deposition layers. However, further parameter studies are still required to optimize the controller parameters e.g. to increase the robustness and to accelerate the control process.

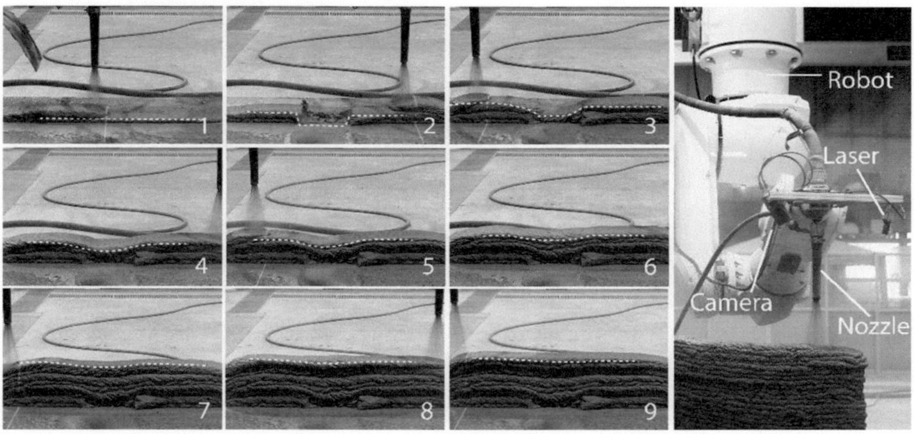

Fig. 6. Online measurement and closed-loop control for manufacturing of a straight wall

4 Case Study: Multidirectional Printing and Introducing of Reinforcement

In order to investigate the full potential of SC3DP a case study has been designed, combining all previous mentioned features. The build-up strategy is a combination of a horizontal layering, vertical application, introduction of reinforcement and a 90° overhang. As first production step, a single wall is produced with horizontal layering, as described in Sect. 2. The single wall element functions as a "backbone" (Fig. 7a) which is used as a base for upcoming vertical applications. One side of the wall element is straight and the other side is curved. The produced wall element had a height of 1 m and a width of 1.40 m. Pre-bend steel rebars have been manually placed in between the application of the horizontal layer (Fig. 7b). Afterwards vertically spraying with an average thickness of approximately 20 mm was used to cover the side of the wall (Fig. 7c). The applied concrete has been sprayed with no additional accelerator, which allowed additional post-processing. Prefabricated reinforcement, in this case a carbon fibre fabric, has been placed in the still workable concrete (Fig. 7d). An additional

20 mm layer of concrete was applied to cover the fabric (Fig. 7e).This final layer has been manually smoothed, in order to create an even surface finish. As a final step, a console has been manufactured by spraying vertical layers, cantilevering from the vertical surface (Fig. 7f). The use of accelerator and the supported by the rebar enabled the production of a cantilever of 17 cm. The results indicate that even bigger overhangs are possible to produce, depending on the shape, curing time and additional support.

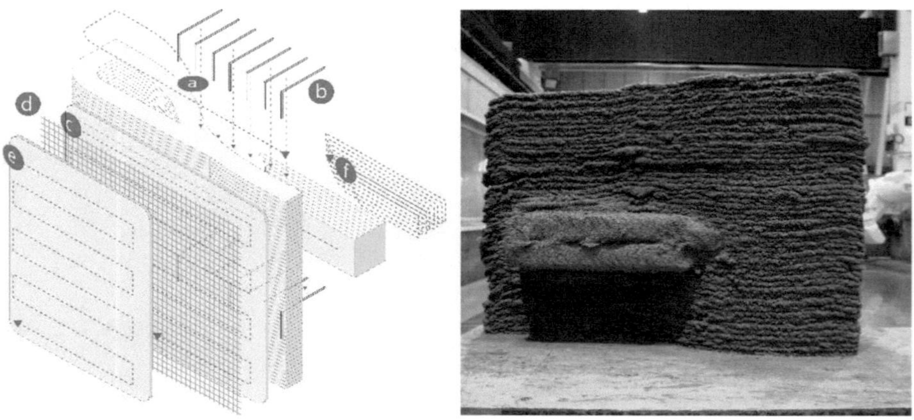

Fig. 7. Offline planning (left) and manufacturing result (right) of a complexes reinforced concrete component produced by Shotcrete 3D Printing

5 Conclusion

The results of the experiments and case studies carried out have shown, for the first time, that complex concrete geometries with large overhangs and integrated reinforcement are possible to produce using Shotcrete 3D Printing. The experimental knowledge from several parameter studies can be used for an efficient offline process planning for the complex production processes. However, the results also led to the insight that high component accuracy is hard to achieve with model-based offline planning alone. Therefore, a multi-loop control concept for SC3DP has been developed that enables compensation of inaccuracies and thus precise production of the concrete components. The developed controller integrates a novel 3D measuring based compensation approach for in-process control of the component geometry. This has significantly increased the robustness of the production process especially with regard to material displacements. Production defects that previously led to wasted parts can now be avoided.

However, further developments are necessary in order to industrialize the manufacturing process. On the one hand, an automated post-process for the surface has to be developed to compensate for the very rough surface quality of the additive process. On the other hand, further parameter studies have to be carried out to examine the limits of the process (e.g. overhangs). Additionally process chains for the automated integration of reinforcement and prefabricated components (e.g. empty conduits) have to be developed. A major disadvantage of the laser triangulation system is currently the use of a single laser line. With the current system, it is only possible to measure before or after the spray jet, which limits the use of the geometry controller for only one direction of movement. Additionally the geometric freedom for robots paths is limited in due to shaded areas in narrow radii (e.g. sharp corners). It is therefore planned to equip the system with three laser lines, which will be positioned in a triangle around the spray jet. In addition, the control algorithm for adjusting the nozzle distance will be extended with a close loop-control for the layer width to further increase the accuracy of the manufacturing process. Since the laser system can also be used to measure the entire component, it is planned to use the 3D data, for generating a robotic paths for further post-processing (e.g. smoothing or subtractive post-processing).

References

1. Neudecker, S., Bruns, C., Gerbers, R., Heyn, J., Dietrich, F., Dröder, K., Raatz, A., Kloft, H.: A new robotic spray technology for generative manufacturing of complex concrete structures. Procedia CIRP **43**, 333–338 (2016)
2. Labonnote, N., Rønnquist, A., Manum, B., Rüther, P.: Additive construction. State-of-the-art, challenges and opportunities—a review. Autom. Constr. **72**, 347–366 (2016)
3. Lowke, D., Dini, E., Perrot, A., Weger, D., Gehlen, C., Dillenburger, B.: Particle-bed 3D printing in concrete construction—possibilities and challenges. Cem. Concr. Res. (2018, Accepted for publication)
4. Khoshnevis, B.: Automated construction by contour crafting-related robotics and information technologies. Autom. Constr. **13**, 5–19 (2004)
5. Lim, S., et al.: Development of a viable concrete printing process. In: Proceedings of the 28th International Symposium on Automation and Robotics in Construction (ISARC 2011), Seoul, South Korea, 29th June–2nd July 2011, pp. 665–670 (2011)
6. Hack, H., Wangler, T., Mata-Falcon, J., Dörfler, K., Walzer, N., Graser, K., Reiter, L., Richner, H., Buchli, J., Kaufmann, W., Flatt, R., Gramazio, F., Kohler, M.: Mesh mould: an on site, robotically fabricated, functional formwork. In: Conference: High Performance concrete and Concrete Innovation Conference, At Tromsø, Norway, Volume: 11th HPC and 2nd CIC (2016)
7. Wangler, T., Lloret, E., Reiter, L., Hack, N., Gramazio, F., Kohler, M., Bernhard, M., Dillenburger, B., Buchli, J., Roussel, N., Flatt, R.J.: Digital concrete: opportunities and challenges. RILEM Tech. Lett. **1**, 67–75 (2016)
8. Melbye, T.A.: Modern advances and applications of sprayed concrete. In: Keynote paper given at the International Conference on Engineering Developments in Shotcrete, Hobart, Tasmania, Australia, 2nd to 4th April (2001)
9. Heralić, A., Christiansson, A.-K., Lennartson, B.: Height control of laser metal-wire deposition based on iterative learning control and 3D scanning. Opt. Lasers Eng. **50**, 1230–1241 (2012). https://doi.org/10.1016/j.optlaseng.2012.03.016
10. Wolfs, R.J.M., Bos, F.P., van Strien, E.C.F., Salet, T.A.M.: A real-time height measurement and feedback system for 3D concrete printing. In: Hordijk, D., Luković, M. (eds.) High Tech Concrete: Where Technology and Engineering Meet. Springer, Cham (2018)

11. Xiong, J., Zhang, G.: Online measurement of bead geometry in GMAW-based additive manufacturing using passive vision. Meas. Sci. Technol. **24**, 5103 (2013). https://doi.org/10.1088/0957-0233/24/11/115103)
12. Holzmond, O., Li, X.: In situ real time defect detection of 3D printed parts. Addit. Manuf. **7**, 135–142 (2017). https://doi.org/10.1016/j.addma.2017.08.003
13. Tapia, G., Elwany, A.: A review on process monitoring and control in metal-based additive manufacturing. J. Manuf. Sci. Eng. **136**, 060801 (2014). https://doi.org/10.1115/1.4028540
14. Ibrahim, S., Olbrich, A., Lindemann, H., Gerbers, R., Kloft, H., Dröder, K., Raatz, A.: Automated additive manufacturing of concrete structures without formwork. In: Concept for Path Planning Conference: Montage, Handhabung und Industrierobotik, At Erlangen, Germany (2018)

Challenges of Real-Scale Production with Smart Dynamic Casting

E. Lloret-Fritschi[1(✉)], F. Scotto[1], F. Gramazio[1], M. Kohler[1], K. Graser[2], T. Wangler[3], L. Reiter[3], R. J. Flatt[3], and J. Mata-Falcón[4]

[1] Chair of Architecture and Digital Fabrication, Department of Architecture, ETH Zürich, Zurich, Switzerland
lloret@arch.ethz.ch
[2] National Centre of Competence in Research Digital Fabrication in Architecture, Zurich, Switzerland
[3] Department of Civil, Environmental and Geomatic Engineering, Institute of Building Materials, ETH Zurich, Zurich, Switzerland
[4] Department of Civil, Environmental and Geomatic Engineering, Institute of Structural Engineering, ETH Zurich, Zurich, Switzerland

Abstract. Digital fabrication with concrete has for more than a decade been of high interest in both research institutions and industries. A particular interest has been set on Contour Crafting, a type of layered extrusion with concrete, which in recent years has been used for the fabrication of single and multi-story buildings. However, these have been done with little proof of systematic integration of reinforcement, which until now still requires tedious post processing to obtain the structural capabilities required.

Smart dynamic casting, a robotic fabrication process for standard and non-standard vertical structures, has recently proven a systematic integration of reinforcement and is thereby the first digital fabrication process world-wide which has unified reinforcement and concreting in a single robotic fabrication process. This paper presents the latest developments and challenges of SDC and introduces the first architectural application in the form of structurally optimised façade mullions that are to be installed in the dfab House at the EMPA premises in Dübendorf, Switzerland.

Keywords: Slipforming · Digital fabrication · Concrete · Processing

1 Introduction

For more than a decade digital technologies and the introduction of computer-aided additive fabrication in architecture have demonstrated strong potential to construct bespoke complex concrete structures [1, 2]. Due to the enormous use of concrete in the construction industry and the high labour cost for formwork, digital fabrication with concrete has become a focus to reduce the economic and environmental impact of concrete structures. The global aim in concrete construction is to obtain more design freedom paired with material and structural optimisation, and in the best case with zero-waste.

T. Wangler and R. J. Flatt (Eds.): DC 2018, RILEM Bookseries 19, pp. 299–310, 2019.
https://doi.org/10.1007/978-3-319-99519-9_28

Several research projects have emerged in digital concrete construction, who all target the same goal, to construct bespoke concrete structures efficiently. The pioneering method is *Contour Crafting,* [3] a layer-by-layer extrusion process with concrete – currently the most investigated additive fabrication process with concrete worldwide, with which several larger-scale structures has been erected. Other strategies have included *Mesh Mould*, which uses robotically constructed steel cages as a lost formwork [4], as well as *stay-in-place knitted formworks for complex concrete structures*, a system which combines the flexibility of textile with moulding and structural properties [5]. Each techniques face challenges, ranging from: control of rheological material properties during processing, optimal surface quality, systematic integration of reinforcement as well as compliance with building codes, just to mention a few. *Smart Dynamic Casting* (SDC), the subject of this paper, has the same types of requirements. In this paper, the challenges encountered in full scale production of 15 bespoke façade mullions are presented, including (1) defining a formwork and actuation system that reduces friction force and enables shaping, (2) defining and scaling down a material processing system, and (3) integrating all systems together with incorporation of reinforcement to produce custom designed mullions tailor made for particular design loads.

2 Fabrication and Challenges

Smart Dynamic Casting (SDC) is a robotic slipforming process in which concrete is shaped by an actuated formwork much smaller than the element it is producing, requiring precise material control and process feedback [6]. SDC's greatest advantage is that to date, it is the only robotic fabrication process to produce bespoke reinforced concrete elements without any post processing.

Smart Dynamic Casting has been developed in several stages, first focusing on the required development of a material system capable of being shaped in the moment when concrete is hard enough to sustain its own weight, but soft enough to still be deformed [6]. In a second stage, the limits of this gravity-driven process were tested by greatly increasing the surface area to volume ratio of the formwork – this was proven in an experiment in which a canoe shell was formed, and continues in the work of examining the possibility of forming thin folded structures [7]. In a third stage, two separate research threads were pursued, one being the automation of the material processing and the second investigating a new shaping concept which enabled the production of structures with significant cross sectional change, a relevant fact for the production of structurally optimized elements [8].

The latest stage and the topic of this paper is related to the production of 15 extremely slim façade mullions for the DFAB House at the Empa NEST Premises, in Dübendorf Switzerland [9]. Overall the aim of the development was to prove that SDC is suitable for mass-customising bespoke structurally optimized columns. A very high quality of surface and size accuracy was required along-side with the integration of reinforcement and installation details. Thus, to achieve this, intense development was required on the following topics: formwork and actuation, processing, reinforcement integration, and structural design.

2.1 Overall Fabrication Concept

The overall fabrication setup used for the mullion production consists of a linear unit onto which a flexible formwork and the feeding system is mounted, depicted in Fig. 1. This setup is controlled by a custom software that synchronizes the process of material filling, cross section change of the formwork and slipping speed.

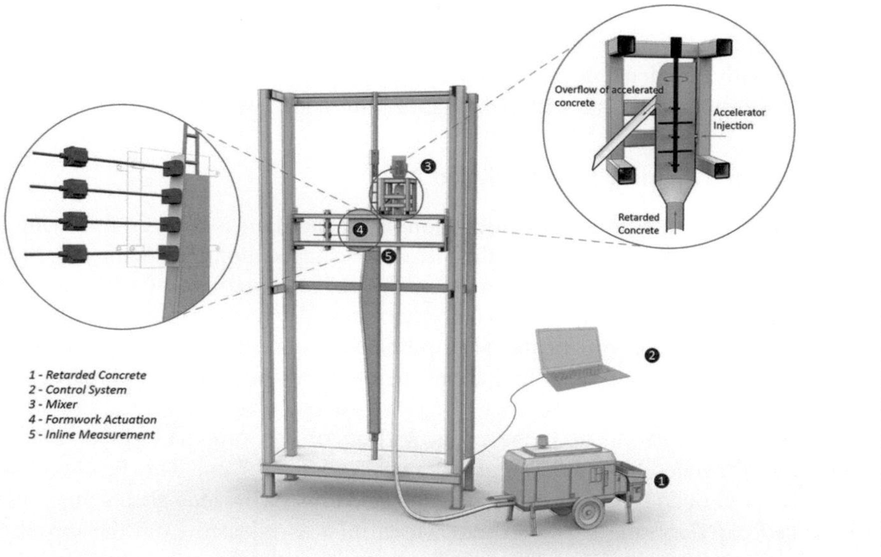

Fig. 1. Diagrammatic representation of SDC fabrication setup

The diagram in Fig. 1 features a schematic representation of the fabrication process. In which initially, a large batch of retarded mortar is prepared and placed into a universal conveying pump (1). A custom software system (2) controls the pumping of small batches of the retarded mix and accelerators into a mixer (3) placed right above the formwork (4). From the mixer small dosages of self-compacting concrete overflows into the formwork. Depending on the dimensions of the formwork, the intervals of pumping automatically vary, to assure that the amount of material inside the formwork remains at a constant height. Furthermore, the software synchronizes the vertical movement of the formwork with the pumping rate to ensure a continuous casting process. The inline feedback system (5) feeds the information on material strength and formwork pressure back to the control software that computes the information, evaluating the property of the material and adapting the fabrication speed and pumping rate accordingly [8].

3 Fabrication Challenges

It was initially believed that simply by scaling down a working process, the goal of 15 slender façade mullions with changing cross section could be achieved, as this had already been achieved with larger cross sections [8]. Scaling down, however, posed new challenges for the SDC system. By scaling down by 50%, friction in the formwork increased significantly as the surface to volume ration increased. Additionally, precision movement of the formwork is affected by decreased cross sectional area. Also, by scaling down, the material processing changed significantly, which impacted material rheology. Finally, implementation of rebar had previously only been performed in an exploratory study, and in this case compliance with codes was necessary.

3.1 Formwork, Actuation and Friction

The formwork system used for the production of the mullions is built up following the concept of 'global shaping', in which the material is shaped along the entire height of a flexible shaping strip, by means of actuators. The range of movement of the actuators and the surface material of the formwork determines the main constraint in the definition of feasible geometries for the SDC fabrication system. In addition, the design also had to negotiate with the limitations defined by the material properties and structural requirements of the façade.

For the mullion production, the cross section of the original 'global shaping' formwork was reduced from a width of 15 cm to a width of 7 cm. This formwork was constructed of plywood plates laminated with scaled polypropylene sheets, making up the so-called capillary oiling system into which oil was injected from the top of the formwork. The shaping strips were equally made of polypropylene scales, with an integrated oiling system mounted onto a strip of spring metal. The previous experiments, using the larger dimensions of the formwork had proven this formwork enabled the production of columns with a smooth surface. However, scaling down the formwork resulted in increased friction, as the surface area to volume ratio was increased by a factor of 3.5. The resulting process caused the mullions to have a rippled surface quality and traces of the shaping strips on the side of the mullions as illustrated in Fig. 2. This was caused primarily by the deviations of the polypropylene lamination capillary system, which consisted of multiple glued layers, secondly, the shaping strips dissolved from the metal during production caused imprecision along the height of the mullion, and thirdly, the oiling from the top of the formwork at times caused an overflow of oil into the still soft material. All these issues compromised the surface quality of the mullions and the late strength properties of the final structure.

Several experiments, not reported in this paper, enabled the definition of a new and simplified formwork system consisting of stainless steel (see Fig. 3). The outer formwork shell was in this case simplified to a U-shaped element (see Fig. 3a) onto which the capillary oiling system could be mounted. The oiling system consisted of a 112 mm high bended steel plate mounted with three screws, and multiple oiling channels assured that oil continuously flowed inside the formwork during the slipping. This system eliminated the deviations caused by the polypropylene scales. Further, as the oil injection points had been moved 100 mm from the top, this assured that the oil

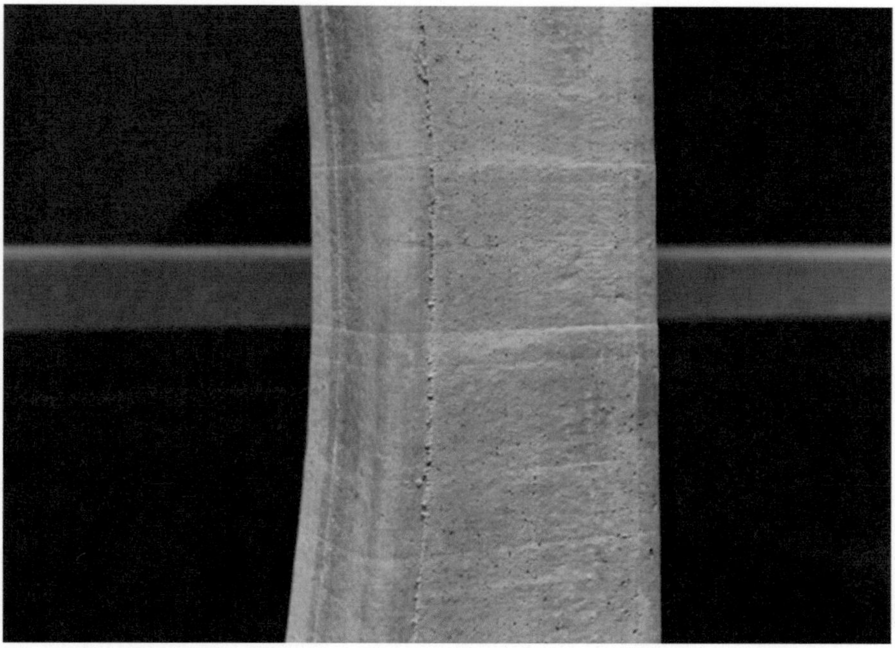

Fig. 2. Result of column produced with wooden formwork with coated with polypropylene scales, using shaping strips of scaled polypropylene. The result of the surface shows ripples and traces from the shaping strips on the left side.

would not penetrate into the still soft concrete. Finally, the shaping strip was constructed of bent steel scales welded onto a spring metal plate (see Fig. 3c), into which a channel for capillary oiling was mounted for continuous oiling along this interface. All these changes resulted in the significant reduction of friction. During the study, however, it was detected that the self-compacting properties of the material filled into the formwork had just as much of an effect on the surface quality. Thus, having a highly precise formwork system could not solve the problem of surface quality alone.

3.2 Material Processing

In regards to material processing, an automated acceleration and feeding system had been developed earlier in SDC [8], suitable for structures with cross sectional areas ranging between 15 × 15 cm up to 15 × 80 cm. The significant downscaling of the formwork required a re-dimensioning of the acceleration system and a re-adjustment of the material mix design to achieve the process requirements throughout the entire process of slipping. Scaling down the process to more narrow elements required a smaller material flowrate, and thus a shorter residence time in the mixing funnel before placement, so a mixer was appropriately sized. It was discovered that this resulted in a material which did not hydrate quickly enough in the formwork to maintain normal

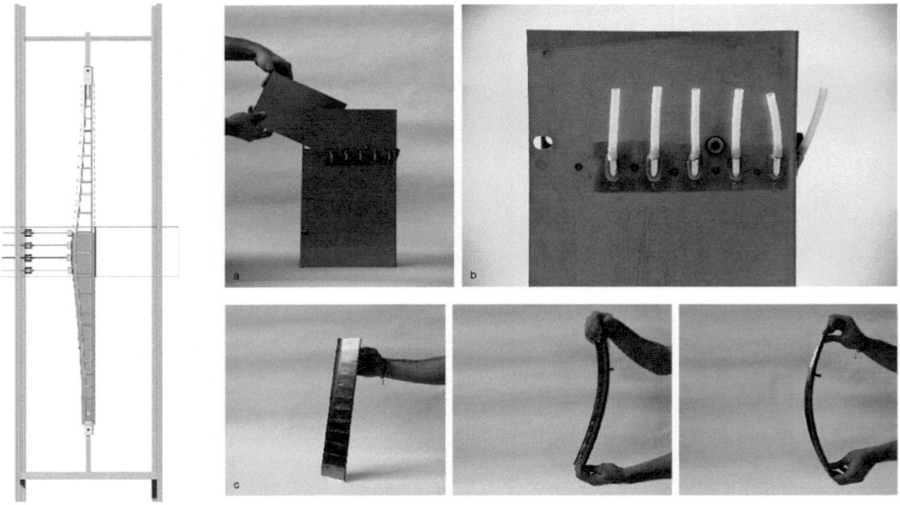

Fig. 3. Left diagrammatic illustration of the formwork and its actuators during a shaping process. Top Right (a) Formwork (height 450 mm × width 350 mm) and insertion of oiling foil, (b) capillary oiling system pipes, assures that the formwork is continuously oiled during the slipping process. (c) shaping strip consisting of multiple overlapping steel foils of 0.5 mm, that are welded onto a piece of spring metal enabling the needed flexibility for shaping.

processing speeds, and thus a higher residence time in the mixing funnel was investigated by adjusting the inlet position of the accelerator, as detailed in Fig. 4. Complicating things further, a new delivery of cement from the cement supplier resulted in a change in hydration behavior in the material, significantly slowing down the hydration in the normal process. To address these issues, the mix design was simply modified by significantly increasing the amount of the accelerator. While this solved the issues of having a stable, self-compacting mix placed into the formwork, it introduced durability issues because of the very high dosage of accelerator (8.5 wt% cement of Sika Rapid C100), which is a calcium nitrate based accelerator. This high accelerator dosage was enough to cause corrosion of the steel reinforcement, which was resolved simply by the use of stainless steel. An additional problem is that this high dosage of calcium nitrate based accelerator could introduce problems of salt crystallization within the concrete. This problem is one that can develop over time, and to address this RH cycles are being performed on select samples that have come from the process, and onsite monitoring of the RH, temperature will be performed on the columns, in addition to periodic visual inspection and ultrasonic measurements. This is to be performed for the duration of the dfab House on the NEST building, which is expected to be five years.

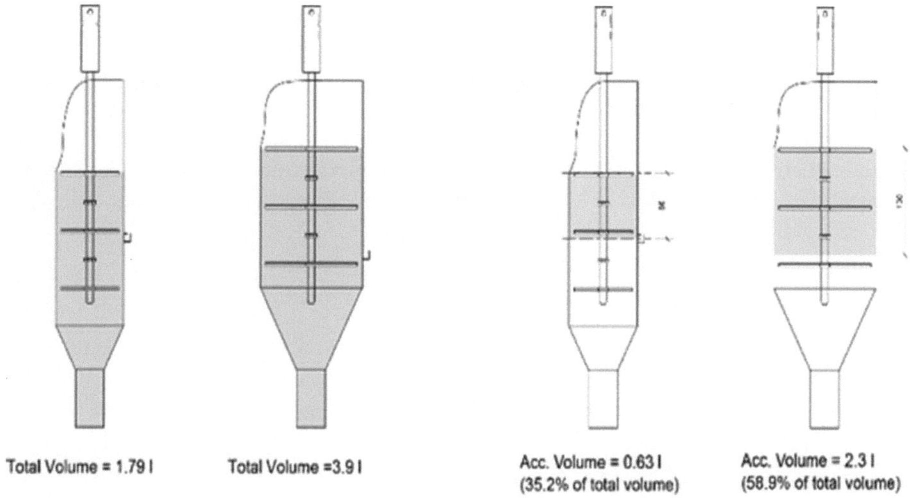

Fig. 4. Schematic of acceleration funnel design of two different sizes, showing total volume (left) and injection volume for mixing of accelerator (right)

4 Design and Production

The mullions function as part of a structural façade system in the dfab House in the NEST building at the Empa premises, and have been designed to optimally support the wind actions acting on the façade. The mullion geometry, which starts from slender to wide and back to slender (Fig. 1), has been defined to adapt to the acting design load case. Due to structural requirements from the above structures, the tributary length of the façade glazing varies between 700 and 1400 mm and consequently the wind loads acting on each mullion are different. Therefore, to correspond to the varying loads, each mullion was designed to respond to the individual load case.

4.1 Material Characterization

SDC is a batch by batch slipping process along the longitudinal axis of the member, in which concrete is processed very differently than in conventional casting processes. The particularities of the process (absence of formwork at a very early age) and of the specific mullion application (minimal section), may lead to a different concrete mechanical behaviour than the typically assumed by building codes. To assess this effect of the SDC process, a set of comparative tests were conducted comparing the SDC mix design in samples produced with the SDC technology versus samples cast conventionally into a formwork. The compressive strength at 28 days in cubic samples produced with the SDC technique (150 × 150 × 150 mm) was 77 MPa (CoV = 0.01), lower than the strength in conventionally produced samples (110 MPa, CoV = 0.01), but still in a high strength range, more than enough for the intended application. The shear strength measured by 500 mm span 3-point bending tests in samples with 70 × 100 mm cross section and Ø12 longitudinal bars, lead to similar values for the SDC and the conventional production. Hence, the shear strength in a diagonal direction to the batch by batch process is not affected by the SDC slipping process.

As a result, based on the conducted mechanical tests, SDC structures behave similarly as conventional structures and can be designed with conventional methodologies when using the proper concrete material properties accounting for the slipping process. The structural design of the dfab House mullions was conducted following SIA 261:2014 and SIA 262:2013 codes [10, 11].

4.2 Structural Design

The SDC elements in the dfab House were conceived as mullions supporting the wind actions acting on the glass façade (maximum design values of 0.84 kN/m^2 and 1.90 kN/m^2 in pressure and suction respectively). The connection system to the top and bottom slabs was designed as a pinned connection (see Fig. 5) that avoids the

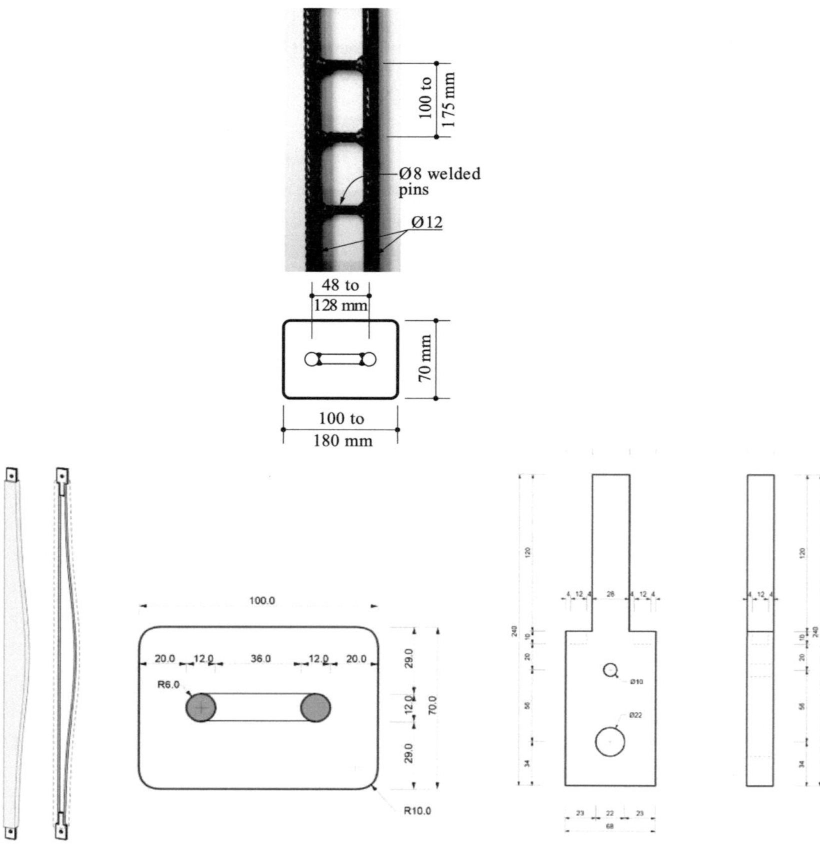

Fig. 5. Reinforcement and connection definition

transmission of vertical loads. Therefore, the mullions structurally behave as single supported elements with distributed pressure or suction loads. Because of fabrication constrains, the minimum cross section was 100 × 70 mm. Based on these requirements, the SDC system was designed in order to guarantee the structural integrity of the application (the mullions themselves, as well as the glass façade which maximum deformation is restricted to 8 mm).

In order to resist the bending actions it was necessary to reinforce longitudinally the mullions with Ø12 conventional deformed steel bars. Different transversal reinforcement alternatives ranging from carbon fiber to deformed steel bars were studied [12] in order to provide sufficient shear capacity. This study showed that the shear strength without transversal reinforcement was sufficient; however, transversal welded deformed steel bars (Fig. 5) were introduced in the final design in order to provide deformation capacity, complying with design codes.

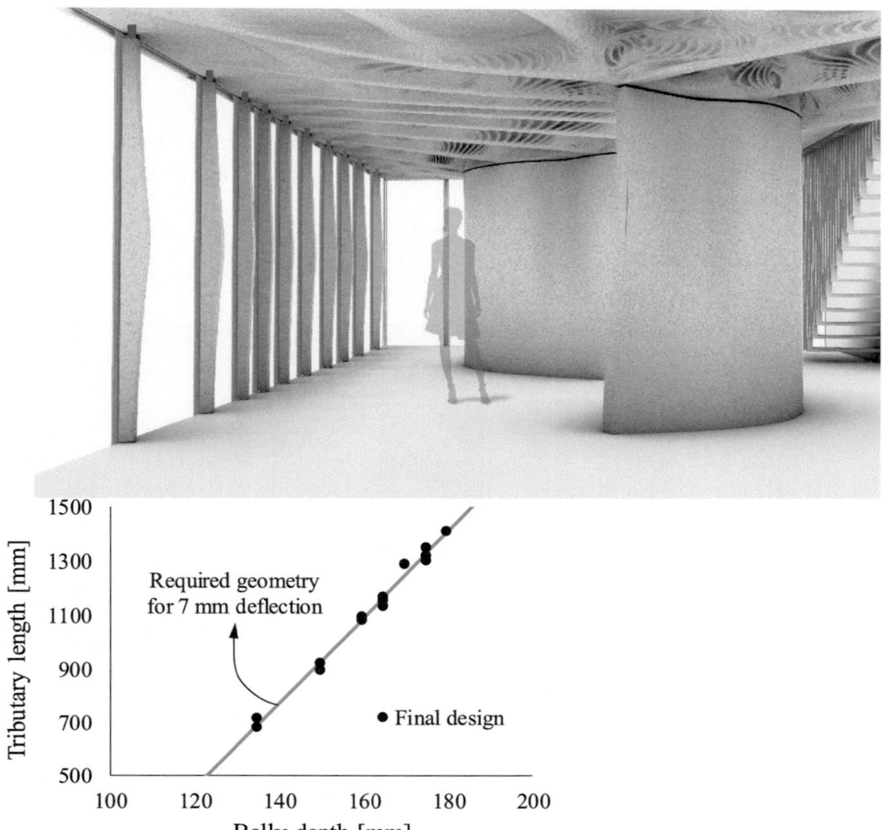

Fig. 6. Geometry of the SDC mullions: (a) overall view, (b) final design fulfilling optimally the requirement of maximum deflection of the façade glazing

Besides the structural integrity of the mullions, guaranteed by the provided reinforcement, the integrity of the façade glazing required limiting the deflections of the mullions below 1/400 of their height (i.e. below 8 mm). This deflection verification was decisive for the final design of the mullions' geometry. The widest cross-section (defined as the "belly") of the 15 mullions was varied from a minimum of 70 mm × 120 mm up to 70 mm × 180 mm in order to optimally fulfil the deflection requirement (see Fig. 6). Thanks to flexibility of the SDC system, it was possible to produce the bespoke optimum mullions without increasing the complexity of the production.

4.3 Material Processing Challenges: Reinforcement Integration

Based on previous studies, it was known that the integration of traditional rebar cages in a very small cross section would be a challenge, due to the significant deviations they usually have. While in traditional casting this issue is typically solved by using spacers inside the formwork, this was not feasible in SDC as (i) there is not a fixed formwork to put the spacers before the production and (ii) the spacers would prevent the material from flowing properly inside the minimal cross sectional area. To solve this issue, computer numerically control (CNC) bent reinforcement bars with reduced tolerances (up to 2 mm) were used. In spite of the tight bent tolerances provided by the CNC process, the welding of transversal pins increased the final tolerances. Therefore, to guarantee the minimum 20 mm required concrete coverage in the reinforcement, the bars were slightly pre-tensioned by means of a tailored made hinge system (see Fig. 7) to assure their straightness during production.

Fig. 7. (left) linear axis prepared for production, (bottom center) pretension attachment of rebar, (top center) rebar tensioned upwards, (right) beginning of slipping process

5 Conclusion

This paper represents a case study in a real, construction scale application for a new digitally controlled slipforming process. Fifteen façade mullions, custom designed to withstand variable loads, have been produced in a single, zero-waste, digitally controlled system. While the production of these mullions has been so far successful, numerous challenges arose as a consequence of this project, detailed below:

1. In a gravity-driven process such as slipforming, low surface-to-volume ratios within the formwork lead to smaller processing windows due to the increased friction, and engineering solutions can be taken up to a certain point.
2. Scaling the process up or down is not straightforward, and the process is highly sensitive to incoming material variations.
3. Certain processing requirements can lead to compromises in other areas, e.g. the required increase in accelerator leads to corrosion susceptibility and salt precipitation risk.
4. The research and development of the mullions turned out to be an iterative process, which demonstrates that to develop a digital fabrication process with concrete, each component involved must be analyzed as part of an integrated process. The final product is a result of several constraints mainly informed primarily by the fabrication process, but indeed also of the building code as in the case of the reinforcement.
5. Specifically, this means that each element is informed by each other, be it the material, the material processing, the formwork or the digital tool. Indeed each element is developed separately, but must always be tested in an integrative process, in which each element informs each other.

Acknowledgments. The work in this study was funded by the National Centre for Competence in Research in Digital Fabrication in Architecture. The authors acknowledge the contributions of Heinz Richner and Andreas Reusser, as well as the support of Hans Flueckiger (Zühlke AG).

References

1. Bos, F., Wolfs, R., Ahmed, Z., Salet, T.: Additive manufacturing of concrete in construction: potentials and challenges of 3D concrete printing. Virtual Phys. Prototyp. **11**(3), 209–225 (2016)
2. Wangler, T., Lloret, E., Reiter, L., Hack, N., Gramazio, F., Kohler, M., Bernhard, M., Dillenburger, B., Buchli, J., Roussel, N., Flatt, R.: Digital concrete: opportunities and challenges. RILEM Tech. Lett. **1**, 67–75 (2016)
3. Khoshnevis, B.: Automated construction by contour crafting – related robotics and information technologies. Autom. Constr. **13**(1), 5–19 (2004)
4. Hack, N., Wangler, T., Mata-Falcon, J., Dörfler, K., Kumar, N., Walzer, A.N., Graser, K., Reiter, L., Richner, H., Buchli, J., Kaufmann, W., Flatt, R.J., Gramazio, F., Kohler, M.: Mesh mould: an on site, robotically fabricated, functional formwork. In: Second Concrete Innovation Conference (2nd CIC), Paper No. 19, Tromsø, Norway (2017)

5. Popescu, M., Reiter, L., Liew, A., Van Mele, T., Flatt, R.J., Block, P.: Building in concrete with an ultra-lightweight knitted stay-in-place formwork: prototype of a concrete shell bridge. Structures **14**, 322–332 (2018)
6. Lloret, E., Shahab, A.R., Linus, M., Flatt, R.J., Gramazio, F., Kohler, M., Langenberg, S.: Complex concrete structures: merging existing casting techniques with digital fabrication. Comput. Aided Des. **60**, 40–49 (2015)
7. Szabo, A., Reiter, L., Lloret-Fritschi, E., Gramazio, F., Kohler, M., Flatt, R.J.: Adapting smart dynamic casting to thin folded geometries. In: Proceedings of the 1st International Conference for Concrete and Digital Fabrication, Zurich, Switzerland (2018)
8. Lloret-Fritschi, E., Reiter, L., Wangler, T., Gramazio, F., Kohler, M., Flatt, R.J.: Smart dynamic casting: slipforming with flexible formwork – inline measurement and control. In: Second Concrete Innovation Conference (2nd CIC), Paper No. 27, Tromsø, Norway (2017)
9. http://dfabhouse.ch/dfab-house/. Accessed 13 June 2018
10. SIA 261: Einwirkungen auf Tragwerke (Actions on Load-Bearing Structures) (2014)
11. SIA 262: Betonbau – Ergänzende Festlegungen (Concrete Constructions – Complementary provisions) (2013)
12. Fuhrimann, L.: Reinforcement strategies for smart dynamic casting. ETH Zurich Bachelor Thesis (2017)

The Tectonics of Digitally Fabricated Concrete. A Case for Robotic Hot Wire Cutting

Pedro Filipe Martins[1](✉), Paulo Fonseca de Campos[2], Sandra Nunes[3], and José Pedro Sousa[1]

[1] Faculty of Architecture, University of Porto, Porto, Portugal
pcarvalho@arq.up.pt
[2] Faculty of Architecture and Urbanism, University of São Paulo, São Paulo, Brazil
[3] Faculty of Engineering, University of Porto, Porto, Portugal

Abstract. In the last decades, digital fabrication technologies have stimulated the materialization of complex and customized solutions in several materials. Recently, the integration of these technologies with such a variable and rich material as concrete has prompted an explosion of possible processes and outcomes for digitally fabricated concrete structures. In this context, this paper examines current digital fabrication strategies for concrete, focusing on their applications and in identifying critical issues for their adoption. From this point, through the presentation of two case studies, we propose and discuss Robotic Hot Wire Cutting as a technically and tectonically relevant digital fabrication technology for customized concrete architecture.

Keywords: Concrete · Digital fabrication · Robotic Hot Wire Cutting

1 Introduction

1.1 Form, Surface and Composition

In the last 30 years, the dissemination of digital design and fabrication technologies in architecture has inspired and enabled the production of increasingly complex and customized design and constructive solutions [1]. Concrete is a particularly relevant material in this context. Its ability to assume virtually any shape has, throughout history, enabled architects and engineers to explore its plasticity in complex free forms, hardly achievable through other materials. In turn, their construction process has been one of the main challenges of concrete architecture, presenting technological, economic and material limitations to the realization of the most complex solutions.

Yet, it was through this construction process that most of the aesthetic and tectonic value of concrete architecture was defined. Not only in the creation of expressive **form**, but also in the character of its **surface** textures and in the properties of its material **composition**. Kenneth Frampton would characterize it as a poetic consistency between a form and its construction [2]. Examples such as the freeform thin shells of Felix Candela, the textured concrete surfaces of Paul Rudolph and the exposed inner

T. Wangler and R. J. Flatt (Eds.): DC 2018, RILEM Bookseries 19, pp. 311–322, 2019.
https://doi.org/10.1007/978-3-319-99519-9_29

compositions and colored concrete of I.M Pei, represent the past relevance of these three tectonic dimensions in concrete.

Therefore, when examining the impact of digital fabrication technologies in concrete architecture, it is important to consider not only the possibility to overcome existing building limitations but also to unveil the emerging possibilities in those three tectonic levels. Within this framework, both technical and tectonic, this work seeks to answer the complex question: How can digital fabrication technologies expand traditional concrete architecture in form, surface and composition, through economically viable, sustainable and tectonically relevant methodologies.

To answer this question, this paper will be developed in two main sections. In the first, the authors perform a critical analysis on existing digital fabrication strategies, focusing on understanding how their application is shaping concrete architecture and concluding on the main challenges facing their widespread adoption. In the second section, the authors put forth a case for Robotic Hot Wire Cutting (RHWC) as a viable and tectonically relevant digital fabrication method capable of overcoming this condition.

2 Digital Fabrication Technologies in Concrete Architecture

Over time, the multifaceted nature of concrete and its complex production process has been a fertile ground for new material expressions, which are framed by existing production technologies and processes. With the recent dissemination of digital design and fabrication technologies, the production of concrete is facing new design opportunities. The most visible impact is related with the possibility for embracing design customization and complexity in concrete architecture. Towards this end, several digital fabrication processes have been developed.

2.1 Evaluating Existing and Emerging Solutions

Contemporary construction methods for complex concrete architecture are today still largely dependent on manually produced, customized formwork or CNC milled, Expanded Polystyrene (EPS) molds. Nevertheless, several other solutions have been developed that take advantage of digital fabrication technologies, either already in use in industrial settings or proposed in academic research prototypes. Such technologies can be broadly distinguished by their application for formwork production or in the direct production of concrete structures. In the following chapters, we provide a critical analysis of the most relevant technologies, focusing on their current and potential applications.

Digital Fabrication of Formwork for Concrete. The production of customized formwork for complex concrete architecture has been one of the main goals for the integration of digital fabrication technologies in concrete construction.

Subtractive Processes. The first applications of digital fabrication in concrete employed the capabilities of CNC (Computer Numeric Control) cutting or milling, to produce complex curved formwork from materials such as wood or EPS. CNC cutting of wood panels was used in works such as the Mercedes Benz Museum (2006) and the Rolex Learning Center (2009) to achieve moderately curved geometries with smooth finishes, while still largely dependent on manual labor for assembly [3, 4]. On the other hand EPS foam molds have been produced through the use of CNC milling technologies or, more recently, with RHWC processes. While the former is capable of producing complex freeform geometries, having been thoroughly explored and optimized for 5-axis robotic technologies in the recent TailorCrete Project [5], RHWC is still mainly suitable for the production of ruled geometry. Nevertheless, it is regarded as a materially optimized and economically viable alternative to CNC milling for the production of EPS molds for concrete [6, 7].

Subtractive digital fabrication strategies for formwork production have been adapted to both in-situ and precast applications. In the first case, generally in the form of monolithic freeform constructions, such as the Spencer Dock Bridge by Amanda Levete Architects [8]. In the second case, oriented towards the production of complex assemblies of discrete elements at smaller scales such as the case of the interior precast panels of the Louisiana Sports Hall of Fame by Trahan Architects [9].

Formative Processes. An alternative approach to subtractive processes in mold production is to use formative processes through digitally controlled, reconfigurable molds with pin actuated systems, also developed in TailorCrete Project and several institutions [10, 11]. While benefitting from high adaptability, reusability and surface quality, these technologies are generally restrained to the production of surfaces similar to 3-axis milling and further limited to specific sizes and perimeters of the mold systems and to limited fabrication times when considering the need for multiple production runs. As such, examples of these systems currently show a tendency towards the prefabrication of double curved and high quality paneling systems [12].

Additive Processes. Although most commonly seen as direct concreting strategies, additive fabrication has also been explored for the production of molds for concrete. In these cases, 3D printing technologies are used to produce large quantities of differentiated molds in materials such as wax, sandstone and plastics [13, 14]. Although also limited to the size of printing systems, these solutions, benefit from the increased geometric freedom of 3D printing when compared to subtractive processes, particularly in the case of sintering or binder jetting. This makes them primarily suited for small-scale prefabrication of highly complex and differentiated components such as those resulting from the design of topological optimized structures [15].

Digital Fabrication of Concrete Structures. The direct production of concrete elements bypassing the need for formwork has been explored mainly through the use of additive technologies. Although some experiments have been developed using subtractive technologies (e.g. Robotic Diamond Wire Cutting) for the production of voussoir assemblies directly on cured concrete, these processes have been more suited for other materials such as stone [16].

Additive Processes. To date, several 3D concrete printing processes have been developed in academic and industrial settings. These cases use either an extrusion process where a digitally controlled nozzle deposits wet concrete in a layered process (e.g. Contour Crafting, Freeform Construction) or a binder jetting process, where a binder solution is deposited on a powdered matrix, curing layer by layer (e.g. D-Shape, Emerging Objects). While binder jetting processes have the ability to produce extremely complex geometries, including voids and cantilever sections, extrusion processes are much more limited in possible shapes, considering the need for each layer to be supported by the previous one. As such, 3D printing of concrete structures has diverging results when considering different scales and methods. These range from small to medium scale assemblies of complex and customized elements, such as the case of the 3d printed beams developed at ETHZ [17], to large scale constructions with constrained geometries, like the prototype houses printed by the Chinese company WinSun [18]. Other strategies have also been developed that use additive technologies in optimized yet highly constrained and design specific processes, such as the robotic slip forming technology developed at the ETHZ for the production of column-like structures [19].

2.2 Expanding Form, Surface and Composition

Although not exhaustive, the previous analysis presents a large picture of what is today technologically feasible. At the same time, the different applications of such contrasting processes also point to emerging tectonic qualities of concrete structures produced with digital fabrication processes. These can be expressed as an expansion of the previously mentioned concepts of form, surface and composition in concrete.

Complex Double-Curved Geometries and Variable Modular Components. When considering the issue of form in digitally fabricated concrete architecture, we can identify two overall themes patent in the already mentioned examples of the Spencer Dock Bridge and the Louisiana Sports Hall of Fame: (a) double-curved surfaces in monolithic structures; (b) small to medium-scale assemblies of complex and customized components. As seen in the previous analysis, while the first is achieved mainly through the use of subtractive processes for formwork production, the second typically concerns prefabrication with additive and formative technologies.

Process-Driven Surface Textures. Although the capabilities of CNC milling and 3D printing have made possible the design and production of virtually any textural effect desired, digital fabrication processes in general have the potential for the creation of new and tectonically relevant process-driven surface textures. Milling paths, hot wire cutting marks and 3D printing layers are recurring traces in digitally fabricated concrete structures that have already been explored as aesthetic expression. For instance, the Building Academy Extension by SOMA Architects, where CNC milling toolpaths were explored as surface textures in cast concrete [20].

Variable Material Compositions. A common denominator in the use of digital fabrication technologies in general and in additive fabrication processes in particular is the relevance of the material composition, usually as a technical consideration. While concrete is traditionally cast in homogenous compositions, the use of digital fabrication processes such as multi-material 3D printing is also empowering the use variable concrete compositions in concrete. Such processes hint at materially optimized structures [21] and interesting new design possibilities that explore gradation in different material properties, such as density and color [22], opening new design spaces for concrete architecture.

2.3 Critical Issues for Digital Fabrication in Concrete

The previous analysis shows a wide range of digital fabrication processes and the possibility expanding traditional concepts of form, surface and composition in innovative solutions for concrete architecture. Nevertheless, digital fabrication technologies still present critical issues that should be addressed to further their implementation into practice.

One major issue is the balance of geometric freedom, with production and material optimization. Being the most used digital fabrication solution for customized concrete architecture, this issue is particularly relevant in the case of CNC milling of EPS molds. Although CNC milling is a useful process for the production of complex and customized formwork solutions, it is also severely limited when considering three key aspects:

1. **high production times** resulting from the high-resolution milling process necessary for smooth finishes;
2. **low material optimization and sustainability**, considering the large quantities of material discarded as waste for each mold surface produced through milling, when compared with the potential absence of waste from hotwire cutting.
3. **high overall production costs,** from materials, labor and fabrication times.

Based on these issues, CNC milling of EPS molds is usually considered a cost-ineffective process only suitable in works of exceptional character and budgets, with particular complex geometries. This issue was apparent in the recent case of the Waal Bridge Extension by Zwart & Jansma, realized with traditional, manually produced wood formwork despite its freeform geometry, considering the high costs inherent to CNC milling [23].

A second relevant issue is the adaptability and integration of digital processes into the construction industry. Processes such as 3D printing of molds and finished structures present substantial improvements in sustainability by eliminating waste production. Although potentially disruptive, these processes require highly technological setups, more suited the controlled environments of prefabrication and laboratory settings. Their application in the construction industry is therefore not easily fulfilled in current industrial contexts, requiring the transformation of existing processes, methodologies and equipment, limiting the possibility for their adoption and widespread use.

Considering these critical issues and the emerging design opportunities of digital fabrication, in the next sections we present a case for RHWC as a viable and tectonically relevant digital fabrication technology for concrete architecture.

3 A Case for Robotic Hot Wire Cutting

3.1 Description

RHWC is a recent evolution of wire cutting processes already found in industrial and pre-industrial settings [24, 25] and used for the fast carving of large volumes of materials such as clay, wood, stone and more recently EPS foam. Later industrial developments have paired this tool with different automation mechanisms such as 3 and 4 axis CNC gantries. In the case of current RHWC systems, a thin heated wire is stretched in an end-effector, attached to a digitally controlled 6-axis robotic arm and used to cut through solid EPS blocks, producing molds for concrete casting. The main features of this process are its overall speed (when compared with CNC milling) and its inherent limitation to the description of ruled surfaces, resulting from the movement of the hot wire through the EPS material. As such, RHWC is applicable mainly in the production of such surfaces or in the approximation of more general double-curved designs which have been rationalized to fit the ruled surface criteria [26].

In the research field of architecture, RHWC has seen several developments in the last decade, with experimental works such as the Periscope Tower, the RDM Vault [6] and the Opticut prototype [16], exploring the design potential of RHWC. More recent developments are focusing on overcoming the geometric limitations of ruled surfaces through the exploration of flexible blades and multiple robot setups, allowing for the direct creation of more general double-curved surfaces [23, 27].

These and other experiments have suggested that RHWC can address the most critical issues described in Sect. 2.3, proposing a fast and cost-effective digital fabrication strategy for the production of formwork for concrete architecture.

To understand the potential of RHWC and explore this technology, two case studies were developed, focusing on different technical and design aspects of RHWC and with a specific interest in the development of mold design and production solutions for prefabricated concrete architecture.

3.2 Ruled Concrete Panels

The Ruled Concrete Panels experiment was developed as a primary exploration into the techniques and processes of RHWC. Its main objectives were to evaluate the limitations and potentials of this technology for the fabrication of customized concrete molds. For this purpose, a set of interlocking elements was designed, based on the geometry of typical precast panels for mechanically stabilized earth systems (MSE). This particular geometry was projected onto a hyperbolic paraboloid surface, resulting in a set of double curved ruled panels, suitable for RHWC. With this arrangement, three challenges were identified and used as specific objectives for the ruled panels case study:

1. **Form** - addressing the geometrical constraints of RHWC;
2. **Surface** - exploring process specific surface textures in RHWC;
3. **Composition** - exploring the possibility of variable concrete composition through mold design.

Case Description. Each panel addressed one of the highlighted challenges. Panel 1 was designed as a control, addressing only geometrical issues. Panel 2 was used for exploring ruled surface textures. Panel 3 employed a variable composition through a layered casting process. With these objectives in mind, the molds were designed specifically for the RHWC process. Each was composed of 3 stacked parts, defining a front surface, a back surface and an interior perimeter (see Fig. 1b). The mold for Panel 3 featured an intermediate subdivision to enable the layered casting required (see Fig. 1c). Using this strategy, the complete molds were cut from 1000 × 500 × 350 mm EPS blocks in a sequence of sweeping cuts, defining each respective surface.

Fig. 1. (a) RHWC fabrication process for textured panel 2; (b) EPS mold for panel 1 (c) EPS mold for panel 3; (d) final assembly of panels 1 and 3, featuring layered concrete composition.

During this process, we identified several key constraints and characteristics of this technology, regarding aspects of speed, precision and geometry. The relation between speed, temperature and the resulting cut width (kerf) was a fundamental parameter. After initial runs, a relatively low speed and temperature were defined to improve accuracy and to avoid excessive melting in specific areas such as ridges and corners which after casting would have compromised the assembly of the finished components. Although a conservative speed was used, each cutting sequence was completed in approximately 7 min, without significant increases in the more complex, textured mold surfaces or the subdivided mold of panel 3.

After assembling and bracing the molds to withstand the concrete pressure, they were cast with a mix of self-compacting concrete, reinforced with polypropylene fibers. For the variable composition panel, different colored mixes were cast in two phases, separated by a 24 h period. After curing, the panels were easily demolded by separating the mold layers. A more thorough description of this experiment and its findings can be found in Martins et al. [28].

Case Conclusions. One of the main advantages proposed by RHWC was that of increased speed and material optimization. As was verified in this case study, the increased cutting area of a line vs. a point (as in milling or 3D printing) allowed for a drastically improved processing time, several orders of magnitude faster than most subtractive or additive processes.

The material optimization inherent to RHWC was also demonstrated with this work. The preservation of material volume after each cut eliminated the production of waste (from transformation) and opened the possibility for the use of successive cuts in the layered mold strategy. This had the effect of halving the required material volume in comparison with typical milled EPS formwork. When combined, these factors greatly decrease production times and costs, increasing the overall economic feasibility and sustainability of this technology.

Regarding surface quality, although the produced surfaces were not smooth, we found that the ruled textural effects produced through RHWC had an interesting aesthetic and tectonic potential, describing a visual relation between the RHWC process and the design of the ruled surfaces.

Finally, the layered mold strategy enabled by the RHWC process also opened the possibility for more complex mold assemblies, allowing for the combination of different material compositions in one single component.

3.3 The CorkCrete Arch

The CorkCrete Arch was developed as an application and expansion of the previously explored strategies and methods, testing them in a full-scale design prototype. Other specific objectives were also addressed in the scope of this project:

1. Testing RHWC processes in large-scale geometries
2. Testing the layered mold strategy in an industrial environment
3. Testing the integration of RHWC with other processes and materials
4. Testing the precision of RHWC in large-scale assemblies

Case Description. To achieve these objectives, the CorkCrete Arch was developed as a composite system, composed from the assemblage of 3 structural Glassfibre Reinforced Concrete (GRC) elements and 18 cork panels. Its geometry was defined to challenge the RHWC process, featuring an interior creased surface resulting from the intersection of 2 ruled surfaces and a lateral face of variable width (see Fig. 2d). These features became the central geometric challenges for the RHWC process.

The design of the molds for the 3 GRC panels followed the same layered strategy employed in the previous case with few exceptions. To overcome the proposed geometric and size challenges, each mold was divided in two halves through their longitudinal axis, prior to fabrication. Each mold half was cut in a sequence of 4 hot wire cuts plus one milling operation, combining the advantages of both processes. Similarly to the previous case study, each cutting sequence was completed in approximately 10 min, illustrating a slight increase in time, consistent with the increase in cutting area. The finalized molds were shipped to a precast facility and used for the production of the final components through a normal GRC spray-up procedure. Once completed, the

Fig. 2. (a) Hot wire cutting of mold with interior ruled surface exposed - center component; (b) GRC spray process after assembled mold – center component; (c) finished GRC center arch component; (d) final assembled Corkcrete Arch.

CorkCrete Arch was manually assembled with success in several different settings, attesting to the accuracy of the fabrication process. The RHWC mold fabrication and GRC casting process for the Corkcrete Arch experiment was fully described in Sousa et al. [29].

Case Conclusions. The layered mold strategy confirmed, once more, the material optimization and speed advantages of RHWC, in this case, in the context of large-scale elements. Furthermore, the mold fabrication strategy proved to be successful in accurately reproducing the complex geometry of all the contact faces between GRC panels as well as GRC-cork interface surfaces.

This case study also revealed that a thoughtful consideration of formwork subdivision can be used to address specific geometrical challenges, increasing the design possibilities of RHWC. Additionally, since these subdivisions became imprinted on the finished surfaces, an interesting aesthetical relation between design and fabrication process was developed.

In terms of adaptability and integration, as already stated, RHWC can be used for the production of molds for prefabrication or in-situ construction by combining formwork sections as was the case in the CorkCrete Arch. In both these contexts, the production process and outputs were demonstrated to be easily integrated in industrial settings, without the need for specific equipment or procedures.

4 Conclusion

This paper presented the question “How can digital fabrication technologies expand traditional concrete architecture, through economically viable, sustainable, and tectonically relevant methodologies?” To address this question, a specific digital fabrication technology was proposed and two case studies were developed.

Through these experiments, RHWC proved to be able to overcome the main critical issues facing digital fabrication in concrete, having a large set of advantages in comparison with other processes, balancing its main geometrical drawback. Most notably, RHWC allowed for a significant decrease in fabrication times, combined with high material optimization and sustainability, making it an optimal solution for the production of formwork. Furthermore, considering the issue of adaptability and integration, RHWC processes and outputs were found to be easily adapted to the industrial context without the need for exceptional considerations. Although clearly limited in terms of geometry, we believe that RHWC still holds potential regarding this subject for three main reasons:

- ruled geometry is still significantly relevant today as an architectural language;
- when compared with traditional means, 6-axis robotic strategies greatly increase the variability and geometric freedom within the realm of ruled surfaces;
- ruled geometry can also be used to rationalize general double curved surfaces, increasing its relevance in architectural practice.

In light of this research, we also believe that Robotic Hot Wire Cutting is a process with great tectonic relevance. Being constrained to an intrinsic formal vocabulary opens the need to consider the fabrication strategy at an early design stage. This fact results in tectonically interesting solutions with a strong relation between the design and fabrication processes. This aspect became clear in the overall design solutions tailored by the RHWC process which impacted in all three tectonic dimensions of form, surface and composition. In addition to the already mentioned ruled vocabulary, the process-driven design of surface textures extracted directly from the RHWC process and the layered mold strategy enabling composite material compositions all contributed to establish specific tectonic qualities of digitally fabricated concrete through Robotic Hot Wire Cutting.

Acknowledgements. The authors would like to thank all the Digital Fabrication Lab (DFL) team, the FEUP/LABEST team and our industrial partner, Mota-Engil for the overall support in these experiments. The works presented were co-financed by the European Regional Development Fund (ERDF) through the COMPETE 2020 – Operational Programme for Competitiveness and Internationalization (POCI) and national funds by the FCT under the POCI-01-0145-FEDER-007744 Project and the previous PTDC/ATP-AQI/5124/2012 research Project.

References

1. Kolarevic, B.: Architecture in the Digital Age: Design and Manufacturing. Spon Press, New York (2003)
2. Frampton, K.: Studies in Tectonic Culture, pp. 3–4. MIT Press, Cambridge (1995)
3. Peck, M. (ed.): Modern Concrete Construction Manual, pp. 109–110. Institut Fur Internationale Architektur-Dokumentation, Munich (2014)
4. Weilandt, A., et al.: Rolex learning center in Lausanne. In: Proceedings of Evolution and Trends in Design, Analysis and Construction of Shell and Spatial Structures. Editorial Universitat Politècnica de València (2009)

5. Andersen, T.J., da Silva, W.R.L., Thrane, L.N.: Lessons from the TailorCrete Project. Concr. Int. **38**(3), 54–61 (2016)
6. Feringa, J., Søndergaard, A.: An integral approach to structural optimization and fabrication. In: Synthetic Digital Ecologies, ACADIA 2012, San Francisco, pp. 491–497 (2012)
7. McGee, W., Feringa, J., Søndergaard, A.: Processes for an architecture of volume: robotic hotwire cutting. In: Brell, S., Braumann, J. (eds.) Robarch 2012: Robotic Fabrication in Architecture, Art and Design, pp. 62–71. Springer, Vienna (2012)
8. Garcia, M.: Amanda levete architects. Archit. Des. **80**(1), 106–113 (2010)
9. In Progress: Louisiana State Museum and Sports Hall of Fame. https://www.archdaily.com/202678/in-progress-louisiana-state-museum-and-sports-hall-of-fame-trahan-architects. Accessed 14 Mar 2018
10. Raun, C., Kristensen, M., Kirkegaard, P.: Dynamic double curvature mould system. In: Computational Design Modelling, Proceedings of the Design Modelling Symposium, Berlin 2011, pp. 291–300. Springer, Berlin (2011)
11. Oesterle, S., Vansteenkiste, A., Mirjan, A.: Zero waste free-form formwork. In: Orr, J., Evernden, M., Darby, A., Ibell, T. (eds.) Proceedings of the Second International Conference on Flexible Formwork, ICFF. CICM and University of Bath, Dept. of Architecture and Civil Engineering (2012)
12. Peters, S., Trummer, A., Amtsberg, F., Parmann, G.: Precast concrete shells: a structural challenge. In: Menges, A., Sheil, B., Glynn, R., Skavara, M. (eds.) Fabricate 2017, pp. 250–257. UCL Press, London (2017)
13. Gardiner, B., Janssen, R.: FreeFab. Development of a construction-scale robotic formwork 3D printer. In: RobArch 2014: Robotic Fabrication in Architecture, Art and Design, pp. 131–146. Springer, Cham (2014)
14. Morel, P., Schwartz, T.: Automated casting systems for spatial concrete lattices. In: Modelling Behaviour, pp. 213–223. Springer, Cham (2015)
15. Aghaei-Meibodi, M., Bernhard, M., Jipa, A., Dillenburger, B.: The smart takes from the strong. In: Fabricate 2017, pp. 210–217. UCL Press, London (2017)
16. Feringa, J., Søndergaard, A.: Fabricating architectural volume: stereotomic investigations in robotic craft. In: Gramazio, F., Kohler, M., Langenberg, S. (eds.) Fabricate 2014: Negotiating, Design and Making, pp. 76–83. UCL Press, London (2014)
17. Digital Building Technologies Homepage. http://dbt.arch.ethz.ch/project/3d-printed-reinforced-beam/. Accessed 10 Mar 2018
18. Wu, P., Wang, J., Wang, X.: A critical review of the use of 3-D printing in the construction industry. Autom. Constr. **68**, 21–31 (2016)
19. Kristensen, E., Gramazio, F., Kohler, M.: Complex concrete constructions: merging existing casting techniques with digital fabrication. In: Proceedings of the 18th International Conference on Computer-Aided Architectural Design Research in Asia, pp. 613–622 (2013)
20. Peck, M. (ed.): Modern Concrete Construction Manual, pp. 111. Institut Fur Internationale Architektur-Dokumentation, Munich (2014)
21. Herrmann, M., Wolf, C., Sobek, W.: Design and manufacturing of optimal structures made from functionally graded concrete. In: Proceedings of IASS Annual Symposia, vol. 2015, no. 5, pp. 1–11. International Association for Shell and Spatial Structures (2015)
22. Oxman, N., Keating, S., Tsai, E.: Functionally graded rapid prototyping. In: Proceedings of VRAP: Advanced Research in Virtual and Rapid Prototyping (2011)
23. Søndergaard, A., et al.: Robotic hot-blade cutting. In: RobArch 2016: Robotic Fabrication in Architecture, Art and Design, pp. 150–164. Springer, Cham (2016)
24. Jones, D., Simons, E.: Story of the Saw, p. 49. Newman Neame, London (1961)

25. Smogorzewska, A.: Technological marks on pottery vessels. Evidence from Tell Arbid, Tell Rad Shaqrah and Tell Jassa el-Gharbi (Northeastern Syria). Polish archaeology in the Mediterranean, 19, pp. 555–564 (2007)
26. Flöry, S., Pottmann, H.: Ruled surfaces for rationalization and design in architecture. In: Proceedings of the Conference of the Association for Computer Aided Design in Architecture (ACADIA) (2010)
27. Rust, R., et al.: Spatial wire cutting – cooperative robotic cutting of non-ruled surface geometries for bespoke building components. In: Living Systems and Micro-Utopias: Towards Continuous Designing, Proceedings of the 21st International Conference of the Association for Computer-Aided Architectural Design Research in Asia (2016)
28. Martins, P., Campos, P., Nunes, S., Sousa, J.: Expanding the material possibilities of lightweight prefabrication in concrete through robotic hot-wire cutting – form, texture and composition. In: Real Time. Proceedings of the 33rd eCAADe Conference, vol. 2, pp. 341–351. Vienna University of Technology, Vienna (2015)
29. Sousa, J., Martins, P.: The robotic production of the GRC panels in the CorkCrete arch project. A stratified strategy for the fabrication of customized molds. In: Complexity & Simplicity, Proceedings of the 34th eCAADe Conference, vol. 2, pp. 153–160. University of Oulu, Oulu (2016)

Compliance, Stress-Based and Multi-physics Topology Optimization for 3D-Printed Concrete Structures

Gieljan Vantyghem[1(✉)], Veerle Boel[1], Wouter De Corte[1], and Marijke Steeman[2]

[1] Department of Structural Engineering, Ghent University, 9000 Ghent, Belgium
Gieljan.Vantyghem@UGent.be

[2] Department of Architecture and Urban Planning, Ghent University, 9000 Ghent, Belgium

Abstract. Recent advancements in Additive Manufacturing (AM) technologies have pushed the limits of manufacturability and have encouraged the design of products with increased complexity. Topology Optimization (TO) algorithms, on the other hand, have provided engineers with a tool for intelligently exploiting this design freedom by efficiently optimizing the shape of engineering structures. In this paper, three important developments of TO that might influence the manufacturing process and design of 3D-printed concrete structures are discussed. The first example shows how general structural TO problems, such as the well-known minimum compliance problem, can help to determine the optimal printing path and can discover the ideal location of the steel reinforcements. Secondly, it is considered how stress-based TO can enhance the shape of fiber-reinforced concrete components where the lack of steel reinforcements introduces a non-negligible strength asymmetry. In a third and last example, traditional structural TO techniques are extended to allow for multi-physics optimization. The thermal transmittance through a construction component is minimized, while the overall material usage is restricted. Results show the generation of very efficient (multi-material) structures that are aesthetically pleasing at the same time. The presented techniques aid in the search for more efficient structural design and might help overcome some of the technological challenges related to large-scale concrete 3D-printing.

Keywords: Multi-physics · Topology optimization · 3D concrete printing

1 Introduction

3D Concrete Printing (3DCP) refers to a recent technology that uses robotic concrete extrusion processes in order to fabricate functional building and construction components (see Fig. 1). In these processes, 3D objects are commonly produced by stacking layers of concrete (or cementitious material) on top of each other [1]. The extrusion nozzle is controlled automatically either by large gantry robots with 3 or 4 Degrees of Freedom (DOF), or articulated robots with 6 or more DOF [2, 3]. This kind of production process is suitable for deployment on-site as well as off-site and it claims to optimize

T. Wangler and R. J. Flatt (Eds.): DC 2018, RILEM Bookseries 19, pp. 323–332, 2019.
https://doi.org/10.1007/978-3-319-99519-9_30

Fig. 1. State-of-the-art in 3DCP: The world's first 3D-printed house (a) and villa (b) by WinSun co, the world's first 3D-printed concrete castle by Andrey Rudenko (c), the world's first 3D-printed hotel suite by Lewis Yakich (d), a 3D-printed army barrack by ERDC (e), and the world's first 3D printed bridge by BAM Infra/TU/e (f).

construction time and manual labor costs [4, 5]. Furthermore, 3DCP offers new possibilities to the way of production, and more importantly, it introduces new design freedoms that were as yet non-existent [6].

In general, Additive Manufacturing (AM) techniques are renowned for their capability to produce complicated and mass-customizable objects where the cost of production does not usually increase with complexity [7, 8]. However, as to date, most examples made by 3DCP experiments have been rectilinear, solid, and their design based on long-established and familiar shapes. It is clear that current 3DCP experiments focus on improving the manufacturing process, and not yet on the design. Innovative design methods could however take advantage of these new design freedoms, and could prove to be beneficial – opening up unexplored architectural possibilities. Interestingly, automated design methods such as topology optimization (TO) and other generative design techniques have been around for quite some time and can generate optimal designs in many applications [9]. TO processes are very efficient on computational resources and have proven to deliver trustworthy results [10]. The continuous process of generating improved design iterations is based on gradient information, finite element simulations, and can be extended to take into account the complex thermo-mechanical phenomena that take place during and after printing. This is especially important, where design iterations, i.e., discarding a defective or unusable print, are very expensive in terms of both material cost and manufacturing time [11].

General TO methods solve a material distribution problem to generate an optimal topology. Based on structural aspects only, the algorithms can achieve cost minimization as well as a certain performance and efficiency maximization. TO technology is currently in active development, and many extensions have been provided. TO results, however, are often very complex and, due to traditional manufacturing constraints, commonly require either simplification following the optimization process, or constraining of the design space to only allow feasible designs [12]. Consequently, the link with AM has never before been stronger and is studied by many [13, 14]. For example, in 3DCP, the maximum overhang angle of the successive layers is limited. In response, some researchers try to actively control the rheology of the fresh concrete during casting and try to trigger early-age stiffening [15]. A totally different approach would be to change the shape of the concrete components so that overhangs are not needed in the first place, or where the maximum angle of overhang is being restricted. For example, in [16, 17], new TO algorithms were developed that were capable of taking this into account.

Another printing obstacle that has been partially solved by TO, is a strength-loss problem caused by the orientation and layer-induced anisotropy of printed elements [18–20]. One approach would be to improve the inter-layer and path-dependent anisotropic properties of 3D-printed infill patterns. Another approach is to consider these anisotropic limitations while searching for the optimal design [21]. Of course, one solution, does not exclude the other, and the best solution is probably to combine both.

In this paper, three important developments of TO are discussed that might directly influence the manufacturing and design process of 3D-printed concrete structures. Additionally, the paper provides recommendations on future opportunities and potential algorithmic limitations.

1. The first example discusses the use of classical structural TO to determine the optimal printing path, and the positioning of steel reinforcements for 3D-printed structural components. It incorporates existing TO methods and their results, and links them with current concrete printing technology. The automated generation of strut-and-tie models by TO is discussed and a methodology to transform the results into printing paths is given.
2. The second example considers how stress-based topology optimization can enhance a concrete design when no steel reinforcements are used. Today, many concrete 3D printers use fast-hardening concrete and cement pastes with small aggregates that have limited tensile strength. This causes a strength asymmetry and weakens the results of a traditionally-optimized design. Adding steel fibers to the material mixture can improve the tensile strength of the material, but it cannot fully eliminate this asymmetry. By introducing stress constraints in the TO algorithms, optimal shapes can be generated that take this problem into account.
3. In the final example, a multi-physics approach in TO is investigated. The material usage (volume fraction) of a large-scale dome structure is restricted, while having the thermal transmittance through the dome wall minimized. Additionally, the structure is subjected to structural constraints, gravity and surface loads. To comply with these conflicting restrictions, the concept of a two-component extrusion technique is discussed. Similar to multi-material plastic printing processes, this should enable the extrusion of both plain concrete and a thermally-efficient substitute.

2 Compliance-Based Topology Optimization

The minimum compliance problem is one of the most well-known TO problems in literature. In this problem, the strain energy (also called compliance) is a global measure of the displacements. By minimizing the strain energy, the stiffness of the structure is maximized [22]. Additionally, a volume constraint is added to act as an opposing restriction, and is comparable to a cost factor in reality. Mathematically, this material distribution problem is solved very efficiently using gradient-based optimizers coupled with adjoint sensitivity analysis. The mathematical formulation reads as follows:

$$\begin{aligned} \text{min:} \quad & c(\mathbf{x}) = \mathbf{U}^{\mathbf{T}}\mathbf{K}\mathbf{U} = \sum_{e=1}^{N} E_e(x_e)\mathbf{u}_e^{\mathbf{T}}\mathbf{k}_e^0\mathbf{u}_e \\ \text{subject to:} \quad & V(\mathbf{x})/V_0 = f \\ & \mathrm{KU} = \mathrm{F} \\ & 0 \leq \mathbf{x} \leq 1 \end{aligned} \tag{1}$$

In this formulation, $c(\mathbf{x})$ is the compliance and $\mathbf{x}$ the element density vector (i.e. the vector containing all design variables). **U** is the global displacement vector, **K** is the global stiffness matrix, and **F** the vector with the mechanical loads. $E_e(x_e)$ is the element's Young's modulus and depends on its density. For this dependency, the modified SIMP approach is used [23]. $V(\mathbf{x})/V_0$ represents the maximum allowable volume fraction (f), and N is equal to the number of elements used to discretize the design domain. Various programming techniques can be applied to optimize the design. However, in this paper, a benchmarked solver called MMA [24] is used. In this solver, the original functions are replaced by "certain approximating functions that are based on gradient information at the current iteration point and some parameters (moving asymptotes)" (Citation from [23]). In essence, the partial derivative of the objective function $c(\mathbf{x})$ and the material volume $V(\mathbf{x})$ with respect to the element densities are required. In the example presented below, the MBB benchmark problem [9] and its optimal shape are presented (see Fig. 2).

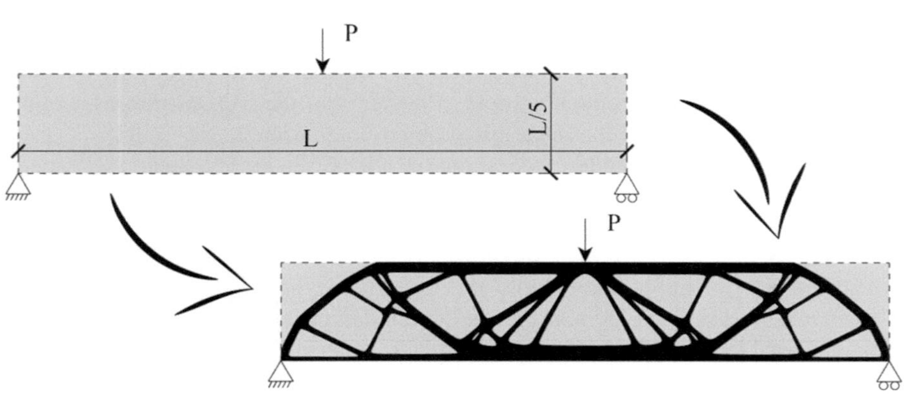

Fig. 2. MBB benchmark problem: Design domain and boundary condition and optimal continuum topology.

Now, the following question may arise: What does an optimized structural shape has to do with 3D concrete printing? The optimization problem does not even take into account the advanced material properties of concrete. The answer to this question is that a link can be made between the generated TO result and the determination of a printing path.

First, a useful application of a minimum compliance design is explained. That is to say, TO algorithms enable the automatic generation of strut-and-tie models for reinforced concrete design [25–27]. These strut-and-tie models are designed to reduce load deformation response, and can help in the dimensioning of steel reinforcements. "In 1980, Schlaich stated that the stiffest truss model is the one that will produce the safest load-deformation response, because limiting truss deflection prevents large plastic deformations in the concrete. However, the engineering judgment required to obtain an accurate truss model was viewed as a drawback of the design approach" (Citation from [27]). Now, maximizing stiffness correlates mathematically to minimizing the reinforcing steel's elastic strain energy. In other word, performing a minimum compliance design was suggested to provide excellent strut-and-tie models. Experiments and nonlinear finite element modelling in [27] confirmed this benefit. Nevertheless, and although better performing, the increase in complexity of the models remained a drawback of the method.

In this example however, the TO results are not used to dimension the reinforced concrete structure, but to print them more efficiently. To start with, the strut-and-tie model can serve as a direct input for the printing path where the generated patterns are in line with the principal elastic stress trajectories. Additionally, the patterns could function as an internal skeleton where the rest of the design domain is restored. For this, lower quality concrete could be used. Finally, the tensile regions (or "ties" from the strut-and-tie model) could be printed with an embedded steel chain. Figure 3 illustrates such a conceptual 3DCP setup.

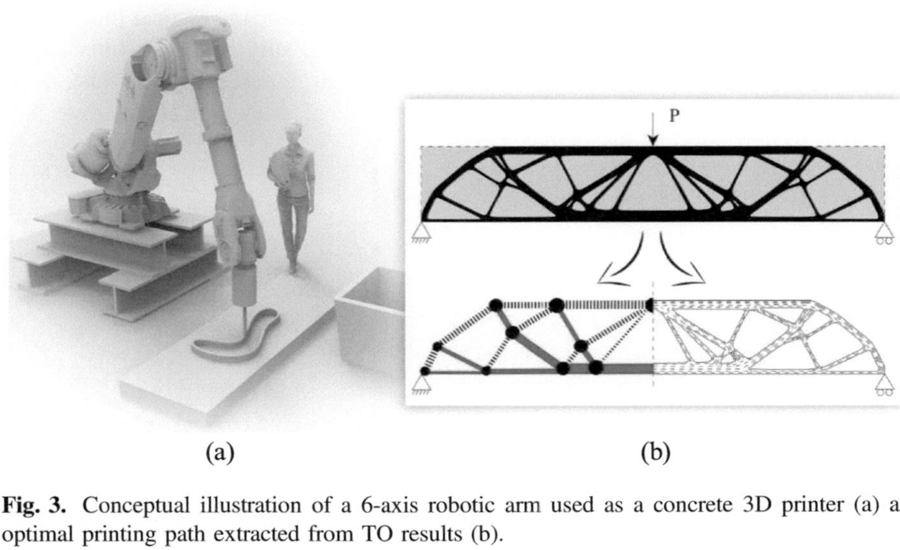

(a) (b)

Fig. 3. Conceptual illustration of a 6-axis robotic arm used as a concrete 3D printer (a) and optimal printing path extracted from TO results (b).

3 Stress-Based Topology Optimization

In the previous section, an optimal printing path and the positioning of the steel (chain) reinforcements was extracted from TO results. However, in today's date, the use of steel reinforcements in not commonly seen in 3D concrete printing processes. Only few experimental setups were found using any form of tensile reinforcement. For example, in 2014, a 3D-printed house by Winsun co reported using high-grade cement paste and glass fiber reinforcements [28]. However, "pictures of the project seem to suggest that the glass fiber was not used as an additive to the mixture, but rather as a separate mesh in between the printed layers" (Citation from [2]). Currently, this lack of use of embedded tensile reinforcements hampers the fabrication of "functional" large-scale building components. A viable solution to this problem, is studied in [29], where carbon, glass and basalt fibers are being added to the cement mix before (or during) extrusion. Study showed that using reinforcing short fibers can result in materials that exhibit much higher flexural (up to 30 MPa) and compressive strength (up to 80 MPa). Additionally, "an alignment of the fibers, caused by the 3D-printing extrusion process is observed, opening up the possibility to use the print path direction as a means to control fiber orientation within the printed structures" (Citation from [29]). In [30], the use of UHPFRC (ultra-high performance fiber-reinforced concrete) was reviewed and their superiority in comparison to conventional fiber-reinforced concrete (FRC) structures proven.

While FRC and UHPFRC improve the tensile and flexural strength of the material, it will always remain lower than that of the compressive strength. Regrettably, this deviation causes a strength asymmetry for which "a minimum compliance design" is not fully optimal. The optimized structures are in fact oversized in the compression zones and would crack or break rather easily in the tensile zones. By introducing stress constraints in the topology optimization algorithm, optimal shapes can be generated which are optimized and take into account this strength asymmetry. In Fig. 4 illustrates this concept and presents a small concrete specimen which was optimized using Drucker–Prager yield criterion. For this case study, the maximum compressive strength was 60 MPa and the maximum tensile strength 15 MPa. The mathematical formulation was based on a study by [31], and optimized using a globally convergent version of MMA (GCMMA) [32]. In post-production, the resulting topology was reverse engineered in Fusion 360, and using Abaqus, a few manual design iterations were performed to further improve the model.

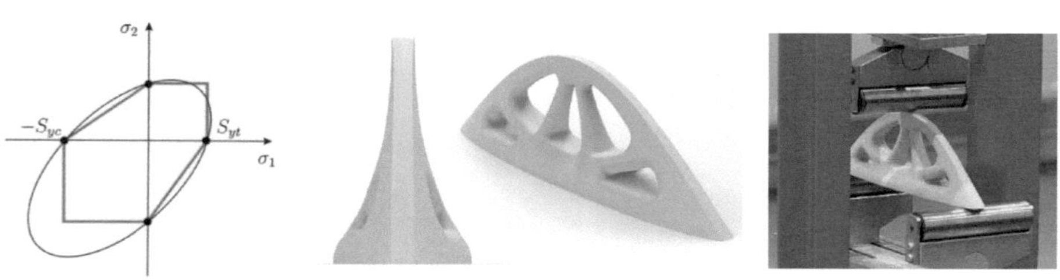

(a) (b) (c)

Fig. 4. Mohr–Coulomb (red) and Drucker–Prager (blue) yield criterion (a), optimized concrete specimen using stress-based TO (b), and an experimental 3-point-bending test (c). (Color figure online)

4 Multi-physics Topology Optimization

In this third and last example, traditional structural topology optimization techniques are extended to include multi-physical requirements. By using such multi-physics approach, an optimal design can be found which not only meets structural requirements, but where also the heat transfer characteristics of the structure are optimized [33, 34]. In the case that is presented here, the thermal transmittance through the dome wall is minimized, the weight or volume of the structure is restricted, and the domain is subjected to gravity and surface loads. Similar to previous examples, the Young's modulus of each element depends on the density. However, in this example, the element's thermal conductivity is also calculated from this density. In contrast to previous studies, the design variables (element densities) have three optimal states. One state symbolizes the surrounding air (where, $x_e = 0$), another state represents the solid structure ($x_e = 1$), and finally, a third optimal state is created ($x_e = 0.5$) which symbolizes a thermally-efficient mesostructure made from an intermediate density. Originally, this idea was inspired by the infill pattern used in plastic 3D printing. However, the principle can be applied to concrete 3D printing as well, where one extrusion nozzle uses plain or fiber-reinforced concrete, and another nozzle could use a thermally-efficient substitute such as "Foamcrete" [35]. Figure 5 for the first picture reference and Fig. 6 for the second.

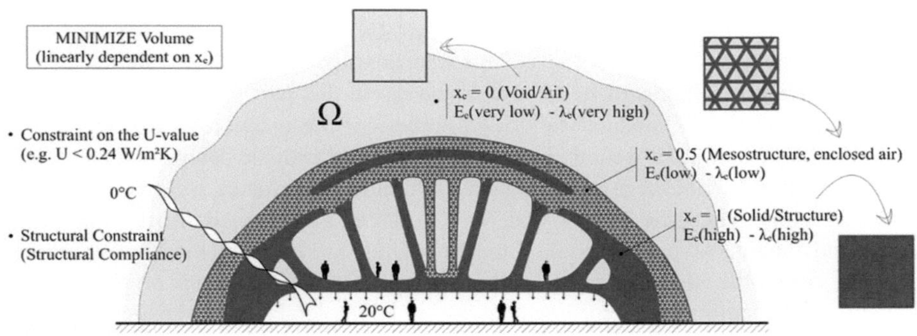

Fig. 5. Multi-physics topology optimization of a dome structure where the element density variable has three optimal states. x_e= 0 (void), x_e= 0.5 (mesostructure), and x_e= 1 (solid).

Additionally, the development of a novel dispenser technology could further enhance the capabilities of this design approach. If, for example, the rheology of the concrete mix could be changed actively inside the extrusion nozzle, the TO algorithms could not only allow three optimal states but a whole range. This potential touches the concept of a functionally-graded concrete studied in [36], and could really open up a new range of design opportunities. The TO algorithm could determine the exact requirements of mixture at every location, allowing for a precise assessment of the weight savings. A strong concrete mixture where it needs to be strong, less strong where it is allowed, and thermally insulating where it is required.

$$\begin{aligned}
\text{max :}\quad & C_t(\mathbf{x}) = \mathbf{T}^{\mathrm{T}}\mathbf{K}_t\mathbf{T} = \sum_{e=1}^{N} \lambda_e(x_e)\boldsymbol{\theta}_e^{\mathrm{T}}\mathbf{k}_{t,e}^{0}\boldsymbol{\theta}_e \\
\text{s.t. :}\quad & C_s(\mathbf{x})/C_{s\max} - 1 \le 0 \\
& V(\mathbf{x})/V_{\max} - 1 \le 0 \\
\text{w. :}\quad & C_s(\mathbf{x}) = \mathbf{U}^{\mathrm{T}}\mathbf{K}_s\mathbf{U} = \sum_{e=1}^{N} E_e(x_e)\mathbf{u}_e^{\mathrm{T}}\mathbf{k}_{s,e}^{0}\mathbf{u}_e \\
& 0 \le \mathbf{x} \le 1
\end{aligned}$$

(a)

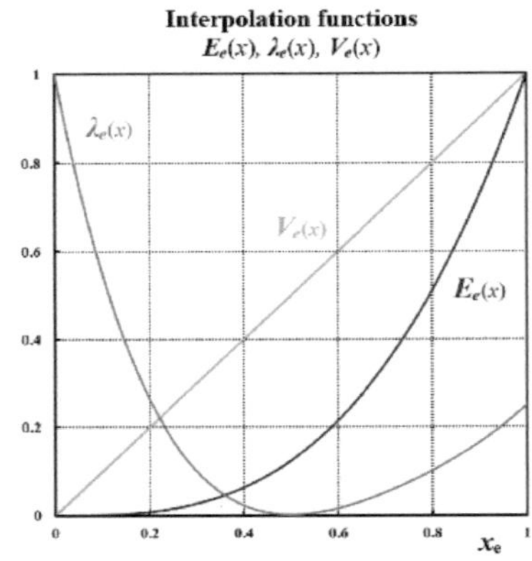

(b)

Fig. 6. Mathematical formulation of the multi-physics topology optimization problem discussed in Sect. 4 (a) and the applied interpolation schemes: $E_e = x_e^3$, $\lambda_e = 1.75\ (0.5 - x_e)^2 + 3\ (0.5 - x_e)^3 + 3\ (0.5 - x_e)^4$, and $V_e = x_e$ (b).

5 Conclusions and Final Remarks

The use of 3D-printing provides access to the production of complex geometries and offers the potential to print in a continuous way. 3DCP has the ability to revolutionize the building industry, however, much research is still needed to provide strong and material efficient structures in future constructions. In this paper TO was suggested as an excellent method to enhance the design process of these concrete structures.

In the first example, it was discussed how TO could provide design engineers with a convenient methodology for obtaining a minimum compliance truss or continuum model. A model where the "struts" and "ties" could be transformed directly to printing paths, and where the pattern could function as an internal skeleton. Lower quality concrete could fill in the voids/gaps, restoring the structure to its original design domain. Additionally, steel reinforcements could be implemented in the tensile regions of the optimized shape. In the second example, stress-based TO was used to further improve the optimized design where the TO algorithm could take into account the advanced non-linear material properties of the concrete mixture. A small concrete beam was optimized using the Drucker–Prager yield criterion, which resulted in a much higher bending-efficiency. Finally, in the last example a multi-physics approach was implemented and the thermal transmittance of the structure was minimized. Especially this last example showed great potential for the future of construction, where not only structural requirements, but other aspects should be taken into consideration. Finally, TO opportunities are largely linked with the current state of manufacturing processes, and should hence be studied together.

Acknowledgements. This research was supported by Ghent University. We thank our colleagues from the Magnel Laboratory and all members of Concre3dLab – Ghent who provided insight and expertise that greatly assisted the research. The authors also thank Krister Svanberg for providing the MMA optimizer code.

References

1. Lim, S., Buswell, R., Le, T., Austin, S., Gibb, A., Thorpe, T.: Developments in construction-scale additive manufacturing processes. Autom. Constr. **21**, 262–268 (2012)
2. Bos, F., Wolfs, R., Ahmed, Z., Salet, T.: Additive manufacturing of concrete in construction: potentials and challenges of 3D concrete printing. Virtual Phys. Prototyp. **11**(3), 209–225 (2016)
3. Duballet, R., Baverel, O., Dirrenberger, J.: Classification of building systems for concrete 3D printing. Autom. Constr. **83**, 247–258 (2017)
4. Perkins, I., Skitmore, M.: Three-dimensional printing in the construction industry: a review. Int. J. Constr. Manag. **15**(1), 1–9 (2015)
5. Tay, Y., Panda, B., Paul, S., Noor Mohamed, N., Tan, M., Leong, K.: 3D printing trends in building and construction industry: a review. Virtual Phys. Prototyp. **12**(3), 261–276 (2017)
6. Gosselin, C., Duballet, R., Roux, P., Gaudillière, N., Dirrenberger, J., Morel, P.: Large-scale 3D printing of ultra-high performance concrete – a new processing route for architects and builders. Mater. Des. **100**, 102–109 (2016)
7. Hague, R., Campbell, I., Dickens, P.: Implications on design of rapid manufacturing. Proc. Inst. Mech. Eng. Part C J. Mech. Eng. Sci. **217**(1), 25–30 (2003)
8. Labonnote, N., Rønnquist, A., Manum, B., Rüther, P.: Additive construction: state-of-the-art, challenges and opportunities. Autom. Constr. **72**, 347–366 (2016)
9. Bendsøe, M., Sigmund, O.: Topology Optimization. Springer, Berlin (2011)
10. Rozvany, G.: Exact analytical solutions for some popular benchmark problems in topology optimization. Struct. Optim. **15**(1), 42–48 (1998)
11. "Simulations in 3D Printing", 3D Hubs (2018). https://www.3dhubs.com/knowledge-base/simulations-3d-printing. Accessed 26 Feb 2018
12. Brackett, D., Ashcroft, I., Hague, R.: Topology optimization for additive manufacturing. In: 22nd Annual International Solid Freeform Fabrication Symposium – An Additive Manufacturing
13. Zegard, T., Paulino, G.: Bridging topology optimization and additive manufacturing. Struct. Multidiscip. Optim. **53**(1), 175–192 (2015)
14. Mhapsekar, K., McConaha, M., Anand, S.: Additive manufacturing constraints in topology optimization for improved manufacturability. J. Manuf. Sci. Eng. **140**, 051017 (2018)
15. ERC Advanced Grant Project "SmartCast" (693755) "Smart casting of concrete structures by active control of rheology (SmartCast)" (2017)
16. Gaynor, A., Guest, J.: Topology optimization considering overhang constraints: eliminating sacrificial support material in additive manufacturing through design. Struct. Multidiscip. Optim. **54**(5), 1157–1172 (2016)
17. Langelaar, M.: Topology optimization of 3D self-supporting structures for additive manufacturing. Addit. Manuf. **12**, 60–70 (2016)
18. Bellini, A., Güçeri, S.: Mechanical characterization of parts fabricated using fused deposition modeling. Rapid Prototyp. J. **9**(4), 252–264 (2003)
19. Lee, C., Kim, S., Kim, H., Ahn, S.: Measurement of anisotropic compressive strength of rapid prototyping parts. J. Mater. Process. Technol. **187–188**, 627–630 (2007)

20. Vega, V., Clements, J., Lam, T., Abad, A., Fritz, B., Ula, N., Es-Said, O.: The effect of layer orientation on the mechanical properties and microstructure of a polymer. J. Mater. Eng. Perform. **20**(6), 978–988 (2010)
21. Mirzendehdel, A., Rankouhi, B., Suresh, K.: Strength-based topology optimization for anisotropic parts. Addit. Manuf. **19**, 104–113 (2018)
22. Christensen, P., Klarbring, A.: An Introduction to Structural Optimization. Springer, Dordrecht (2009)
23. Andreassen, E., Clausen, A., Schevenels, M., Lazarov, B., Sigmund, O.: Efficient topology optimization in MATLAB using 88 lines of code. Struct. Multidiscip. Optim. **43**(1), 1–16 (2010)
24. Svanberg, K.: The method of moving asymptotes—a new method for structural optimization. Int. J. Numer. Methods Eng. **24**(2), 359–373 (1987)
25. Victoria, M., Querin, O., Martí, P.: Generation of strut-and-tie models by topology design using different material properties in tension and compression. Struct. Multidiscip. Optim. **44** (2), 247–258 (2011)
26. Almeida, V., Simonetti, H., Oliveira Neto, L.: The strut-and-tie models in reinforced concrete structures analysed by a numerical technique. Rev. IBRACON Estrut. Mater. **6**(1), 139–157 (2013)
27. Gaynor, A., Guest, J., Moen, C.: Reinforced concrete force visualization and design using bilinear truss-continuum topology optimization. J. Struct. Eng. **139**(4), 607–618 (2013)
28. Wu, P., Wang, J., Wang, X.: A critical review of the use of 3-D printing in the construction industry. Autom. Constr. **68**, 21–31 (2016)
29. Hambach, M., Volkmer, D.: Properties of 3D-printed fiber-reinforced Portland cement paste. Cem. Concr. Compos. **79**, 62–70 (2017)
30. Yoo, D., Yoon, Y.: A review on structural behavior, design, and application of ultra-high-performance fiber-reinforced concrete. Int. J. Concr. Struct. Mater. **10**(2), 125–142 (2016)
31. Bruggi, M., Duysinx, P.: Topology optimization for minimum weight with compliance and stress constraints. Struct. Multidiscip. Optim. **46**(3), 369–384 (2012)
32. Svanberg, K.: A class of globally convergent optimization methods based on conservative convex separable approximations. SIAM J. Optim. **12**(2), 555–573 (2002)
33. Bruggi, M., Taliercio, A.: Design of masonry blocks with enhanced thermomechanical performances by topology optimization. Constr. Build. Mater. **48**, 424–433 (2013)
34. Vantyghem, G., Steeman, M., De Corte, W., Boel, V.: Design of Cellular materials and mesostructures with improved structural and thermal performances. In: Advances in Structural and Multidisciplinary Optimization, pp. 1983–1996 (2017)
35. Othuman Mydin, M.: An experimental investigation on thermal conductivity of lightweight Foamcrete for thermal insulation. Jurnal Teknologi **63**(1) (2013)
36. Herrmann, M., Sobek, W.: Functionally graded concrete: numerical design methods and experimental tests of mass-optimized structural components. Struct. Concr. **18**(1), 54–66 (2017)

Author Index

T. Wangler and R. J. Flatt (Eds.): DC 2018, RILEM Bookseries 19, pp. 333–334, 2019.
https://doi.org/10.1007/978-3-319-99519-9

MIX
Papier aus verantwortungsvollen Quellen
Paper from responsible sources
FSC® C105338

If you have any concerns about our products,
you can contact us on
ProductSafety@springernature.com

In case Publisher is established outside the EU,
the EU authorized representative is:
Springer Nature Customer Service Center GmbH
Europaplatz 3, 69115 Heidelberg, Germany

Printed by Libri Plureos GmbH
in Hamburg, Germany